Challenges and Advances in Computational Chemistry and Physics

Volume 37

Series Editor

Jerzy Leszczynski, Department of Chemistry and Biochemistry, Jackson State University, Jackson, MS, USA

This book series provides reviews on the most recent developments in computational chemistry and physics. It covers both the method developments and their applications. Each volume consists of chapters devoted to the one research area. The series highlights the most notable advances in applications of the computational methods. The volumes include nanotechnology, material sciences, molecular biology, structures and bonding in molecular complexes, and atmospheric chemistry. The authors are recruited from among the most prominent researchers in their research areas. As computational chemistry and physics is one of the most rapidly advancing scientific areas such timely overviews are desired by chemists, physicists, molecular biologists and material scientists. The books are intended for graduate students and researchers.

All contributions to edited volumes should undergo standard peer review to ensure high scientific quality, while monographs should be reviewed by at least two experts in the field. Submitted manuscripts will be reviewed and decided by the series editor, Prof. Jerzy Leszczynski.

Manoj Shukla · Elizabeth Ferguson ·
Jerzy Leszczynski
Editors

Emerging Materials and Environment

Springer

Editors
Manoj Shukla
Environmental Laboratory
US Army Engineer Research
and Development Center
Vicksburg, MS, USA

Elizabeth Ferguson
Environmental Laboratory
US Army Engineer Research
and Development Center
Vicksburg, MS, USA

Jerzy Leszczynski
Department of Chemistry and Biochemistry
Jackson State University
Jackson, MS, USA

ISSN 2542-4491 ISSN 2542-4483 (electronic)
Challenges and Advances in Computational Chemistry and Physics
ISBN 978-3-031-39469-0 ISBN 978-3-031-39470-6 (eBook)
https://doi.org/10.1007/978-3-031-39470-6

This Springer imprint is published by the registered company Springer Nature Switzerland AG
The registered company address is: Gewerbestrasse 11, 6330 Cham, Switzerland

Paper in this product is recyclable.

Preface

It is with an immense pleasure we take this opportunity to bring this volume "Emerging Materials and Environment" under the "Challenges and Advances in Computational Chemistry and Physics" book series. Every day, humans are exposed to natural and manufactured chemicals in different forms. Human population will keep growing and this growth with exert asymmetric pressure on the natural resources. Obviously, natural resources are limited and will not be able to satisfy our ever-growing need. Human population will more depend on the synthetic materials. Necessity is the mother of invention, and thus, we may develop new classes of materials in future ahead. This volume brings theoreticians and experimentalists on a common platform integrating reviews of recent work in area of emerging materials and their potential for release in the environment distributed in nine chapters.

Chapter 1 describes a brief overview of some emerging technologies and potential for release of novel materials in the environment. This chapter contributed by Shukla et al. discusses different emerging technologies such as 3D printing, nanoreinforced composite materials, bio- and cellulosic materials, microplastics, life-saving antibiotics, and volatile organic compounds from various technological advances and how these technologies may contribute to the release of new or traditional materials in the environment. Mechanochemistry is an area where mechanical forces are utilized to modify chemical reaction pathways and reaction rates. In Chap. 2, Jha and Subramanian reveal application of Generalized Force-Modified Potential Energy Surface (G-FMPES) formulation to study chemical reactions through mechanochemical simulations. The chemicals released due to widespread use of pharmaceuticals, personal care products (PPCPs), and agrochemicals are classified under contaminants of emerging concerns (CEC) and/or environmental pollutants (EPs). The chemometric-based modeling and machine learning (ML) models that can be used to study adsorption of CECs/EPs are described in Chap. 3 by Kar and Leszczynski. Ionic liquids (ILs) are mostly organic salts, while halloysite nanotubes (HNTs) are naturally occurring clays with nanotubular structures. ILs and HNTs find increased application in different area of technology. Nelson discusses whether increased use of these compounds would be of environmental concern in Chap. 4. There is a need for the development of synthetic approach for functional heterostructured nanocomposites

(HNCs) which is chemically green, simple, and scalable and shows low product variability. In Chap. 5, Mukherjee reveals a synthetic approach "Laser Ablation Synthesis in Solution-Galvanic Replacement Reaction (LASiS-GRR)" for the development of such materials. Ceria, a prototypical reducible oxide, mimics enzyme-like catalytic activities and finds potential applications in areas such as biochemistry, medicine, and environment remediation. Bhasker-Ranganath and Xu review reactivity and mode of action of ceria in Chap. 6. It is a challenge to provide a safe drinking water to world population even in the modern time. In Chap. 7, Gates et al. discuss the application of 2D materials for water purification. Per- and polyfluoroalkyl substances (PFAS) are man-made synthetic compounds that were designed to possess novel properties and have been used as firefighting agents. These compounds are ubiquitous and remain stable in the environments and thus nicknamed as "forever chemicals". These compounds are now known to have contaminated many regions around the globe. Various aspects of PFAS are discussed in the last two chapters of the current volume. In Chap. 8, Kolel-Veetil and Ganjigunteramaswamy review different potential technologies and approaches for the capture, removal, and degradation of PFAS compounds from the contaminated media. In the last Chap. 9, Carre-Burritt and Vyas discuss life-cycle aspects of PFAS compounds including applications, environmental release and uptake by plants and animals, and remediation technologies.

We take this opportunity to thank all contributors for devoting their time and hard work to make this project a success. Of course, many thanks go to our family and friends. Without their support, the development and completion of the book would not have been possible.

Vicksburg, MS, USA Manoj Shukla
Vicksburg, MS, USA Elizabeth Ferguson
Jackson, MS, USA Jerzy Leszczynski

Contents

Contributors

Suman Bhasker-Ranganath SUNCAT Center for Interface Science and Catalysis, SLAC National Accelerator Laboratory, Menlo Park, CA, USA;
SUNCAT Center for Interface Science and Catalysis, Stanford University, Stanford, CA, USA

Caitlin G. Bresnahan US Army Engineer Research and Development Center, Environmental Laboratory, Vicksburg, MS, USA

Asa E. Carre-Burritt Department of Chemistry, Colorado School of Mines, Golden, CO, USA

Brian D. Etz US Army Engineer Research and Development Center, Environmental Laboratory, Vicksburg, MS, USA;
Oak Ridge Institute for Science and Education, Oak Ridge, TN, USA;
Simetri, Inc., Winter Park, FL, USA

Swathi Iyer Ganjigunteramaswamy Chemistry Division, Naval Research Laboratory, Washington, DC, USA;
Nova Research, Inc., Alexandria, VA, USA

Ye Gao Department of Chemistry and Biochemistry, Jackson State University, Jackson, MS, USA

Kaelin Gates Department of Chemistry and Biochemistry, Jackson State University, Jackson, MS, USA

Glen R. Jenness US Army Engineer Research and Development Center, Environmental Laboratory, Vicksburg, MS, USA

Sanjiv K. Jha Mathematics and Science Department, GateWay Community College, Maricopa Community Colleges, Phoenix, AZ, USA

Supratik Kar Chemometrics & Molecular Modeling Laboratory, Department of Chemistry, Kean University, Union, New Jersey, USA

Manoj Kolel-Veetil Chemistry Division, Naval Research Laboratory, Washington, DC, USA

Ashlyn M. Koval US Army Engineer Research and Development Center, Environmental Laboratory, Vicksburg, MS, USA;
Oak Ridge Institute for Science and Education, Oak Ridge, TN, USA;
Simetri, Inc., Winter Park, FL, USA

Robert W. Lamb US Army Engineer Research and Development Center, Environmental Laboratory, Vicksburg, MS, USA;
Oak Ridge Institute for Science and Education, Oak Ridge, TN, USA;
Simetri, Inc., Winter Park, FL, USA

Jerzy Leszczynski Interdisciplinary Center for Nanotoxicity, Department of Chemistry, Physics and Atmospheric Sciences, Jackson State University, Jackson, MS, USA

Charles M. Luft US Army Engineer Research and Development Center, Environmental Laboratory, Vicksburg, MS, USA;
Oak Ridge Institute for Science and Education, Oak Ridge, TN, USA;
Simetri, Inc., Winter Park, FL, USA

Levi A. Lystrom US Army Engineer Research and Development Center, Environmental Laboratory, Vicksburg, MS, USA;
Oak Ridge Institute for Science and Education, Oak Ridge, TN, USA;
Los Alamos National Laboratory, Los Alamos, NM, USA

Harley R. McAlexander US Army Engineer Research and Development Center, Environmental Laboratory, Vicksburg, MS, USA

Dibyendu Mukherjee Department of Chemical and Biomolecular Engineering, Nano-Bio Materials Laboratory for Energy, Energetics and Environment (nbml-E3), University of Tennessee, Knoxville, TN, USA

William M. Nelson Engineer Research and Development Center, Environmental Laboratory, US Army Corps of Engineers, Vicksburg, MS, USA

Shamily Patibandla Department of Chemistry and Biochemistry, Jackson State University, Jackson, MS, USA

William A. Pisani US Army Engineer Research and Development Center, Environmental Laboratory, Vicksburg, MS, USA;
Oak Ridge Institute for Science and Education, Oak Ridge, TN, USA;
Simetri, Inc., Winter Park, FL, USA

Katarina M. Pittman US Army Engineer Research and Development Center, Environmental Laboratory, Vicksburg, MS, USA;
Oak Ridge Institute for Science and Education, Oak Ridge, TN, USA;
University of Mississippi Medical Center, Jackson, MS, USA

Avijit Pramanik Department of Chemistry and Biochemistry, Jackson State University, Jackson, MS, USA

Paresh Chandra Ray Department of Chemistry and Biochemistry, Jackson State University, Jackson, MS, USA

Michael R. Roth US Army Engineer Research and Development Center, Environmental Laboratory, Vicksburg, MS, USA

Timothy C. Schutt US Army Engineer Research and Development Center, Environmental Laboratory, Vicksburg, MS, USA

Manoj K. Shukla US Army Engineer Research and Development Center, Environmental Laboratory, Vicksburg, MS, USA

Gopinath Subramanian X-Computational Physics Division, Los Alamos National Laboratory, Los Alamos, NM, USA

Shubham Vyas Department of Chemistry, Colorado School of Mines, Golden, CO, USA

Ye Xu Cain Department of Chemical Engineering, Louisiana State University, Baton Rouge, LA, USA

Chapter 1
Emerging Materials and Environment: A Brief Introduction

Manoj K. Shukla, Charles M. Luft, Ashlyn M. Koval, William A. Pisani, Robert W. Lamb, Levi A. Lystrom, Brian D. Etz, Katarina M. Pittman, Michael R. Roth, Caitlin G. Bresnahan, Timothy C. Schutt, Glen R. Jenness, and Harley R. McAlexander

Abstract This chapter provides a brief overview of emerging materials that either have the potential to be or have already been identified as problematic for the environment. The growing population, estimated to be ≈10 billion people by the year 2050, will create asymmetric pressure on available resources, leading to the need for novel materials to drive technological advancement and alleviate the burden on

M. K. Shukla (✉) · C. M. Luft · A. M. Koval · W. A. Pisani · R. W. Lamb · L. A. Lystrom · B. D. Etz · K. M. Pittman · M. R. Roth · C. G. Bresnahan · T. C. Schutt · G. R. Jenness · H. R. McAlexander
US Army Engineer Research and Development Center, Environmental Laboratory, Vicksburg, MS 39180, USA
e-mail: manoj.k.shukla@usace.army.mil

C. M. Luft
e-mail: charles.m.luft@usace.army.mil

A. M. Koval
e-mail: ashley.m.koval@usace.army.mil

W. A. Pisani
e-mail: william.a.pisani@usace.army.mil

R. W. Lamb
e-mail: robert.w.lamb@usace.army.mil

L. A. Lystrom
e-mail: llystrom@lanl.gov

B. D. Etz
e-mail: brian.d.etz@erdc.dren.mil

K. M. Pittman
e-mail: kmpittman@umc.edu

M. R. Roth
e-mail: michael.m.roth@usace.army.mil

C. G. Bresnahan
e-mail: caitlin.g.bresnahan@usace.army.mil

T. C. Schutt
e-mail: timothy.c.schutt@usace.army.mil

M. Shukla et al. (eds.), *Emerging Materials and Environment*,
Challenges and Advances in Computational Chemistry and Physics 37,
https://doi.org/10.1007/978-3-031-39470-6_1

natural resources. Development and implementation of new materials and technologies driven by necessity may exacerbate environmental contamination as these new materials are rushed into use without forethought into their environmental impacts. In this chapter, various aspects of 3D printing, nanocomposites, electronic waste (E-waste), biomaterials, cellulosic materials, volatile organic compounds (VOC), microplastics, and antibiotics have been discussed in terms of their current or potential environmental relevance. For example, Ag- and TiO_2-nanoparticles (NPs) have potential for antibacterial, and UV protection applications, respectively, and are used in textiles, medical devices, dental fillings, etc. However, these NPs can pose a threat if released into the environment, which may occur through leaching mechanisms or through textile laundering. The annual global E-waste production is projected to increase to 74.7 Mt by the year 2030, thus increasing the potential for environmental contamination unless efficient recovery and remediation technologies are developed. Photovoltaic panels (PVs) are one such example that have emerged as significant E-waste contaminants. These devices have only recently been classified as E-waste by the European Commission, and their volumes are anticipated to increase rapidly. During pyrolysis processes (combustion, biomass conversion, etc.), a significant quantity of VOCs such as benzene, toluene, and phenol are expected to be released and pose both carcinogenic and noncarcinogenic health hazards to the workforce, neighboring general population, and environment. Furthermore, the tracking and detection of increased antibiotic resistance, and accumulation of microplastics leading to organic and metallic pollutants will be highlighted. One major environmental contaminant relevant in today's society, per- and polyfluoroalkyl substances (PFAS), will not be discussed in this chapter as this topic is discussed in two separate chapters in the book. Material types, pros and cons, and modes of release into the environment will be discussed. The topics reviewed in this chapter will support parallel research on environmental impacts of next-generation materials as new technologies are developed and implemented in society.

G. R. Jenness
e-mail: glen.r.jenness@usace.army.mil

H. R. McAlexander
e-mail: harley.r.mcalexander@usace.army.mil

C. M. Luft · A. M. Koval · W. A. Pisani · R. W. Lamb · L. A. Lystrom · B. D. Etz · K. M. Pittman
Oak Ridge Institute for Science and Education, Oak Ridge, TN, USA

C. M. Luft · A. M. Koval · W. A. Pisani · R. W. Lamb · B. D. Etz
Simetri, Inc., Winter Park, FL 32792, USA

L. A. Lystrom
Los Alamos National Laboratory, Los Alamos, NM 87545, USA

K. M. Pittman
University of Mississippi Medical Center, Jackson, MS, USA

1.1 Introduction

Currently, there are approximately 7.5 billion people living on the earth. This number continues to grow and is estimated to surpass the 10 billion mark by the year 2050. Obviously, there will be tremendous asymmetrical pressure on resources and on the environment to provide food, water, and energy to the large population. Necessity will drive novel invention and innovation to address these pressures as they develop. Such inventions lead to technological advancements, including development of novel materials that are useful for every aspect of life. However, the use of these materials may also have disadvantages; when unmanaged, they may become a threat to the environment as well as to human health. For example, asbestos—which is an excellent insulator and highly heat resistant, exhibits excellent tensile strength, and is resistant to chemical degradation—has been used many years in building materials [1, 2]. It is now well known that prolonged exposure to asbestos can lead to severe health issues [1, 2]. Currently, several countries have a complete or partial ban on the use of asbestos [3]. Similarly, owing to daily use in various aspects of our lives (including life-saving devices), the production and utilization of plastic have skyrocketed in recent decades [4]. The resistance to degradation and thus longer longevity of plastic is posing tremendous challenges to waste management. Moreover, over 40% of produced plastics are employed for single-use packaging and thus generate a huge amount of plastic waste, thereby causing environmental and human health concerns. The presence of microplastics in the Great Pacific garbage patch is well documented [5]. It is estimated that humans consume 39,000 to 52,000 microplastic particles per year from food and beverages alone [6]. Moreover, these particles can also carry additives from the manufacturing process, chemicals adsorbed on the plastics, and pathogens or parasites that might be on the plastics. In this way, different contaminants are introduced to the food chain via microplastics in the environment [7]. Recent research also suggests the evidence of microplastics in human placenta [8].

Per- and polyfluoroalkyl substances (PFAS) are a large family of manufactured industrial chemicals developed in the 1940s [9]. Due to novel features such as stability at high temperature, strong chemical resistance, and water and grease repellant properties, this class of chemicals has found applications in numerous settings—from the fast food industry to military use. PFAS have been used in the military within aqueous film forming foams (AFFF) for fire training and emergency response purposes. PFAS stability relies on carbon–fluorine (C–F) bonds which are the strongest and most stable covalent bonds in organic chemistry. Not surprisingly, these chemicals are nicknamed "forever chemicals" because of their omnipresence and resistance to natural degradation. As of August 2023, there were 3,186 sites in 50 states, the District of Columbia and two territories of United States of America (USA) that are affected by the PFAS contamination [10]. The ubiquity of PFAS in the environment has been investigated in numerous studies, for example, recent research suggests widespread contamination of PFAS in freshwater fish in the USA [11]. For six PFAS compounds (perfluorooctanoic acid (PFOA), perfluorooctane sulfonic acid (PFOS), perfluorononanoic acid (PFNA), hexafluoropropylene oxide dimer acid

(HFPO-DA, commonly known as GenX Chemicals), perfluorohexane sulfonic acid (PFHxS), and perfluorobutane sulfonic acid (PFBS)), the US Environmental Protection Agency (EPA) recently announced the proposed National Primary Drinking Water Regulation (NPDWR) with maximum contamination level (MCL) of 4.0 parts per trillion (ppt) for PFOA and PFOS [12, 13]. Due to human health concerns, manufacturers in North America and Europe in the early 2000s started to phase out some long-chain PFAS, replacing them with short-chain molecules. However, short-chain PFAS have since been found to have an adverse effect on human health as well and are more mobile than their long-chain cousins [14]. Because PFAS are so chemically stable, it is not surprising that conventional methods are generally not suitable for the isolation, degradation, and destruction of these compounds, except with only limited success. Recently, several review articles have been published summarizing and comparing these efforts [15, 16]. It has been found that a majority of these efforts have been done at the laboratory scale, with some at field trials which achieved only limited success or resulted in new issues for concern. For example, the incineration of PFAS with other wastes has the potential to generate harmful compounds such as dioxins and furans. Moreover, combustion of PFAS produced byproducts such as tetrafluoromethane and hexafluoroethane which are active greenhouse gases with global-warming potentials ranging from 5,300–7,400 and 9,200–14,000 with long lifetimes of 50,000 and 10,000 years, respectively [17]. Thus, we must take care in treatment approaches that such technologies will not produce byproducts that will have long-lasting environmental and/or health impact.

Therefore, it is not surprising that innovative materials and novel technologies are under continuous development. For example, advances in additive manufacturing, multifunctional materials, remediation technologies, nanocomposite materials, and multiscale modeling for materials development are making phenomenal progress in recent times. The COVID-19 pandemic has added another dimension to the ever-growing demand for the well-being of the human population. This pandemic has shown our limitations and need to develop technologies rapidly enough to contain such a pandemic and to develop reliable medical remedies. The recent push on climate change and renewable energy will also drive development of better materials and new technologies to reduce dependency on traditional resources. As technologies mature some of these emerging materials will find their way into the environment and thus have the potential to mix with soil and water bodies depending upon their ability to be environmentally persistent. In the following sections we will discuss some environmental implications of emerging materials which currently trend toward increased use and have potential for environmental release. It is our hope that the extent of future environmental challenges can be curtailed by the awareness and predictive capability we develop in the present time for addressing potential environmental impacts of emerging materials.

1.2 The Many Facets of Three-Dimensional Printing

Additive manufacturing via three-dimensional (3D) printing has been a revolutionary technology that has seen an increased interest in research over the last few decades. Proponents of the technology report increased economic efficiency due to a reduction in scrap material, less waste, and streamlined logistics on storage and shipping of components that can be printed locally. This methodology has branched across several fields with a variety of applications, each allowing for highly customizable materials and pieces to be formed. As the use of this technology continues to grow, it is essential to know the potential applications, contaminant release, and impacts of the materials as they are printed with respect to the environment and to human health. To this end, we review the types of 3D-printing methods and applications, and then focus on fused deposition modeling of polylactic acid and acrylonitrile–butadiene–styrene.

1.2.1 3D Printing: Applications and Methods

Three-dimensional printing (3DP) technologies have existed since the 1980s, and continue to be a fast-growing field of research. The basic concept behind the technology is that a computer-aided design (CAD) is created and saved to an ".stl" file, which lists the data required to describe the design. The object can be printed layer by layer to the given specifications. In this way, highly complex geometries can be created and printed in a single object.

The spectrum of applications for 3DP is profound. The architecture and building community are working to print buildings and have found that construction equipment may easily be printed [18, 19]. Dentistry is already seeing applications in generating models to use for study, building custom implants, custom orthodontics such as braces, and materials to act as drill guides while performing surgery [20–23]. The medical field alone has a vast array of potential applications for 3DP including: models for surgical training, surgical tools, drug design and delivery, hearing aids which can be custom designed based on the ear canal, medical implants, tissues such as heart valves, and cartilage [19, 20, 24–33].

Fields such as aerospace are in a desirable position to use 3DP technologies due to 3DP's aptitude for generation of complex components that are also light weight [19, 34–36]. While extensive use of 3DP parts in aerospace has not yet been realized, there is an example where an environmental control system was able to be printed, reducing the number of required parts down from 16 to 1 [20]. The automotive industry finds use of 3DP through prototyping, engine part fabrication, and generation of tools for assembly [19, 20]. Indeed, even the fashion industry and the food industry have seen the application of 3DP [19, 37–39].

The wide variety of potential applications extends the field of 3DP research as the printing and material requirements vary greatly for different uses. For instance, in medical and dental implementations it is essential that the materials are biocompatible, and to this end, much research has been performed to determine how to reduce inflammation effects and toxicity [25, 29, 40–43]. The aerospace industry needs lightweight materials, and continues research into developing lightweight alloys for 3DP applications [36]. Custom 3DP "inks" are continuously developed in order to maximize conductive properties, minimize structural defects, and/or optimize polymers to achieve ideal physical and mechanical properties. Unfortunately it becomes difficult to determine the detailed impact of the materials used, as a plethora of the inks, thermoplastics, resins, and powder mixtures are subjects of proprietary information.

Several different 3DP approaches have emerged in order to support the vast array of applications. Stereolithography, laser sintering, and fused deposition modeling are a few popular methods. No approach is superior to the other; rather, they are each designed to fulfill different needs and functions. Stereolithography is a form of vat photopolymerization in which a liquid photopolymer is irradiated with a light source in order to allow polymerization to occur at the focal point of the light source [19, 21, 24, 29]. The polymer hardens to a certain thickness onto the platform where the light is directed. Once the layer is complete, the platform is moved to allow more liquid polymer on the site, and the process repeats for another layer. Following the completion of printing and cleaning, the material often requires further curing to transform any remaining unreacted polymer.

The selective laser sintering procedure is similar in design [19, 21, 28, 44]. In this approach, a powder is placed upon a platform, and a laser sinters the polymer as directed by the CAD design. More powder is placed upon the build site, and the process repeats in a layer-by-layer fashion transforming the powder into a solid shape and forming a 3D object.

Perhaps the most well-known approach, due to its commercial availability and use, is the fused deposition modeling (FDM) method. This approach is commonly used for commercial desktop 3D printers. In this method, a filament is heated so that it liquefies and is then released through an extruder nozzle. The polymer material then cools and hardens on the building platform. The platform is subsequently lowered incrementally, and the next layer is deposited. This layer cools, and the sequence continues resulting in an object built in a bottom up fashion, layer by layer, as dictated by the CAD [19–21, 28]. The FDM desktop 3DP, its filaments, recyclability, and emissions will be discussed in further detail in subsequent sections.

1.2.2 Advantages and Disadvantages of 3DP

Additive manufacturing, by nature, involves many advantages over its subtractive manufacturing counterpart, which has been historically favored in industry [19, 20]. In theory, waste is minimized as the material is being printed, rather than cut/trimmed with the scraps being discarded. It is thought that carbon emissions will be reduced as products can be designed around the world, described by a digital .stl file, and then sent to the locality where it is to be printed and used. Increased local manufacturing (via 3DP) would significantly reduce transportation costs over the current manufacturing paradigm. Indeed, 3DP may alter the manufacturing process by minimizing the amount of human interaction and oversight required to build an object. The 3DP approach finds its niche in its ability to generate very complex geometries that may not be available to current manufacturing techniques and may also require fewer parts to obtain a final object. Moreover, material recyclability is advantageous in that it may reduce energy outputs and reduce the amount of single-use plastics.

While the potential advantages of 3DP are extensive, several disadvantages must be taken into consideration as this emerging technique gains traction around the world [19, 20]. The same design which makes it possible to transform the manufacturing industry by creating .stl files that can be downloaded anywhere in the world for printing may be vulnerable to piracy attacks. Similarly, the technique can be used to create weapons such as knives and guns, which allows the population to potentially circumvent weapon laws. Additionally, widespread adoption of 3DP may decrease the number of traditional manufacturing jobs, while increased local production may negatively impact countries dependent on the export of manufactured goods. Lastly, polymer and resin formulations tend to be proprietary, which has several implications to different arenas of the working world. For instance, the lack of detailed compositional data makes it difficult to have transferability of feedstocks, reproducibility between experiments using different brands of filament, prediction of the effects on the environment, and the emissions profile released during the printing process (which may affect the safety of the workers). Furthermore, the lack of compositional data may also affect the efficacy of the printed product.

1.2.3 Fused Deposition Modeling (FDM)

As mentioned earlier, FDM is a form of 3DP where a filament is heated and is extruded through a nozzle. After the hot filament is deposited on the platform bed, it cools and hardens. The platform is lowered, the next layer is applied, and the cycle continues until the object has been completed. While this method allows for complex geometries to be formed, it comes at a cost. For instance, common desktop 3DP processes require an hour to form a cube of 1.5 inches, which is unquestionably slower than typical injection molding [20]. However, in the case of microreactors, the complex geometry can be directly printed, leaving behind more arduous multi-step manufacturing processes [45].

Several parameters must be taken into account when using FDM such as the heat of the nozzle, the temperature of the deposit bed, flow rates, and the diameter of the nozzle [45]. These parameters are often rigorously tested and provided by the 3DP manufacturer for the baseline use-case; however, even slight changes to the material composition or printing needs will require careful modification of the parameters.

1.2.3.1 Filaments

Many different materials and polymers can be used as filament feedstock. However, for the standard desktop 3DP the polymers tend to consist of polylactic acid, acrylonitrile–butadiene–styrene, nylon, high-impact polystyrene, and polycarbonate [46]. The most commonly used filaments tend to be polylactic acid (PLA) and acrylonitrile–butadiene–styrene (ABS). While the base compositions of each filament are generally similar (i.e., two different ABS materials from different companies are similar), there are additives such as coloring agents, plasticizers, stabilizers, and deviations in both the molecular weight and copolymer distribution that affect the desired properties [47]. Given the popularity of the PLA and ABS filaments, the remainder of this discussion will focus on common degradation of both thermoplastics, studies into the emissions given throughout the printing process, and the potential for recycling these filaments.

1.2.3.2 Polylactic Acid (PLA)

PLA is a favorite in the 3DP world due to its ecofriendliness; the filament material exhibits biodegradability and is made from renewable source materials like cornstarch [48]. Typical suggested operating parameters for the PLA filament in an FDM printer include an extrusion temperature that ranges from 180 to 210 °C [49]. The thermal degradation of PLA in oxygen-free environments does not occur until temperatures exceed 230 °C [50]. Furthermore, it is biocompatible and has a low impact on the inflammation system, which makes its application accessible for use in the medical field. It is expected to achieve merit for biomaterials such as cartilage repair, tissue regeneration, bone screws [48, 51, 52]. Further research has examined using the polymer as scaffolding, determining its degradation in vitro as well as its biodegradation pathway [48, 51]. Molecular dynamics in conjunction with experimental techniques were deployed to study biodegradation of PLA with an emphasis on chain scission and solvation [48]. They found that both effects play a role; the solvent water causes chain scission via hydrolysis, and water participates in interactions with the acidic moieties at the end of the chains. Likewise, a previous study on thermal-oxidative degradation (tested at constant temperatures between 70 and 105 °C) found that PLA has random chain scission in its degradation process [53].

PLA's degradability into naturally abundant compounds is advantageous from the perspective of low environmental hazard; however, it can be a disadvantage when looking at waste material. Methods of degradation include photo-oxidative degradation, thermal-oxidative degradation, and hydrolysis [50]. Hydrolysis requires a minimum temperature of 30 °C in the presence of moisture. On the other hand, another disadvantage of the PLA polymer is that the material has a low melting point, meaning it may deform easily [49]. This may not be ideal depending on the application of the printed part.

1.2.3.3 Acrylonitrile–Butadiene–Styrene (ABS)

ABS polymers are also commonly used for 3DP. The extrusion temperature for ABS is recommended to be between 230 and 250 °C [49]. A 1986 review on the combustion products of ABS noted that 27 compounds are released from this process [54]. However, it should be noted that two samples of ABS resulted in different levels of combustion products, highlighting the difficulty in studying expected emissions given an unknown molecular weight distribution of a standard polymer. The main chemicals emitted from the combustion studies consistently include carbon monoxide, carbon dioxide, and hydrogen cyanide, with varying concentrations depending on the experiment [54]. The release of carbon monoxide and hydrogen cyanide is cause for concern as they are each toxic. Nonetheless, these are combustion studies, and ideal usage of 3DP would not burn the filaments. The combustion products are merely presented to illustrate potential hazards of misusing the filament as well as the consequences of end products being incinerated.

Studies have been performed to investigate the thermal degradation at a temperature of greater than 80 °C over a period of time [55]. It was found that the color of the polymer did change; however, the degradation appeared to be localized to the surface of the object. The degradation process needs oxygen to occur, and after a certain depth the oxygen cannot penetrate to the internal polymer pieces. That being said, the thermo-oxidative process begins when a hydrogen abstraction occurs to the poly-butadiene, allowing for radical formation, which assists in cross-linking reactions [55], and excessive crosslinking makes the material brittle.

1.2.4 Emission Studies

Plastic polymers are far from being a new technology, and as such, much research has been done in order to determine their health and environmental risks. ABS terpolymer was ranked with a higher hazard score than most other plastics [56]. In fact, styrene itself may have an impact on the endocrine system. On the other hand, the same ranking study has no classifications for PLA, in line with the idea that PLA is a relatively safe polymer.

3DP works at temperatures high enough to melt thermoplastics and several studies have looked into the emissions released throughout the desktop printing process. Studies on airborne emissions of thermal processing of plastics found that the largest factor in thermoforming emissions depends on temperature [57]. In general, research regarding emissions from 3DP has focused on the size of particles or the presence of volatile organic compounds. Both subjects are discussed below.

1.2.4.1 Particulate Formation

One of the first reports to detail size and concentration data from 3DP was from Stephens et al. [58] in 2013. In this work they operated a 3DP in an office space to determine particle concentrations of emitted material, and used both PLA and ABS filament. They found an increase in particles over 20 nm when the printers were running versus the background before printer operation. They also found that the concentration increased for the first fifteen minutes and then leveled off. Ultrafine particle (UFP) concentrations rose to about three times that of the background measurements. The study determined that ABS had higher rates of UFP emissions but stated that both PLA and ABS should be considered high emitters of UFP. A 2015 article from Kim et al. [59] confirmed that ABS had higher rates of particle emission and determined the emission rates of ABS to be two orders of magnitude higher than PLA. They also corroborated that the concentration of particles significantly increased during 3DP. Zontek et al. [60] set out to examine emissions using ABS and PLA filament. It was established that in well-ventilated setups, a majority of the room was unaffected by the 3DP (around 75% of the room did not experience a significant increase in particle concentrations) when the feedstock was PLA. They also used ABS in a poorly ventilated environment and found that the room experienced an increase in particle concentration demonstrating the importance of the ventilation and use of an enclosure while printing. Their study also ascertained that 68% of particles fell within the UFP range and that compounds released from ABS included cyclohexane, n-decane, isocyanic acid, 1-decanol, and ethylene–propylene–diene terpolymer [60].

Expanding on the work done on PLA and ABS filaments, Kwon et al. [61] looked at emissions from PLA, high-impact polystyrene, nylon, poly(vinyl alcohol), ABS, and laywood. In all cases, they confirmed what previous studies had seen, which is that the amount of particles increases for the first few minutes of printing (between 5 and 15 min) and decreased to a lower level, still higher than that of the background, for the duration of printing. They found that when operating all filaments by the given manufacturer recommendation, their list from high to low emitters went as follows: high-impact polystyrene, nylon, ABS, PLA, poly(vinyl alcohol), and laywood. Interestingly, they also tested all the filaments at the same temperature (265 °C for the extruder and 90 °C for the bed), and the results indicated higher emissions when compared to the ideal settings determined by the manufacturer. This highlights the importance of following the manufacturers recommendations while using a 3DP as mishandling the machinery can lead to an increase in emissions. Mendes et al. [49] set out to conduct a size resolved study of the materials released while printing. Similar to the other studies, they found that emissions initiated upon printing and minimized after a brief window of time, and that most particles fell within the range of nanoparticles. Furthermore, increasing the temperature led to more particle emission. Mendes et al. [49] found that the size of the particle was dependent on the temperature at which the filament was extruded and reported mean particle sizes of 7.6 to 10 nm, significantly smaller than the other studies. The authors believe that the previous experiments had larger particles as they ran studies in typical air settings (like an office), where some of their experiments were executed in an emission test chamber.

It should be noted that investigations like those performed by Vance et al. [62], who found that emissions likely lead to agglomeration over time, have suggested that the emissions from 3DP is similar in size and concentration to typical home and office appliances such as a laser printer.

All of these studies together conclude that in a real-world setting, PLA filaments do not significantly increase particle concentrations over the original background. ABS emits more particles than PLA, and all extruder temperatures should follow the manufacturing recommendations. Emissions peak within the first 15 min and then decrease. It is recommended that printing should occur in a well-ventilated room. Excitingly, Zontek et al. [63] were able to develop a method using Monte Carlo analysis and the eddy diffusion model to determine number concentration ranges during 3DP events.

1.2.4.2 Volatile Organic Compounds (VOC)

In addition to UFP released during printing, which are known to cause respiratory issues, knowledge of the composition of the gaseous emissions would go a long way into helping determine regulatory standards for 3DP operations. To address this, several studies have been performed to achieve a deeper understanding of the content of emissions.

The study by Kim et al. [59] discussed in the previous section was one of the first publications to attempt to determine the composition of the emitted materials. In the case of the ABS filament and one of the PLA filaments, they were able to identify an increase in toluene, ethylbenzene, m-xylene, and p-xylene. Formaldehyde, acetalehyde, and isovaleraldehyde concentrations increased during printing operations for all three studied filaments (ABS and two PLA filaments).

Azimi et al. [64] attempted to quantify the degradation gases using nine different filaments which included the most popularly used filaments (ABS and PLA) as well as high-impact polystyrene, nylon, polycarbonate and other polymers. While several combinations of filaments and printer type were examined, it was found that the filament type has the most influence in determining which VOCs will be released. ABS and high-impact polystyrene were found to have styrene as the largest emitted compound, with emission rates ranging from 12 to 113 μg/min [64]. As expected, PLA had low amounts of emissions, with 4–5 μg/min of lactide being produced.

Rather than studying the emissions directly released from a 3DP, another research group heated up filaments directly and measured the gases released with a focus on ABS, PLA, polyethylene terephthalate and nylon [65]. They found that the main VOCs emitted from ABS consist of styrene and butadiene. PLA's main emission was methyl-methacrylate, while polyethylene terephthalate and nylon produced relatively few VOCs. Stefaniak et al. [66] also looked at how the printer and filament type affect the VOCs, while additionally making the point to use different color filaments to determine the role this variable played. Similar to Azimi et al. [64] they found that ABS had higher emissions than PLA. The study found that within a filament type, the color had a relatively small impact on the presence of VOCs. The study recorded acetaldehyde, ethanol, acetone, and isopropyl alcohol emitted for both ABS and PLA filaments. While 13 VOCs were identified for ABS, only 9 were detected for the PLA filaments.

The investigations outlined so far have focused on typical ABS and PLA filaments. However, additives to achieve ideal polymeric, mechanical, electronic, and other physical properties encompass a large area of modern research. Potter et al. [47] investigated the effect that carbon nanotubes, a now common additive, had on VOC emission. They found that VOCs slightly decreased when carbon nanotubes were integrated into the filament matrix. While styrene was still present, the concentrations were lower compared to the neat ABS sample. The ratios of the VOCs produced also differed between the neat ABS and the carbon nanotube ABS. The authors believe the carbon nanotubes may have surface interactions with some of the VOCs, which consequently leads to a reduction in emitted VOCs. Their findings illustrate the importance of gaining a deep understanding about the role additives in a filament have on their emission profile.

1.2.5 Health Effects of 3DP

As determined from the studies outlined during the previous sections, there are UFPs and VOCs released upon the operation of a 3DP. While it was determined that the amount of particles released are similar to other typical household products and machinery, very few studies have investigated the direct impact of the VOCs on health.

In 2017, a case study was released discussing the potential for asthma to have been caused by 3DP operations. A man started a business that operated 19 FDM 3DPs, using ABS filament [67]. Within 10 days of operation that man developed asthmatic symptoms such as shortness of breath, and coughing. Eventually the man reduced the amount of printers used, switched filaments to PLA, and introduced air purifying resources leading to an improvement of his symptoms. Five other employees at the business did not show any asthma-like symptoms throughout this process. It should be noted that the previous section discussed several studies which found that ABS released more particles than PLA. This case study may be an example of this phenomena in practice; however, the addition of a purifying system and reduction of printers does not allow for a direct comparison.

A study from Stefaniak et al. [68] investigated the effect 3DP emissions had on the cardiovascular system. To this end, they exposed rats to 3DP emissions (nose only exposure) using ABS filament. Compared to unexposed rats, the mean arterial pressure increased by 28% when exposed to 3DP emissions. It was found that while 3DP had no impact on endothelium-independent arteriolar dilation, there was an impact on endothelium-dependent arteriolar dilation, meaning that the emissions did alter microvascular relaxation.

Another issue of concern is whether 3DP parts are toxic, and if so how to detoxify them. In order to understand the relationship between 3DP-printed products and toxicity, Oskui et al. [69] performed a study where zebrafish embryos were exposed to several 3D printed parts. They used an ABS part generated from the FDM printing process. They also used two parts generated through stereolithography (STL). The composition of the material was proprietary and therefore not known. However, they used a second STL part which they treated with UV light in order to detoxify the resin. There was a slight decrease in embryo hatching when exposed to the FDM part compared to the control. The STL/UV part had a decrease of hatching larger in magnitude than that of the FDM exposed hatchings; however, the untreated STL experiment saw the largest decrease in hatchings. The hatchlings exposed to the STL part had the highest rate of malformations, and the FDM exposed hatchlings had the least, although still more than the control group. The hatchlings exposed to the STL-UV part performed better than the untreated STL part, showing the UV treatment did partially detoxify the material.

Inoue et al. [70] examined if thermal treatment could be used to detoxify parts generated from microstereolithography. After printing an object using a photo-curable polymer, they heated the objects at 175, 220, 225, and 250 °C. Their work showed that temperatures over 225 °C reduced the cytotoxicity.

1.2.6 Filament Recycling

As the use of 3DP continues to expand, it is essential to determine a route for handling the used materials. Several studies have looked at the possibility of recycling filament to reduce production costs. In one study, recycled 3DP material was gathered from the 3D printing laboratories at Taylor's University in Malaysia [71]. From the recycled materials, the white PLA waste was extracted and then shredded to a size of 3–5 mm, or roughly the size of a virgin PLA pellet. The recycled printed pieces were then compared to virgin PLA-printed materials. Within the first cycle, they found a decrease in tensile strength (50%), tensile modulus (48%), and elongation properties. The first recycled run also suffered from printing issues due to the inhomogeneity of the recycled pieces through the extruder. The second cycle saw an increase in these issues. The authors speculate that the decrease in physical properties may be related to the printing problems they experienced, in relation to extruder clogging from the recycled filament not melting properly. In addition, they found the melting temperature to slightly increase. They concluded that only one round of recycling is appropriate for PLA filament.

On the other hand, a study by Woern et al. [72] looked at the recyclability of several filaments. They used fused particle fabrication with virgin PLA pellets and four recycled polymers, which were ground up for use and included PLA, ABS, polyethylene terephthalate, and polyporpolyne. It is important that the material feed well into the extruder as issues with printing decrease the physical properties of the piece being printed (as speculated in the previous study). They found that reground polypropylene and PLA had the largest variation in size, but they still fed into the extruder well. Despite the mix of sizes and colors, the authors found that recycled PLA, extruded with the same parameters as virgin PLA, resulted in similar tensile strengths; the tensile strength only differed by 2.5%. Similarly, they found that pieces printed with recycled ABS had a tensile strength that falls within the correct range for FDM printing. Furthermore, there was little change in the nature of the recycled polyethylene terephthalate, with a tensile strength of 40 MPa.

Mohammed et al. [73] investigated recycling ABS for multiple cycles, starting with virgin ABS. They examined blends of one time recycled ABS with virgin ABS. The melt flow decreased when recycled filament was used, and so higher temperatures had to be applied while extruding. The authors believe that thermal-oxidative stress is degrading the polymer, and as such, requires higher temperatures when reusing. In the blends it was found that a mixture of 10% recycled ABS can be used and keep the same flow properties as the virgin ABS.

Mohammed et al. [74] also investigated the possibility of recycling ABS from electronic waste. They separated out ABS plastics from electronics that had been discarded and used it through a melt extrusion device constructed in-house. Their work showed that the recycled ABS was successfully used to print a working pipe connector despite some surface defects. Another study on recycling electronic waste plastics was performed by Gaikwad et al. [75] They performed three extrusion cycles using the material gathered from electronic waste. The first recycled filament extrusion was successful as they found the breaking strength of the recycled object to be 76% that of the virgin ABS. The tensile strength of the recycled 3D-printed product was 83% compared to the virgin ABS. These properties drastically decreased after the second recycling.

In order to rectify some issues that come about when using recycled filament, Zhao et al. [46] examined the effect of an adhesion promoter used with recycled PLA. To this end, they dipped recycled PLA into polydopamine. While the thermal stability decreased due to the coating, the tensile strength showed improvement. The potential for recycling 3DP filament and other thermoplastics is high. As illustrated by the investigations detailed above, this is an active area of research and is finding several potential solutions which are being explored. The degradation of these materials makes recycling difficult, and the unknown impact of the additives on large-scale batch recycling poses further challenges. While the work discussed is very promising, the field merits continued study.

1.3 Release of Nanoparticles from Nanocomposites

The inclusion of a myriad of nanoparticles in various matrices has been rapidly increasing across the aerospace, defense, automotive, medical, and consumer product industries in the last two decades. Nanoparticles are 0D, 1D, 2D, and 3D materials with at least one dimension less than 100 nm, and they tend to possess remarkable electrical, thermal, and mechanical properties. For example, graphene, a 2D nanomaterial consisting of a single sheet of carbon atoms arranged in a hexagonal pattern, is the strongest material on Earth with a tensile strength of 130 GPa and a stiffness of 1 TPa. Unfortunately, structures cannot be built solely out of nanoparticles, so there has been tremendous interest in including these nanomaterials as reinforcements in polymer, metal, and ceramic matrices to improve the thermal, electrical, and mechanical properties of the matrix.

Tremendous time, effort, money, and resources have gone into using these nanoparticles to create vastly more capable materials than ever before. However, nanoparticles can cause unforeseen problems if they are released into the various environmental matrices (air, soil, water) due to their own toxicity and the transformation of their toxicity via processes like photochemistry, oxidation, biotransformation, adsorption of natural organic matter, and abrasive forces [76].

As the usage of nanoparticles increases, so does the risk of potential release of nanoparticles from their host matrices. To more effectively design composite materials and reduce potential contamination, it is necessary to determine the mechanisms of release of nanoparticles from composites. This section focuses on a review of the literature pertaining to the release of carbon nanotubes, graphene, TiO_2, and ZnO from nanocomposites due to the following release mechanisms: machining, weathering, disposal, and the effect of solvents. These mechanisms are split up into their own sections for ease of perusal by the reader.

1.3.1 Release of Reinforcement by Machining

Machining refers to the various mechanical processes that the nanocomposites are exposed to during/post use such as cutting, sanding, grinding, and drilling. Each of these processes has different effects on the stability of the nanocomposite and the potential release mechanisms of the nanoparticles. As the nanoparticles are released, the nanocomposites inevitably become less stable until failure occurs. Factors such as brittleness, surface topology, and hardness all affect the nanocomposites when exposed to these mechanical processes [77]. Understanding how the mechanical processes affect the release of the nanoparticles can help researchers predict which nanocomposites are best suited for the goal of their project and help reduce potential contamination through a reduction in released nanoparticles. Studies on release from these mechanical processes are explored more below.

1.3.1.1 Cutting

Bello et al. [78] also investigated release of carbon nanotubes (CNTs) from a polymer matrix laminate and an alumina fiber cloth due to machine cutting with a dry band-saw and a wet rotary cutting wheel. Four sets of samples were made: (1) epoxy/CNT laminate, (2) epoxy laminate (control), (3) alumina fiber cloth with CNT, and (4) alumina fiber cloth (control). Varying thickness samples were manufactured to investigate the effect of thickness on release. On a per sample basis, 9–15 cuts/release measurements were made. According to the results, ultrafine particles were always released, no matter the type of composite nor whether it was the CNT-filled or the control. The thicker samples generated more released particles than the thinner samples. The study does not report a matrix-dependence on release rate; that is, the authors do not mention if the alumina fiber cloth or the epoxy laminate released more particles than the other. Dry cutting samples with the band saw tended to release more particles than by wet cutting. Sizes of released, aerosolized particles were as follows: 71–89% were around 110 microns, 6–25% were around 0.1–1 micron, and 1–10% were less than 100 nm. The authors used scanning electron microscopy (SEM) and transmission electron microscopy (TEM) to determine the structure of released particles; they did not find any CNTs in the released particles, but they did observe submicron fibers.

1.3.1.2 Drilling

Bello et al. [79] also investigated release of CNTs due to high-speed dry drilling from two types of laminates: one made of alumina and epoxy and another made of epoxy and carbon fibers, both reinforced with aligned CNTs. For the alumina/epoxy/CNT composite, the CNTs were aligned radially from the surface of the alumina fibers. For the epoxy/carbon fiber/CNT composite, the CNTs were aligned in the laminate thickness direction (traditionally labeled "z") and placed in the middle of the laminate. Compared to Bello's cutting experiment [78], drilling created fewer breathable particles, but more released particles overall (3.9×10^6 to 1×10^7 particles/cm^3 for drilling vs. 2×10^4 to 6×10^6 particles/cm^3 for cutting). Sample thickness did not change release dynamics when drilled. Unlike the cutting study, CNT aggregates were detected among the drilling-released particles. The contrast of these two studies shows how changing the release scenario can have a significant impact on release dynamics.

1.3.1.3 Grinding

Ogura et al. [80] investigated release of CNTs from a polystyrene matrix due to grinding. The study used condensation particle counting (CPC) to detect release of particles from the polystyrene/CNT (5 wt% CNT) composite and the polystyrene control. Both the polystyrene/CNT and polystyrene control released a "considerable" amount of particles. It was found via SEM that the particles released were micron and nanoscale fragments of polymer matrix with protruding CNTs embedded; no free CNTs were observed.

1.3.1.4 Sanding/Abrasion

Wohlleben et al. [81, 82] investigated release of CNTs from cement and thermoplastic composites (controls included) due to sanding. Their experimental setup was designed so that the only particles detected would be from the sanding of the composites. All sanded composites released nanoscale debris. For the cement composites, both with and without CNTs, most of the released debris was less than 100 nm (as determined by scanning mobility particle sizer or SMPS). The thermoplastic composites released particles around 2 microns. SEM images of the released particles from both the CNT-containing composites and the controls showed no free CNTs.

Cena and Peters [83] investigated release of 2 wt% multi-walled CNTs (MWCNTs) from epoxy due to sanding (3M 220 grit). In contrast to Wohlleben et al.'s [81, 82] confined test apparatus, Cena and Peters had an operator manually sand sticks of epoxy/MWCNT composite with measurements of released particles (CPC and optical particle counter or OPC) taken in the vicinity of the sander and the air close to the operator. The difference in release between the two measurement locations was insignificant; background levels of particle concentrations were similar to the measurement locations. TEM results show that release did actually occur: MWCNT-like structures embedded and protruding from particles larger than 300 nm. However, no free CNTs were detected.

Huang et al. [84] also investigated release of MWCNTs from sticks of epoxy due to sanding (medium grit, which is considered somewhere between 60 and 100 grit). All sticks exhibited release of nanoscale debris in a bi-modal distribution: less than 100 nm and greater than 500 nm. TEM showed evidence of free MWCNTs as well as MWCNTs embedded in and protruding from the epoxy matrix.

Schlagenhauf et al. [85] investigated release of MWCNTs from an epoxy/-MWCNT composite due to abrasion by a Taber Abraser. Samples with 0, 0.1, and 1 wt% MWCNT (balance being epoxy, Epikote 828LVEL or diglycidyl ether of bisphenol A, and hardener, Epikure 3402) were made. When abraded with the Taber Abraser (wheel H-18), all samples produced nanoscale debris; four modes of release were identified, but less than 100 nm was not one of them. The control sample released smaller debris than the 0.1 or 1 wt% MWCNT samples; the debris released by the composites was 70–90 nm larger than the control. Though TEM revealed free MWCNTs and MWCNT bundles in the 1 wt% MWCNT/epoxy debris, free and bundled MWCNTs were not observed in the 0.1 wt% MWCNT/epoxy debris. It was found in another study that increases in abrasion speed and applied force causes increases in nanoscale debris and that the amount of release was shown to be dependent on grit size [86].

1.3.2 Release of Reinforcement by Weathering

1.3.2.1 UV Exposure

Wohlleben and Neubauer [87] determined that the matrix of a weathered nanocomposite affects release of reinforcement more than the type of reinforcement; the type of matrix changed the release rate by a factor of 100,000 while the specific type of reinforcement within the same matrix changed the release rate by a factor of 10. These factors, attributed to particles (reinforcement) distributed throughout another different and solid material (matrix), will only be able to escape if that material undergoes significant degradation. The authors speculated as to the possible mechanism by which the reinforcement changes the release rate. One possible mechanism is that damage to the matrix could be avoided by the reinforcement absorbing UV light. However, if the reinforcement is a photocatalyst (as in the case of ZnO), photodegradation of the matrix could be accelerated [87].

Nguyen et al. [88] investigated release of MWCNT due to UV exposure from an epoxy/CNT (0.75 wt% CNT) nanocomposite. They discovered that MWCNTs may be able to shield the epoxy matrix from further UV damage by forming a dense, entangled layer on the surface of the composite. Photodegradation causes epoxy molecules to be removed from the surface of the composites, exposing the MWCNTs. The MWCNTs eventually form a dense, entangled layer after enough polymer matrix has been removed via UV radiation. This process may be applicable to all high aspect ratio CNTs. Nguyen et al. [88] did not observe any release of MWCNTs throughout their experiment. Similar results were found by Wohlleben et al. [81] for thermoplastic polyurethane (TPU)/CNT (3 wt% CNT) composites subjected to both dry and wet weathering; the authors report no release of CNTs, but there were CNTs protruding from the composite surface which could be released from the matrix by shear stresses [89].

Bernard et al. [90] investigated the behavior of graphene oxide (GO) in a polyurethane (PU)/GO (2 wt% GO) nanocomposite due to UV exposure. The authors found that GO behaved in a similar manner as the MWCNT in Nguyen et al.'s [88] study: as the polyurethane matrix photodegrades, the GO forms an increasingly dense, entangled network on the new surface of the composite. Determining release of GO was not attempted in this study.

TiO_2 and ZnO degrade when exposed to UV, enhancing their toxicity through different mechanisms. TiO_2 creates hydroxyl radicals ($^{\cdot}OH$) on the surface of the nanocomposite that can react with the surrounding environment and degrade the matrix of the composite [91]. The doped Ti can enter cells when released from the matrix and cause apoptosis and necrosis [92]. ZnO composites exposed to UV result in the formation of Zn^{2+} ions which degrade the matrices holding the nanoparticles and can also react with the surrounding biota [93]. Zn^{2+} ions accumulate in aquatic vertebrates, plants, and microbes, which is the main source of its reported toxicity. Further, recent research has shown that lab-tested composite degradation with exposure to UV underestimates the actual rate of ion release in Zn and Ti composites where researchers found that the rate was increased when exposed to sunlight rather than lab UV [94].

1.3.2.2 Weathering and Mechanical Stress

Hennig et al. [95] investigated the effect of UV light and mechanical stress on the release rate of MWCNTs from epoxy. It has been shown in the literature that photodegradation due to UV radiation accelerates the degradation of polymer matrices and can cause embedded nanoparticles such as single-walled CNTs (SWCNTs) and MWCNTs to be released [77, 89, 96–101]. Mechanical stresses in addition to UV radiation are expected to increase the release rate of CNTs further. Hennig et al. [95] manufactured two sets of MWCNT/epoxy samples and exposed them to different weathering and mechanical processes. The first set (+SSR, simulated sunlight radiation) was exposed to UV radiation for 30 days, then mechanically stressed in three different tests, then exposed to UV radiation for a further 60 days, and then mechanically stressed in the same three tests once again. The weathering process consisted of exposing only one side of the samples to an irradiance of 50 W/m^2 at wavelengths of 300–400 nm; irradiation under these conditions for 30 and 60 days is equivalent to exposure to natural sunlight at 50° northern latitude in Europe for 4.3 and 8.7 months, respectively. The second set of samples (−SSR, no simulated sunlight radiation) was mechanically stressed in three different tests once and then mechanically stressed in the same three tests once again. After each mechanical stress test, released material was gathered. Hennig et al. [95] found that release of MWCNTs from the +SSR samples due to both irradiation and mechanical stress was 22 times higher than for the -SSR samples which only underwent mechanical stress. Under SEM, the unlabeled, nonirradiated epoxy/MWCNT plates showed smooth surfaces with few indentations and protrusions. The SEM images of the plates irradiated for 30 and 60 days showed increasing signs of degradation of the matrix with round structures up to 5 microns in diameter and with visible protrusions. Plates irradiated for 30 days and then mechanically stressed with all three tests showed more evidence of damage than the plates irradiated for 90 days with no mechanical stresses. This suggests that composites in service under mechanical stress and UV radiation may release more nanoparticles than those composites exposed to UV radiation alone. TEM images of released epoxy/MWCNT fragments (E, F, G) showed that MWCNTs were both embedded and protruding. The longest protruding MWCNT (measured from the start of the protrusion) was 200 nm long.

Hirth et al. [89] investigated the combined effect of weathering and mechanical stress on the release of CNTs from a TPU/CNT nanocomposite. TPU/CNT samples were dry- and wet-weathered (humidity of 50 ± 10 %) the equivalent of nine months in Europe at 50° northern latitude, then shaken in a container of surfactant solution at 5 Hz, 1 cm amplitude for 24 h, and finally ultrasonicated for 30 s. The surfactant solution (after being shaken and sonicated) changed colors based on whether the control (yellow) or the nanocomposite (blackish gray) was submerged. Carbon reinforcements are black which means that CNTs were released into the surfactant solution.

Samples of the dried solution were analyzed by SEM, and the results show that not only were embedded, protruding CNTs released, but also free CNTs were released from the nanocomposite. Release of free CNTs and particles less than 150 nm in size from the weathered nanocomposites (wet or dry) is directly proportional to shear stress: as shear stress increases, release increases.

1.3.2.3 Aqueous Weathering

The presence of TiO_2 and ZnO nanocomposites in aquatic environments and the potential toxicological effects have been well documented in literature. These composites slowly release nanoparticles and ions into the surrounding environment where they are absorbed by the local biota. This release negatively affects multiple facets of the habitat, resulting in cytosis, mutation, and many other negative effects on the biota. The degradation rate depended on multiple environmental factors such as pH, electrolyte valence, and humic acid. In general, byproducts decreased with an increase in pH or zeta potential [102]. The degradation rate of ZnO nanocomposites was mostly determined by the pH of the area with electrolytes and salts having lesser effects [103]. Ultrafiltration has been shown to effectively remove nanoparticles from solution [104]; however, this is too expensive for large-scale use.

While there is an understanding of what these nanocomposites release into an aquatic environment and the negative effects that can result, there is a significant knowledge gap involving the actual dissolution of the nanocomposites. Multiple researchers cite difficulties experimentally simulating the complex aquatic matrices that exist in nature and instead test one factor, such as pressure, flow, pH and ignore the effects that different animals, plants, and bacteria have on the release and absorption of the contaminants.

Zanna et al. [105] investigated the release of silver nanoparticles (Ag-NPs) due to immersion in a saline solution from an organosilicon/Ag-NPs coating of stainless steel pipes. They tested two different coatings: one with a low amount of Ag-NPs (7.4%) and one with a relatively high amount of Ag-NPs (20.3%). These two coatings were applied to stainless steel pipes and then immersed in a 30 °C, stirred 0.15 M NaCl saline solution for 2, 3, 4, 8, 18, and 60 days to age the coatings. The retention of coating thickness in a saline solution appears to be dependent on the Ag-NPs loading: the 7.4% Ag-NPs-loaded coating retained its thickness throughout the aging process while the 20.3% Ag-NPs-loaded coating lost most of its thickness by day 18 and then remained constant. Most of the Ag-NPs in surface layers of the low Ag-NPs coating released within the first three days and release stopped by day 18; the inner layers of the low Ag-NPs-loaded coating still had Ag-NPs. The high Ag-NPs-loaded coating experienced significant Ag-NPs release and matrix degradation in the first few days, and then continued until day 18. By day 18, most of the coating thickness was gone, leaving a paltry amount of Ag-NPs remaining in the vastly reduced coating.

Kaegi et al. [106] investigated release of Ag-NPs due to rain from exterior paint applied to expanded polystyrene panels. Ag-NP-imbued white acrylic paint was applied to expanded polystyrene panels and then placed on the west-facing side of a model house; the panels were left on the model house for one year. There were three main conclusions: (1) of the total Ag-NPs in the paint, 30% were released; (2) of those 30% Ag-NPs released, 80% were released in the first two months; and (3) released Ag-NPs analyzed via TEM appeared to still be bound to the paint matrix.

1.3.2.4 Release Due to Contact and Washing

Ag-NPs and TiO_2-NPs are often included in composites that contact the human body in some way (textiles, medical devices like catheters, dental fillings, etc.) either because of their potent antibacterial properties [107, 108] (Ag-NPs) or their UV protection (TiO_2). Contact between human sweat and textiles with embedded Ag-NPs can cause release of Ag-NPs via a leaching mechanism. This leaching mechanism appears to be dependent on a variety of factors including initial loading of Ag-NPs, textile quality, and pH, as shown by Kulthong et al.'s [108] study using artificial sweat formulations. The released Ag-NPs may be in particulate form or in dissolved form. TiO_2-NPs have been shown to leach into artificial sweat solutions as particulates and not in an ionic form [109].

As previously mentioned, Ag-NPs possess an antibacterial property and are often used in textiles to protect against bad odors. Unfortunately, it has been widely shown that Ag-NPs, in both particulate and dissolved ion form, wash out of the textiles they are embedded in when laundered [110–114]. TiO_2 has also been shown to wash out of textiles, but not to the extent that Ag-NPs do and only as particulate matter [115]. The amount and type (particulate, ionic, salt) of Ag-NPs released per wash are dependent on several factors including the textile manufacturing process, the specific textile (cotton vs. wool vs. nylon, etc.), washing solution (tap or distilled water, detergent, bleach), pH, and washing history [111, 114].

1.3.3 Release Due to Disposal

1.3.3.1 Waste

Nanocomposites exposed to waste conditions (landfills, sewers, etc.) have various release dependencies contingent on the pH, concentration, and ions present in the environment [116, 117]. On average, TiO_2 leaches into the environment over 12–14 h before reaching a steady state of equilibrium [118]. The Ti clusters remained on the surface even when exposed to different pH conditions and ionic strengths. ZnO composites retained 80–90% of the original nanoparticles over a similar period of time [119].

The closer the environment is to the isoelectric point of the nanoparticle, the more aggregated the nanoparticle would remain and fewer contaminants would be released into the environment [120]. Optimal pH for Ti and graphene are 6 and 2 respectively [121, 122]. An increase or decrease in the pH from the isoelectric point resulted in an increase of the rate of dissolution of the nanocomposite. The concentration of the nanoparticle in relation to the matrix also affects the dissolution and release of the nanoparticles into the environment. Researchers observed that the more concentrated samples of nanoparticles were more likely to aggregate than dissolve into the environment [123]. As the concentration decreased, the equilibrium shifted, and it became more likely that Ti^{2+} ions were released rather than the formation of TiO_2 aggregates.

The presence of ions also affects the dissolution rate of the nanocomposites. Researchers observed that the presence of Na^+ and Cl^- increased the rate of aggregation of the nanoparticles up to a maximum concentration where that rate would plateau [124]. If a stronger electrolyte was used, the rate could be further enhanced. This enhancement is attributed to the Zeta potential on the surface of the composites that can facilitate the movement of ions across the surface of the nanocomposite [121].

1.3.3.2 Incineration

Bouillard et al. [125] investigated the release of MWCNTs from an acrylonitrile–butadiene–styrene (ABS)/CNT (3 wt% CNT) composite due to incineration. At an incineration temperature of 400 °C, they found free MWCNTs as well as bundles of MWCNTs released in the combustion gas. The size of the released MWCNTs (12 nm in diameter, length of 600 nm) is close to the size they were (10 nm in diameter, length of 1–10 micron) when the composite was manufactured. The presence of MWCNTs in the combustion gas is likely due to a low incineration temperature. Sotiriou et al. [126] report that, at incineration temperatures of 500 and 800 °C, no CNTs (0.09 wt% CNT in a PU/CNT composite) were present in the combustion gas. However, they do go on to say that their equipment may not be accurate enough to detect such a low CNT concentration in the gas phase.

A measure of the effectiveness of incineration is through measuring the oxidative potential of the combustion byproducts which indicates the capacity of the byproducts to oxidize specific molecules after exposure [103]. If the presence of the nanocomposites in question does not increase the oxidative potential (OP) of the byproducts, then incineration can be considered as a potential method for removal of nanocomposites from waste and other products. The incineration of composites containing Ag and TiO_2 significantly affected the OP of the byproducts [127]. Ag nanocomposites resulted in the formation of Ag-NPs and silver sulfide (Ag_2S)-NPs in the affluent after incineration. Meier et al. [128] showed that Ag(0) was the major component in the ashes post incineration. The byproducts of TiO_2 incineration remain unclear. However, current procedures associated with incineration including filtration, pressure drops, and ash collection result in the removal of the majority of the NPs present in the waste. This results in incineration being one of the best methods for removing NPs from the environment at this time [103].

1.3.4 Release Due to Solvents

Different solvents have various effects on the release mechanisms of the nanocomposites. Researchers observed that organic solvents encouraged aggregation of TiO_2 and ZnO nanoparticles. Particularly, solvents with increasing amounts of aromaticity resulted in increased rates of aggregation of the released nanoparticle [129]. Hydrophobicity had a similar effect on the reduction in toxicity of these nanoparticles. The effect on aggregation is attributed to the quenching of the reactive oxygen species in the case of TiO_2 and the reduction in ion release in ZnO.

As mentioned earlier, solvents containing ions also improve the aggregation of the released nanoparticles through the increase in the zeta potential on the surface of the composite. The presence of Cl^- and other anions increases the zeta potential of the composite surface, reducing any potential ion release of the composite. With reduced ion release, most released nanoparticles will remain as TiO_2 or ZnO and aggregate within the solution.

1.3.5 Other Important Nanoparticles of Note

Aside from the nanoparticles mentioned in this section, there are numerous other nanoparticles in use, including SiO, clay, boron nitride nanotubes (BNNTs), CeO, CuO, and many more. Some of these nanoparticles have been partially studied but each has their own unique purpose. For example, BNNTs have unique potential in aerospace applications due to the ability of Boron to absorb neutron radiation [130]. However, multiple toxicity studies have shown BNNT residue to be toxic to biota, resulting in cytosis and mutations in cells [131]. Some research has shown that the size of the nanotubes have a role in the toxicity of BNNTs but limited research exists otherwise, with little to none about release mechanisms of the nanoparticle from the nanocomposite. Further, these composites mentioned in this section are usually stable and exist in the environment for a significant amount of time while undergoing degradation and ion release, continually leaking contaminants into the environment through the mechanisms described in this section. There are some hypothesized methods for the removal of the nanoparticles including pH tuning, ion addition, and biomolecules such as kaolin clays [132] but many questions remain about the fundamentals of nanoparticle release before efficient recovery methods can be determined.

1.4 Biomaterials and Environmental Release

A biomaterial is any material used to make devices to replace a part or a function of the body in a safe, reliable, economic, and physiologically acceptable manner [133]. Indeed, this a is large and varied class of materials, as we will demonstrate through the entirety of this section. Biomaterials are in extensive use in the modern world including sutures, prosthetics, catheters, implants, and tooth fillings to name a few examples. However their utility does not stop with such well-known examples that we may encounter in our daily lives. In fact, with so many potential uses for biomaterials, the scope is nearly infinite, as clever materials and applications are developed and evolving in real time.

A good biomaterial must satisfy several conditions depending upon the type of biological substrate it is chosen to interface with. First and foremost, a good choice of biomaterial must satisfy the biocompatibility requirements of a given system. Consideration must be made to the potential toxic breakdown products of the biomaterial as to not harm the organism it is meant to aid. The material must not induce any detrimental side effects, such as an inflammatory immune response that could harm both the system as well as the biomaterial.

Assuming the aforementioned conditions are met, the biomaterial must also satisfy several physical properties to accomplish its desired goal. Biomaterials such as bone and joint replacements are often considered based on their mechanical strength. For these applications metrics such as the modulus of elasticity, compression modulus, tensile strength, and corrosion resistance are all properties taken under consideration. Polymeric materials are often desired for a given molecular weight, or appropriate relative viscosity and porosity. Fine tuning of such mechanical properties is often achieved by modifying well-known biomaterials or combining existing materials. These derivatives and combinations lead to an immeasurable number of possibilities for biomaterials; several general classes of biomaterials will be discussed in the following subsections.

1.4.1 Bioceramics

Ceramics have been used extensively in the medical field. Given the porosity of ceramics as well as the relative inertness of the materials in contact with tissues, they are often used as artificial bone grafts [134]. Ceramic materials used in tissue implantation are referred to more specifically as bioceramics. Common bioceramic materials include aluminum oxides, bioactive glasses, hydroxyapatite, calcium sulfate, and tricalcium phosphate [135].

While these materials are useful on their own, recent research has focused on composite ceramic materials. Motivation to manufacture less costly materials as well as fine tuning of mechanical properties has driven much of the recent innovation. Prabha et al. [136] developed strontium substituted bioceramic scaffolds as a good load bearing application for orthopedics and dentistry. Bioactive glasses have been in use for decades due to their ability to bind to both bone and soft tissue [137] as well as pronounced biocompatibility [138]. Commercial materials like Bioglass 45S5 composed of calcium sodium phosphosilicate are available for bone grafting applications [139]. Interestingly, Oonishi et al. [140] found that the rate of bonding to bone can be controlled by modifying the chemical composition of the bioglass. For an excellent review of the current state of ceramic materials used in dental implants, see Li et al. [141].

Bioceramics in addition to being biocompatible have the benefit of biodegradability. This property has led to potential uses as a drug delivery system [142]. Mesoporous silica nanoparticles (MSNs) are unique due to the porous structure which can be tuned in size to between 2 and 50 nm [143]. In principle, a smart delivery system could be realized by loading the material with active therapeutic agents and systematically releasing them at their target via some external stimuli. Zhu et al. [144] investigated an MSN polyelectrolyte multilayer (PEM) aptamer in which the aptamer could target cancer cells with high recognition. This proof of concept has been further extended with the use of MSNs to deliver other drug molecules such as antimicrobial peptides [145].

A growing class of bioceramic materials make up a class of composite biomaterials. While metallic alloys, commonplace in human bone and joint replacement, are prized for their structural strength, the functionality can be further improved in conjunction with ceramics. Generally speaking, conventional alloys such as stainless steels, alumina, zirconia, and titanium to name a few are considered bioinert. As a result, they tend to form layers of fibrous tissue at the material–bone interface [146]. Additionally, medical devices and prostheses undergo wear in the body resulting in particulate matter which will trigger osteolysis (bone degradation) and ultimately aseptic loosening of the prosthetic. Ceramic coatings aim to mitigate these issues by enhancing the binding of the prosthetic to the bone tissue as well as enhancing the durability of the metal. A variety of techniques have been developed in recent years to apply bioceramic materials to alloys. Some of these methods include laser surface engineering [147], ion beam sputter coating [148], and thermal spraying [149].

1.4.2 Metals

The malleability, density, ductility, and strength of metals make them ideal biomaterials for a variety of applications. Metal components are often used as medical instruments, medical devices, joint replacements, and implants. Some of the most common metals used to manufacture implants are iron, chromium, cobalt, nickel, titanium, tantalum, molybdenum, vanadium, and tungsten [150]. While these metals are used individually, they are much more commonly deployed as alloys. Nonetheless, while metals have tremendously advantageous properties as outlined above, the principal downside is corrosion in the body which will produce physiological side effects and ultimately failure of the implant in the body.

Alloys make for attractive biomaterials due to the tunability of the material based on the composition. Historically, stainless steel was one of the first alloys utilized for biomedical implantation and was shown to have better corrosion resistance than elemental metals [151]. Precautions must be taken with stainless steel to maintain low nickel content, as this is hazardous to the body. Titanium and its alloys make for better alternatives as these materials show both a greater strength-to-weight ratio and greater biocompatibility. Other common alloys include cobalt-based materials, particularly Co–Cr–Mo has demonstrated high elastic modulus and greater density and stiffness [152]. Another biocompatible metal, tantalum, has found utility as a biomaterial. Wang et al. [153] analyzed tantalum oxide nanotube films and found them to increase protein adhesion and adsorption of rabbit bone mesenchymal stem cells. Porous tantalum-based implants have been used for shoulder arthropasty due to the nature of the material functioning well for proper osseo-integration [154].

Metallic alloys are important biomaterials not just for rigid prostheses and joint replacements. Shape memory alloys are materials described by their ability to return to their predefined dimensions upon heating, induced by deformation training [155]. This shape memory effect is present in a number of alloys including, Au–Cd, Cu–Zn, NiTi, InTl, NiAl, FePt, FePd, MnCu, and FeMnSi. Nickel–titanium alloy is the most widely used shape memory alloy due to its pseudoelasticity as well as its excellent biocompatibility [156]. These properties have made NiTi an ideal material for use as a vascular stent.

Recent advances involve more exotic materials as researchers look to optimize the many factors that make an ideal biomaterial. High-entropy alloys (HEA) are generally defined as an alloy formed by five or more metals and have also been investigated as potential biomaterials [157]. Todai et al. [158] examined the biocompatibility of a novel titanium-based HEA (TiNbTaZrMo) and found it to be superior to its elemental counterpart. Yuan et al. [159] investigated an HEA composed of TiZrHfNbTa and found the material had a unique combination of low elastic modulus and magnetic susceptibility while showing good biocompatibility. Due to the large number of potential HEA materials, it is likely these compounds will have an expanded role in biotechnology as the cost of manufacturing goes down and better fabrication techniques are discovered.

Yet another emerging class of metallic biomaterials are the so-called bulk metal glasses (BMGs). BMGs are defined as "amorphous alloys that exhibit a glass transition, from which derives their unique properties of extreme strength at low temperature and high flexibility at high temperature along with thermodynamic and physical properties that change abruptly at the glass transition temperature (T_g)" [160]. The advantages BMGs hold over crystalline metal structures are numerous and include greater corrosion resistance, greater flexibility, and greater control over the elemental composition of the metal. While this is a newer area of materials, preliminary results show these materials have good biocompatibility. For instance, Szyba et al. [161] looked at the break down of $Ca_{57.5}Mg_{15}Zn_{27.5}$ under physiological conditions and found the break down products to be harmless to the human body. Similarly, a commercially available BMG $Zr_{44}Ti_{11}Cu_{10}Ni_{11}Be_{25}$ investigated by Sawyer et al. [162] showed that by utilizing ceramic conversion treatment the material could be modified into biologically inert layers at the surface.

Control of the degradation of metallic implants is central to the idea of a biodegradable or bioresorbable metal. The challenge with this approach is finding materials that meet mechanical specifications while remaining bioinert in the body. Currently magnesium alloys have been patented for use as heart valve stents [163], as the material will safely absorb in the body making removal no longer a problem. A critical issue with this approach is control of adsorption rates. Kirkland et al. [164] concluded that magnesium alloys could be produced that controlled dissolution rates.

1.4.3 Synthetic Polymers

Polymers composed of repeated monomeric units offer unique functionality for a variety of biomaterial applications. Ultimately, the desired end use of the polymeric biomaterial will dictate the choice of the polymer composition. As usual, the material must be biocompatible, nontoxic, and noninflammatory. Polymeric materials have found many uses as biomaterials including scaffolds for tissue engineering, prostheses, and medical devices [165].

Many of the commonly available manufactured plastics have been incorporated into biomedical devices. Polyethylene (PE) is the world's most produced plastic due to its low cost of manufacturing and versatility, ranging from ultrahigh molecular weight to medium and low-density PE [166]. These terms refer to a range of density for the given material and are controlled by the amount of branching and cross-linking present in the polymer. Low-density PE has a lower tensile strength and can be found in sterile medical packaging and films. High-density polyethylene (HDPE) has a high density-to-strength ratio and has been deployed as a wear-bearing surface of hip and knee arthroplasty and total joint replacement [167].

Polypropylene (PP), similar to PE, can be altered in density to control for the strength of the material. PP is widely used as a surgical mesh scaffold to grow fibro-collagenous tissues as well as a barrier to treat stress urinary incontinence and pelvic organ prolapse [168]. While PP makes for a potentially excellent biomaterial, questions remain about the materials biocompatibility. For instance, implants fabricated from PP induced an inflammatory response when grafted to mice [169].

A number of acrylic polymers are manufactured. Poly(methyl methacrylate) (PMMA) known commercially as Plexiglass is one of these materials. PMMA is an important biomaterial with several uses. PMMA bone cement is used in surgical procedures to fix prosthetics to bone. Traditionally, eye glass lenses have been made out of the material as well as hard contact lenses. Within the body, PMMA is the material of choice for artificial teeth and dentures [170]. PMMA is under consideration for a variety of microfluidic devices, the so-called lab-on-a-chip devices. Microfluidic devices take advantage of liquids and gases chemical properties at the microscale. Recently, PMMA demonstrated proof of concept on a microfluidic H-filter and droplet generator [171]. The cost of the material coupled with improved fabrication methods should increase the number of PMMA devices in use.

Fluorocarbon polymers are a large number of organic polymers composed of a chain of carbon atoms with fluorine atoms appended. Due to the presence of many highly stable carbon-fluorine bonds, these compounds are extremely resilient to degradation by light and chemically inert. Polytetrafluoroethylene (PTFE), known by the trade name Teflon, is the most widely used of these compounds. Other important materials in this family include polyvinyl fluoride and polyvinylidene fluoride (PVDF) as well as the chloride substituted polymer polychlorotrifluoroethylene (PCTFE). The hydrophobicity and low coefficient of friction for these materials make them excellent for nonstick coatings, and PTFE is frequently deployed as a catheter coating [172]. PTFE is also widely used as a surgical arterial graft particularly for hemodialysis access operations [173]. PVDF has attracted attention as a potential medical textile, a scaffold used to reinforce tissue. The polymer shows good biocompatibility, and both the porosity and elasticity can be controlled by thermal treatments [174]. PCTFE while not used as a direct biomaterial is frequently used in pharmaceutical packaging as a moisture barrier [175].

Other important synthetic polymers include the polyamines such as nylon. Nylon in particular is a thermoplastic silky material composed of aliphatic or semi-aromatic polyamides. Nylon is used as a wound suture material sold under the name Ethilon [176]. Additionally, nylon with its excellent biocompatibility and biodegradability make it an excellent material for tissue engineering applications. Nylon has a high tensile strength which has made the material ideal for balloon catheters for angioplasty [177]. Recent research has taken advantage of these properties and seen nylon used as a composite biomaterial. Some examples include titanium/nylon hybrids [178], nylon–chitosan-blended membranes for tissue engineering [179], and nylon–silk blends to produce nanofibers [180].

Biostable polyesters are the final synthetic polymer worth discussing. There are two main biostable polyesters containing aromatic groups which are polycarbonates and poly(ethylene terepththalate). These materials are typically found in membrane filaments and meshes [181]. Polyesters composed of aliphatic glycolic acid or lactic acid make up more biodegradable polymers. Some examples include poly(glycolic acid) (PGA), poly(L-lactic acid) (PLLA), and poly(D-lactic acid) (PDLA). These polymers undergo hydrolytic or enzymatic degradation making them suitably biodegradable within the body [182, 183].

1.4.4 Hydrogels

It is widely regarded that hydrogels were the first biomaterials developed for human use [184]. Hydrogels come in a wide variety of compositions, but a general definition is any polymer gel that is highly hydrated (typically over 30% water by weight). Hydrogels contain an abundance of polymeric interactions including cross-linking. These cross-links can consist of both covalent bonds as well as noncovalent interactions such as hydrogen bonds, chain entanglements, and hydrophobic interactions. Hydrogels encapsulate an extremely broad range of materials. Yannas et al. [185] developed cross-linked collagen–glycosaminoglycan copolymers for use as a scaffold to regenerate the epidermis. Since that time, the use of hydrogels for tissue engineering has greatly expanded. Methods have been developed for the generations of hydrogels produced in situ via photopolymerization [186]. Hydrogels are also vital for the technique of cellular microencapsulation, the process where cells are immobilized within a polymeric semipermeable membrane. Microencapsulation can be extended for use in drug discovery. Hydrogels have been investigated as potential materials for stem cell microencapsulation [187]. This approach has been used specifically for cardiac tissue engineering [188, 189]. The ability to formulate hydrogels that are similar to the extracellular matrix as well as being highly biocompatible make them ideal for a variety of tissue engineering applications [190]. A hydrogel glue has been used to repair lung air leak in mice [191].

The long lifetime of hydrogel research has led to many currently widespread applications of the material. Perhaps most widely used as a contact lens material, hydrogels with their high water content and flexibility make the material ideal for close contact with the human eye. Silicone hydrogel soft contact lenses have been available since 1999 and offer excellent oxygen permeability to the eye [192]. Hydrogels also make for excellent absorbing materials. They have realized utility as sanitary napkins and disposable diapers. Polyvinyl alcohol-based hydrogels have been used as wound dressings [193].

1.4.5 Biopolymers

Polymeric materials produced by living organisms are referred to as biopolymers. These materials have numerous advantages and can be prepared from renewable plant and agricultural sources [194]. Biopolymeric materials can be derived from a diverse set of feedstocks including polysaccharides, proteins, lipids, polyphenols, and polymers, while the organisms used to make these materials span a broad range from bacteria and fungi to plants and animals [195].

We begin our discussion with protein-based biopolymers. Collagen is perhaps the most ubiquitous protein as it is the main structural protein of the extracellular matrix and in the connective tissues of the body. Collagen is prevalent throughout the animal kingdom and is often used in tissue engineering [196]. Moreover, the material may be prepared in a variety of ways including gels [197]. Collagen has been utilized as a biomaterial for millennia and continues to serve for drug delivery, hydrogels, bone filling materials, sutures, and wound dressings [198–200]. The irreversibly hydrolyzed form of collagen is gelatin and has biomedical utility as well. Gelatin is frequently used for both drug delivery systems as well as tissue engineering [201].

Much like collagen, the protein elastin is a main component of the extracellular matrix and is found throughout connective tissue. The main feature of interest with elastin is its highly elastic nature which allows for organs such as blood vessels to stretch and relax billions of times over a lifetime [202]. Elastin is used as a tissue scaffold and has been investigated for in vivo wound healing and nerve repair [203]. Elastomeric proteins are also present in insects, specifically resilin, which is present in *Drosophilia melanogaster* (fruit fly). Resilin is notable for being autofluorescent which makes it a potential biosensor material [204].

Biomaterial proteins extend to the plant kingdom as well. Zein is a so-called prolamine protein (named as such for its high content of the amino acid proline) found in corn. Zein is often found in food products and used as a coating on bakery products. Additionally, zein has been used as a film coating for tissue scaffolding [205]. Soy protein is another plant protein with extensive use in food processing. Soybean products are sold as textured vegetable protein and are the main ingredients in meat and dairy substitutes [206].

Another noteworthy fibrous protein that has been used for centuries is silk produced by arachnids, worms, mites, butterflies, and moths [207]. Silk is mostly composed of the proteins sericin and fibroin which forms beta-pleated sheets. This secondary structure is responsible for the high tensile strength of the material. Silk has been used as a medical suture since 131–211 A.D. and continues to be utilized in present day [208].

1.4.6 Polysaccharide Biomaterials

Much like protein, polysaccharides are biopolymers readily produced by a wide array of organisms. Composed of repeating monosaccharides and/or disaccharides, polysaccharides are complex carbohydrates. By convention, polysaccharides are classified in one of two groups—gums and mucilages [209]. Gums are generally considered to be pathological products formed following injury or unfavorable conditions in the organism. Mucilages are natural products that are byproducts of normal metabolism within the cell [210]. In contrast to synthetic polymers, polysaccharides offer certain advantages. Because polysaccharides are composed of monomeric sugars they are inherently nontoxic. Utilizing natural sources to produce the material will very likely be less costly than any synthetic alternatives.

We begin our discussion with some of the prominent plant-based polysaccharide biomaterials. Although polysaccharides are harvested from diverse organisms, they share key features. Plant-based polysaccharides have emerged in the field of 3D bioprinting. Bioprinting is the process of combining cells, biomaterials, and growth factors to fabricate materials that naturally mimic natural tissues [211]. Several biopolymers readily undergo gelation which allows for control of their viscoelastic properties [212]. Materials such as alignate, a polymer isolated from the cell walls of several specials of brown algae [213], nanocellulose, agarose (derived from red seaweed), carrageenan, pectin, and starch have been manufactured into bioinks [214–216]. The gelation of these materials makes them potential hydrogels in addition to the potential applications mentioned earlier. Wound dressings composed of plant-based polysaccharides have been investigated along with drug delivery systems [217–219].

Polysaccharides are not limited to the plant kingdom as animal-based materials have been used widely as biomaterials as well. In animal tissues, the extracellular matrix is composed of an interlocking meshwork of heteropolysaccharides and fibrous proteins [220]. Some of the most prominent animal-based polysaccharides include hyaluronic acid (HA), a glycosaminoglycan (GAG) found in the extracellular matrix. HA is involved in several key processes in the body including angiogenesis, reactive oxygen species, chondrocytes, cancer, lung injury, immune regulation, and skin [221]. HA has been investigated as a scaffold for bone regeneration and cartilage tissue engineering [222]. HA is thought to play a role in the process of skin aging. The HA molecule has a unique capacity to retain moisture which has made it a key component for novel treatments for skin aging [223].

Other biopolymers of interest include the highly studied polysaccharide chitosan. Isolated from the exoskeletons of certain crustaceans and the cell walls of fungi, chitosan is produced by the deacetylation of chitin. Chitin is readily broken down by the enzyme chitinase, making the material both biocompatible and biodegradable. Chitin can be further modified to obtain one of many potential derivatives with biomedical applications [224].

1.5 Cellulosic Materials and the Environment

Environmental concerns stemming from increased greenhouse gases, air pollution, and toxic soil and water contaminants motivate scientific research toward sustainable materials with low environmental impact. In this regard, cellulose has emerged as one such sustainable material. Research on cellulosic materials has grown in the past 25 years with remarkable advancements for implementing cellulosic technology into a wide range of applications [225–227]. Cellulose, first described in 1838 by French chemist, Anselme Payen, is the primary component of plant cell walls and is widely considered as the most abundant organic polymer in nature [225]. Cellulose is a polysaccharide, and its chemical structure consists of repeating β-(1,4)-D-glucopyranose units. Cellulose, extracted from plants, algae, tunicates, and some bacteria, has been used as raw material for the production of cellulosic nanomaterials (CNM) for a number of applications and the production of biofuels and bioderived chemicals in biorefineries due to its natural accessibility and its beneficial environmental and biocompatible characteristics [226, 228, 229]. While cellulose is well understood and considered safe for human and environment interaction, much less is known about the biological safety of nanomaterials, which are at the forefront of the emergence of cellulosic materials [230, 231].

Processing of cellulose generates two main classes of CNMs (also known as nanocellulosic materials (NCM)). The first class represents nanostructured materials and encompasses cellulose microcrystals (CMC) and microfibrils (CMF), which structurally are defined with nanoscale diameters and lengths on the microscale. The second class of CNMs represents cellulose nanofibers, such as cellulose nanocrystals (CNC) and cellulose nanofibrils (CNF). These nanofibers are characterized by their nanoscale diameter and length dimensionalities. The type of CNM obtained is heavily dependent on the cellulose source and the extraction and processing techniques used in the preparation of the materials [232]. Due to the wide array of available CNMs, numerous beneficial properties are achievable and tunable making this emerging material enticing for applied research spanning several scientific communities. Such chemical properties obtained for CNMs are high strength mechanical properties, high surface area, low density, surface tunability as well as their high abundancy and renewability [225, 227, 228, 232]. Over the years, CNMs have been identified as promising components for a wide range of applications including packaging, insulation, clothing and apparel, production of biofuels, synthesis of biopolymers, information and energy storage devices, intelligent electronic devices, and biomedical applications, among others [233–241].

Due to the ever-growing demand for new, environmentally friendly materials with optimized characteristics, cellulosic materials could be at the forefront of many revolutionary technologies. Many reviews of the various types of cellulosic materials, extraction, processing, and production techniques, as well as chemical properties, modification, and applications have been demonstrated and further support the emergence of cellulosic materials [225–227, 232, 233, 242–246]. However, it is important to consider the environmental impact new materials can have prior to full industrial commercialization and consumer utilization. When considering the environmental impacts of new materials, it is important to not only look at the end product but also the raw materials, processing, transportation, consumer use, and disposal of the product, as all aspects of a material's life cycle can contribute to the possible environmental implications. This section briefly highlights and presents important environmental implications of the increasing use of cellulosic materials. In particular, we will highlight environmental implications of producing cellulosic materials based on life cycle assessment studies as well as discuss the impact of cellulosic materials on biological health.

1.5.1 CNM Production and Associated Environmental Impacts

As CNMs continue to be utilized, it is important to understand the production techniques and possible environmental impacts associated with them. The production of CNMs is dependent on the raw material, desired properties, and the material application [245]. Therefore, there have been many techniques established for producing CNMs, such as physical/mechanical, chemical, enzymatic, or a combination of techniques, each with its own associated environmental impacts [225, 226]. Impacts include water, chemical, and energy usage, as well as the production of various waste streams and greenhouse gas emissions [232]. Several reviews discussing the production techniques of CNMs have been published; however, we will briefly highlight some prominent techniques and evaluate the life cycle assessment literature of CNM production and the associated environmental impacts.

Two primary steps occur during cellulose processing, pretreatment, and refinement. Pretreatment is typically performed to remove noncellulosic components from the raw material (e.g., lignin and hemicellulose in the case of biomass) and to modify the cellulose surface for easier refinement processing [225, 226, 232]. Pretreatment techniques typically involve chemical processes, such as TEMPO-oxidation (2,2,6,6-Tetramethylpiperidine-1-oxyl), carboxymethylation, or acetylation, or involve enzymatic hydrolysis. These treatments are typically performed to modify the cellulose surface lowering the energetic penalty required for mechanical production during the refinement stage. After pretreatment, there are many chemical, enzymatic, and physical refinement processing methods. Further chemical processing includes acid hydrolysis, subsequent oxidation, use of supercritical solvents, transition metal hydrolysis, and the use of ionic liquids. Ionic liquids are typically perceived to be recyclable and nontoxic and therefore have received increased attention in the literature for sustainable cellulose processing [247, 248]. In terms of physical refinement methods, the prominent processes include homogenization, microfluidization, grinding, ball milling, cryocrushing, and sonication. While resources are invested heavily at the lab scale to improve the production of cellulosic materials, the most common industrial scale processes include enzymatic or chemical pretreatment followed by mechanical homogenization and grinding refinement [226]. Assessment of the environmental impacts of these processing techniques is crucial for further expansion and use of cellulose nanomaterials.

Life cycle assessments (LCAs) have been performed on a wide range of materials and technologies and are a valuable tool for evaluating the environmental impact of existing and emerging products [249]. The typical scope for LCA studies involve "cradle-to-gate" or "cradle-to-grave", where the assessment begins from the raw material (cradle) and ends at the factory (gate, i.e., does not reach consumer) or disposal of a product (grave). This scope allows for different stages of the production (extraction, manufacture, transportation, use, and disposal) to be evaluated and generally facilitates comparisons to help identify the most optimal raw materials and processing technologies for improving cost, environmental footprint, and waste generation. Numerous LCA studies have been performed to evaluate the environmental impact of producing cellulose nanomaterials as well as their incorporation into final products, such as reinforced paper, plastics, and production of ethanol [250–254]. The following discussion will briefly review the LCA literature on CNM production.

LCA studies evaluating the production of CNMs have primarily focused on cellulose raw material and the impact of different processing techniques. Raw materials such as wood, cotton, other plants, and food waste can have varying amounts of cellulose and noncellulosic matter. The varying ratios of constituents in different materials result in the need for alternative processing techniques [225, 226]. The typical raw material in industrial cellulosic applications is wood pulp; however, other sources investigated in LCA studies are cotton, coconut, vegetable waste, and recently, bacteria. Studies by Li et al. [255] and Arvidsson et al. [256] investigated various production techniques including the use of chemical or enzymatic pretreatments, no pretreatment, and mechanical processing using wood pulp as the cellulose raw material. The results presented by Li et al. [255] showed that chemical pretreatment (using TEMPO oxidation) followed by mechanical homogenization had the lowest environmental impact; however, this strategy still required a large footprint.

Li et al. [255] and Arvidsson et al. [256] identified that the etherification strategies (chloroacetic acid treatment and carboxymethylation treatment, respectively) contained the highest environmental impact due to high energy demands and need for large amounts of solvents and chemicals. Arvidsson further showed that processing with either no pretreatment methods or enzymatic pretreatment resulted in similar environmental impacts. They determined that the technique with the least environmental impact was TEMPO-oxidation followed by homogenization as it requires less energy and chemicals than etherification and sonication methods. de Figueirdo et al. [257] compared CNMs prepared from cotton and unripe coconut fibers and identified that cotton was a more environmentally friendly raw material to process than unripe coconut because the production of CNMs required less energy and water and produced fewer potential pollutants. However, do Nascimento et al. [258] later showed that CNM production from coconut was possible and environmentally friendly with proper recycling technology. Piccinno et al. [259] further investigated the usage of vegetable waste (specifically carrot waste) using enzymatic and mechanical (homogenizer) processing methods and determined that CNMs could be produced in a more environmentally friendly way than using unripe coconut fibers and was comparable to the processing impact observed by de Figuerido et al. [257] when using cotton fibers. This is due to the lack of chemicals used in the processing allowing for more water recycling to occur and lowering the amount of waste generated. Additionally, the impacts found for this enzymatic production process is very similar to the environmental impacts determined by Li et al. [255] for the TEMPO-oxidation/homogenizer processing. The primary difference between the two methods is a tradeoff between higher energy requirements (Piccinno et al. [259]) or the need for more chemicals (Li et al. [255]).

The results for the LCA studies discussed indicate that environmental impacts such as energy demand, water usage, and waste generation, for processing, CNMs are affected by the starting material, pretreatment, and processing techniques. Current literature surrounding the environmental impacts of CNMs production is limited due to lack of data. Generally, the processing of CNMs exhibits less environmental burden than the processing of other nanomaterials (e.g., carbon nanotubes) [255, 256]. However, further research is needed to fill in crucial knowledge gaps when it comes to scaling up to commercialization and incorporating consumer usage and disposal [260]. With improved overall evaluation of the life cycle of CNMs, emerging materials containing CNMs could be a viable option for the replacement of nonsustainable materials.

1.5.2 Biological Concerns of CNMs

The incorporation of cellulosic materials into new technologies is encouraging for the replacement of harmful petroleum-derived chemicals and other environmentally persistent contaminants. However, the rise in use also brings uncertainty to how cellulosic nanomaterials (CNMs) will affect environmental and human health. CNMs can be released into the environment through several routes, including release during production, consumer use, disposal, and recycling of the material [261, 262]. Several studies have investigated the toxicity of cellulosic materials on humans; however, few studies have focused on the fate of CNMs in the environment, presumably due to the ubiquitous nature of the environmentally derived cellulosic raw materials. Reports by Endes et al. [263], Shatkin and Kim [261], and Ventura et al. [264] have highlighted the need for further investigations into how CNMs affect environmental and human health before further commercialization and utilization in emerging applications [265].

Exposure to CNMs can occur during several stages of a products life cycle, and therefore, the wide dispersity in exposure scenarios makes it incredibly difficult to estimate the harmful impact CNMs will have on humans [263]. Hohenthal et al. [266] identified the most likely exposure routes to consist of inhalation and skin contact of CNMs during the production and application of CNMs in coated board and wet-laid nonwoven materials. They concluded that the estimated overall exposure to humans during production and application of CNMs to be low. Shatkin and Kim [260] performed a systematic LCA to identify potential risks associated with CNMs incorporated into food packaging. Their analysis corroborated the report by Hohenthal et al. [266] and determined the highest risk exposure of CNMs to be via inhalation during handling of dry CNM powder. Vartiainen et al. [267] further identified the exposure to airborne cellulose particles after grinding and spray drying of birch cellulose. Their results showed that workers could be readily exposed to CNM particles during various production stages. While inhalation is a prominent form of exposure for the CNM production stage, other exposure routes, based on application and consumer use, (e.g., textiles, healthcare products, or cosmetics), need further assessment to identify human health concerns.

The study of human cellular effects from interaction and uptake of CNMs is of importance to identify the potential for harmful exposure from CNMs. Once exposed, measures important for understanding CNM effects on humans include cellular uptake; persistence; inflammatory response; immuno-, cycto-, and genotoxicity; and excretion. Endes et al. [263] reviewed a wide body of literature and determined mixed results. They identified that some CNMs have low cellular uptake, low toxicity, and were recommended for biomedical applications. On the other hand, several studies identified CNMs to have varying levels of toxicity depending on the material size/dimensions, surface modifications, exposed cell type, and CNM concentration. Thus, this review highlighted how the diverse physicochemical nature of CNMs makes identifying toxicity of CNM exposure to humans difficult. A recent review of CNM toxicity to humans by Ventura et al. [264] identified similar conclusions. Studies found that CNC had higher cellular uptake than CMC and was mainly related to the size of the cellulosic material. Additionally, while overall toxicity for CNMs was low, a few studies observed inflammatory responses that were dependent on the physicochemical properties of the CNM. Some immuno-, cycto-, and genotoxicity was observed for CNMs; however, all cases were considered mild in comparison to toxicity studies for carbon nanotubes and other metal-containing nanomaterials [263–265]. Therefore, it is clear that the physicochemical properties of the CNMs, such as, size, surface charge/functionality and resulting surface chemistry are largely responsible for cellular uptake and toxicity in humans. Identifying how to lower exposure routes and rationally design CNMs to reduce human toxicity will be highly desirable to further promote the emergence of cellulosic materials.

Similar conclusions discussed for CNM impact on humans are expected when considering the environmental effects of CNMs. Several important measures to establish a materials environmental impact are bioaccumulation, ecotoxicity, persistence, mobility, and degradation [261]. Bioaccumulation and toxicity have been minimally investigated in the literature. Stoudmann et al. [268] modeled the environmental exposure and hazard of CNMs in Europe based on estimation methods used for other prominent nanomaterials. They determined that CNMs had a low-risk characterization ratio based on the predicted no effect concentration (PNEC) and predicted environmental concentration (PEC) of CNMs in surface water. The estimated mean CNM PNEC was 7.8 mg/l and PECs for 2015 and 2025 were 0.23 μg/l and 2.37 μg/l, respectively. Similarly, Kovacs et al. [269] modeled a worst-case scenario spill where roughly 1000 tons of CNMs was lost from a production facility to the neighboring waterways and the estimated concentration of CNMs in the water was 0.24 mg/l. The results of these predictive studies suggest, presently and in the future, that CNMs have a low environmental risk for accumulation in waterways and will not reach toxic levels even in the case of a large-scale production leakage [268, 269]. This conclusion has been supported by several individual studies. Kovacs et al. [269] and Harper et al. [270] investigated the ecotoxicity of CNMs on aquatic species. Kovacs et al. [269] identified that reproduction inhibition was observed for the fathead minnow at a CNM inhibition concentration (IC_{25}) of 0.29g/l. This concentration is 100 times higher than the worst-case scenario spill estimate performed by Kovacs et al. [269]. Harper et al. [270] investigated the uptake and toxicity of CNMs containing physicochemical modifications in zebra fish. The results showed low overall toxicity to developing zebrafish and the changes in surface chemistry did not significantly alter the toxicity of the investigated CNMs. Furthermore, lethal concentrations of CNMs on aquatic species were found to be much higher than the estimated worst-case scenario, indicating low CNM toxicity for aquatic species. Vartiainen et al. [267] and Kovacs et al. [269] studied the effects of exposure of CNMs to the bacteria *Vibrio fischeri*. Their results showed IC_{25} values of $>$2500 μg/ml (Vartiainen [267]) and $>$10,000 μg/ml (Kovacs [269]) indicating the tested bacteria were not acutely toxic to environmentally relevant concentration of CNMs. In contrast, a study performed by Pereira et al. [271] evaluated the toxicological effects of cotton CNMs on freshwater microalgae and observed that CNMs effected the algae growth and cell viability. Based on the limited studies of the environmental impacts of CNM, it is clear that CNMs contain low toxicity in relevant environmental concentrations while CNMs are toxic to aquatic species at much higher concentrations.

The persistence and degradation of cellulosic materials have also been investigated. Kummerer et al. [272] investigated the biodegradability of organic nanoparticles in aqueous environments and found that both macroscopic cellulose fibers and cellulose nanoparticles degrade over the course of a 28 day period reaching 45%

and 54% degradation, respectively. In comparison, carbon nanotubes and fullerenes were investigated and showed no degradation after the 28 day period. Interestingly, the smaller particles (cellulose nanoparticles) were found to have faster degradation kinetics than the macroscopic fibers, which is attributed to higher surface area. Likewise, Hohenthal et al. [266] determined that natural nanofibrillated cellulose and surface modified nanofibrillated cellulose reached 90% degradation during a 70 day test period. While surface modification did affect the kinetics, the overall degradation of the modified and unmodified surface CNM was the same. Furthermore, the researchers showed that coated packaging board containing nanofibrillated cellulose reached a biodegradability rate of 98% in less than 60 days. Lastly, Li et al. [273] showed that nanoporous cellulose gels incorporated into polymer plastics facilitated the degradation rate of the fabricated nanocomposite material. Therefore, CNMs have been identified as materials with low environmental persistence and high biodegradation; however, physicochemical factors of varying materials are likely to affect these properties.

Overall, CNMs have been shown to have low biological toxicity. Several reviews have highlighted the cumulative literature and identified that cellular uptake and toxicity to humans are heavily dependent on the physicochemical properties of the CNM [261, 263–265]. CNM size, surface charge/functionality, resulting surface chemistry, CNM concentration, and specific cell type are all factors responsible for the observed toxicity in humans. Overall, environmental exposure of CNMs to aquatic species, bacteria, and algae showed little to no impact; however, there is a need for more scientific studies for evaluating the effects of CNM in the environment. Lastly, various natural and surface-modified CNMs and CFMs, as well as other prepared cellulose composites have been shown to be easily biodegradable in standard test conditions. Therefore, based on the overall low toxicity and biodegradable nature, it is assumed that CNM contaminants in the environment will not be persistent and harmful to biological species. However, this conclusion would benefit from increased research on CNM structure–property relationships.

1.6 Volatile Organic Compounds: Daily, Hazardous Exposures

In modern life, our environment could become a threat to our health from many sources such as industrial waste [274, 275], polluted water [276], severe storm activity [277], background radiation (both naturally and man-made) [278, 279], smog/air pollution [280, 281], curing of plastics [282], nanoparticles that automotives generate [283, 284], and disposal of electronics (E-waste) [285]. Here we will focus on common exposures in current times from volatile organic compounds (VOCs) released during our everyday life [286], nanoparticles/chemicals released by wear and tear when using automobiles [283, 284], and hazards from disposal of used electronics [285].

1.6.1 Volatile Organic Compounds

Volatile organic compound (VOC) is a general term for organic compounds with low boiling points (<250 °C) at ambient pressure (1 atm) [287–289]. Typically, humans' sense of smell is the first indication that they are being exposed to VOCs because they have a range of smells and effects on the human body [290, 291]. A prime example of being exposed to VOCs is the powerful odor of permanent markers. The strong odor that the nervous system detects is solvent evaporating from the permanent markers [290]. The solvents used in permanent markers vary greatly from manufacturer to manufacturer, but markers with strong odors tend to have solvents such as xylenes and toluenes. Exposure to xylene from normal use of permanent markers can result in nausea/headaches (100–200 ppm) to feeling "high" (200–500 ppm) [292]. These effects are similar to VOCs originating from the use of shellacs/paints [293] and hobby supplies (glues [294], paints [293], and aerosol sprays [289, 295]). For products that release toxic or hazardous VOCs, proper precautions should be followed such as sufficient ventilation and limited exposure.

Odors from VOCs are not necessarily "bad", VOCs can smell "good" and can be used to deodorize or mask unwanted odors. These products include perfume, laundry products, soaps, hand sanitizers, lotions, cleaning supplies, and air fresheners. However, the incorporation of "good" smelling VOCs into such products can have unforeseen side effects. A study of 25 fragrant products—including laundry, personal care, cleaning, and air freshening products—found a total of 133 VOCs with 24 of these compounds being classified as toxic or hazardous compounds under U.S. federal laws as of 2010. In this work, all 25 fragrant products had at least one of the toxic or hazardous compounds with the most common of these being α-pinene [296]. Removing α-pinene is particularly important in the lumber industry, and research has shown that biofilters at the laboratory scale are highly effective at removing α-pinene [297]. In addition to the toxic or hazardous compounds found in the 25 products, some of the VOCs can undergo reactions with ozone-producing secondary pollutants which could result in unforeseen health complications [298]. The other 109 VOCs found in the study are dominated by essential oils, ethers, and alcohols [296]. One such essential oil is limonene, an essential oil found in the peel of citrus fruit, which is present in 23 of the 25 products studied [296]. The incorporation of limonene in these products provides a citrus scent, but limonene also has health benefits when ingested, particularly for gastritis and gastric disorders [299]. The use of limonene is not the first case of essential oils being used both for their fragrance and herbal medicinal properties. The ancient Egyptians documented their use of medical properties of herbs as early as 1500 BC in the Ebers papyrus [300].

Finally, we will discuss the release of VOCs in plastic curing. This topic is highly relevant to modern life because we are currently in the "Plastic Age" according to Nick Smith [301]. Modern life would not be possible without plastic because plastic is in nearly every product we use including building supplies, packing materials, electronic devices, and automobiles. The scent of new automobiles is caused by the curing of plastics, which release VOCs. Sadly, these VOCs do not have health benefits [302]. One source of VOCs in new automobiles are carpets [286], but recent studies have shown that VOCs are released long after the plastics (carpets and other surfaces) have cured and that used automobile VOC levels could be indistinguishable compared to new automobiles [303]. The constant release of VOCs was rationalized by the photodegradation of plastics exposed to sunlight [304]. However, VOCs are not the only hazardous contaminants generated by automobiles. In the next section, the release of nanoparticles and chemicals released during the operation of automobiles will be discussed.

1.6.2 Harmful Impact of Automobiles Usage

The impact of automobiles on our environment has been primarily focused on the generation of greenhouse gases and smog-generating pollutants [305]. The generation of smog-related chemicals from automobiles has been at the forefront of environmentalist work for decades. In 1965, Ralph Nader published "Unsafe at Any Speed" [306]. In this work, Nader showcased how automobiles were, in-part, responsible for smog in the Los Angeles area. The high levels of smog in the Los Angeles valley during this time resulted in the Clean Air Act of 1970, which prohibited the use of fuels or fuel additives that endanger public health [307]. This act effectively banned the use of tetraethyllead as a fuel additive within the USA [308]. Unfortunately, the ban of tetraethyllead did not remove all smog-related emissions from automobiles, largely because ambient air contains 78% N_2 (nitrogen) and 21% O_2 (oxygen). During the combustion process, the temperature within the engine can exceed 2000 °C [309], which is sufficient to convert N_2 and O_2 into NO_x (nitrogen oxides) [310–312]. The generation of NO_x causes secondary reactions producing ozone, carbon monoxide, and sulfur oxides. These secondary products in turn cause numerous environmental problems of their own which can result in the environment being detrimental to human health [313]. Monitoring the levels of NO_x in 50 urban areas across the USA from 2005 to 2014 showed an approximate decrease of 45% in a variety of quantifiers of NO_x levels; this was accomplished by regulating both the power and transportation sectors [313].

While reducing the amount of NO_x produced when burning fossil fuels and the release of VOCs from the plastics used in construction are important factors to reduce the threat of the environment on our health, other factors related to automobile operation are important to understand. Research has shown that both nano- and microsized particles are released during deceleration of the automobile (braking). The size of the particles ranges from 10 nm to 20 μm. The quantity of nanoparticles significantly increased as the temperature of the brakes increased [284]. The generation of nanoparticles is alarming because the side effects of these nano- and microparticles are largely unknown. In the area of semiconducting light-emitting diodes (LEDs), it is well known that the surface of the nanoparticle is highly reactive as the ratio between the surface area and volume becomes larger [314, 315]. The cytotoxicity of metal oxide nanoparticles has been studied via Quantitative Structure Activity

Relationship (QSAR) methods which could be used to predict the toxicity of the particles generated from braking [316, 317]. The implementation of QSAR to predict the toxicity of the nanoparticles generated by brakes would require additional modeling with MoS_2, SnS, ZnS, and graphite as these are the primary compounds found within the brake material [284].

Studying how brakes deteriorate during the operation of automobiles and the release of VOCs related to automobiles is relatively straightforward because brake pads and rotors must be replaced due to wear and VOCs cause the "new" automobile smell. However, automobiles release other toxic chemicals into the environment that could be potential health hazards. The chemicals released from catalytic converters are potential health hazards because every new automobile sold in the USA is required by law to have a catalytic converter. Catalytic converters are designed to convert gas molecules that have the potential to cause pollution, smog, and/or health issues of living organisms into a less hazardous form. This is accomplished primarily by converting NO_x, CO, and HC into N_2, CO_2, H_2O, and H_2 [318]. Catalytic converters were first implemented in the 1970s following the Clean Air Act. The early implementations contained Pt, Pd, and Rh which resulted in increased prices for these noble metals. After 30 years of using Pt group metals as the catalytic surface in catalytic converters, research has found that Pt group metals are also emitted when using automobiles. Moreover, USA interstates can have upward of 30 times the concentration of Pt group metals compared to typical conditions [283]. These levels could pose significant health hazards not only to humans but also to vegetation and marine life [319, 320]. For humans, the health hazards originate from the high concentration of chloride in human lungs, which could react with the Pt complexes generating cisplatin derivatives. The generation of cisplatin compounds can lead to cellular damage [321]. The health hazards created by the use of automobiles is not the only everyday source of hazards in modern life, the use of electronics is ubiquitous in our society. Shifting the automobile industry toward electrified vehicles (EVs) in the form of hybrid or fully electric cars is a clear avenue toward reducing the emission of these heavy metals from catalytic converters, in addition to their merits in reducing greenhouse gas emisison. However, greater electrification of vehicles could present its own challenges as these vehicles effectively become rolling computers. In the next section, we will focus on electronic waste from disposing of old electronics.

1.6.3 Electronic Waste Disposal

E-waste is a generic term used to cover all manners of waste electrical and electronic equipment (EEE or WEEE) and electronic parts that are discarded by the owner without the intent of re-use. Collection of E-waste offers the prospect to recover a variety of recyclable metals such as gold, silver, platinum. However, this process is complicated by the presence of potentially toxic substances (cadmium, mercury, lead, chromium, and other heavy metals) [322] as well as materials used to construct the devices (e.g., circuit boards) which forces the recycling facility to prioritize which materials will ultimately be recovered and the methods used to recover the desired materials. Further complicating the matter is the scale at which E-waste is generated. In 2014, the Global E-waste Monitor noted that 44.4 million metric tons (Mt) of E-waste was generated globally; in 2019, that figure increased to 53.6 Mt and by the year 2030 that number is projected to grow to 74.7 Mt [323]. Moreover, these increasing quantities of E-waste are not being effectively recycled. Of the 53.6 Mt produced in 2019, only 17% (9 Mt) was documented and properly recycled, while the remaining 83% (45 Mt) was discarded, traded, or recycled in nonenvironmentally sound ways. Unfortunately, the recycling of E-waste is not keeping pace with the global generation of E-waste; due to this, several of the aforementioned hazardous materials will likely find their way into the environment, potentially causing damage to ecosystems and human health.

Several governing bodies have taken action to help curb improper disposal and recycling practices: Resource Conservation and Recovery Act (RCRA) in the USA in 1976, the Basel Convention in 1989, the WEEE Directive in 2003 in the EU, and the Nairobi Declaration in 2006, to name a few [324]. Over time, developed countries have increased regulations on E-waste disposal and recycling which has led to an increased cost of disposal, whereas in developing countries the disposal costs remained low due to the lack of these restrictions [324–326]. The cost disparities between developed and developing countries has led to an increase in E-waste trading and export to these less developed countries. Furthermore, in developing countries there is likely a lack of appropriate infrastructure resulting in treatment/recycling practices of E-waste which expose workers to unsafe levels of hazardous materials. Additionally, the same lower regulations allow for increased runoff pollution further exposing the human population to the hazardous materials [327, 328]. In the following subsections, common E-waste will be discussed focusing on the hazardous compounds that can be released, the impact of these hazardous compounds on the environment and potential health impacts.

1.6.3.1 Lithium Ion Batteries

A large portion of E-waste originates from portable electronic devices such as laptops and cell phones. Most of these devices utilize lithium (Li) ion batteries due to the high charge-to-mass ratio of the Li cations. The Li ion batteries are composed of several materials typically: a graphite anode, an electrolyte solution of ethylene carbonate or diethyl carbonate, and a Li (I) salt cathode. The cathode materials can vary greatly, but $LiCoO_2$ [329], $LiFePO_4$ [330], and $LiMnPO_4$ [331] are common cathode materials. When the battery's casing is compromised, phosphates from $LiFePO_4$ [330] and $LiMnPO_4$ [331] can leach into the soil or wastewater streams. While the phosphates themselves are not particularly toxic to humans, their presence in water streams can lead to accumulation and contribute to harmful algal blooms [332]. The harmful impact on the environment due to runoff of phosphates by E-waste is likely overshadowed by the runoff of excess phosphates in the agricultural industry; however, the contributing phosphates from E-waste could become more important as the amount of E-waste increases. Li ion batteries contain other chemicals that could make their way into the environment; metal ions of Co, Ni, and Mn for instance could pose a health hazard. The data relating Co to health hazards is limited; however, the limited data suggest that Co compounds pose negligible health risks [333, 334]. Studies exposing animals to injections of Co compounds only resulted in sarcomas at the injection-site, but only with high dosages of Co [333]. Ni ions could be generated by Li ion battery E-waste, which can cause problems for local plant-life surrounding recycling facilities because excess Ni ions disrupt Ni-based enzymes in plants [335–338] leading to inhibition of enzymatic activity, photosynthesis, and chlorophyll production in plants [339]. Unlike plants, Ni has no inherent nutritional value in humans. There is evidence that exposure to excess Ni via food, water, or even inhalation can behave as both a neurotoxin and carcinogen [340–347], although the mechanisms of which are not yet clear. Unlike Ni and Co, excess Mn is known to be toxic to humans. Mn is an essential nutrient for normal brain function, but excess Mn can lead to neurological conditions. Numerous studies have shown that Mn compounds can accumulate in the brain via an olfactory route (inhalation) [348, 349] and can result in cognitive, behavioral, and motor abnormalities [349–353]. While the brain has homeostatic mechanisms to maintain appropriate levels of Mn, and thus acute Mn exposure can be controlled, there is some evidence that prolonged Mn exposure can cause permanent structural abnormalities [354–357]. It should be noted that most of these studies are related to Mn exposure from welding processes, where Mn is an essential component for providing strength to the weld.

Informal recycling facilities, particularly in developing countries, may employ a workforce that is generally untrained, tend to use riskier practices, and may not have appropriate conditions or equipment for proper safety [358–360]. Under these conditions, both the workforce and nearby general population can be exposed to the gaseous byproducts from smelting (for recycling) or burning (disposal) processes [327]. In this manner, both Ni and Mn exposure can become a problem for the workforce and the nearby general population.

1.6.3.2 Printed Circuit Boards

Printed circuit boards (PCBs) have found their way into nearly every type of electronic device such as: personal computers, cellphones, smart devices, and vehicles. PCBs have become an important component in automation throughout our lives. However, the vast quantity of circuit boards created will likely lead to larger numbers of boards disposed of or recycled. Unfortunately, many of these devices employ a variety of materials (i.e., plastics and metal alloys) and are quite compact in their construction, as in the case for mobile electronic devices. These attributes make it incredibly difficult to fully recycle all of the components from PCBs [361–363]. Several techniques have been developed to aid in the process of recovering both base metals (Cu, Pb, Zn, and Cd) and precious metals (Au and Ag) [364]. However, recycling centers in developing countries may not have access to all of the equipment needed to fully recycle all components of PCBs [365]. Even worse, material may be recovered at the expense of creating additional pollution by means of particulate matter released into the air (pyrometallurgy), sludge in runoff (electrometallurgy), as well as the large amount of energy consumed during the recycling process [364]. These techniques can therefore lead to release of some of these heavy metals into the environment. In particular, chronic exposure to Cd via the water supply, even at low doses, has been linked to tubular impairment in the kidneys resulting in a loss in reuptake of other essential nutrients, such as Zn, glucose, amino acids, phosphate, and Ca [366]. Losses of these nutrients can lead to additional maladies (i.e., osteoporosis or nerve issues due to lack of Ca). Other heavy metals, such as Pb which is common in the solder used to connect components to the PCB and each other, also present a myriad of well-documented problems for both humans and the rest of the environment [367]. However, more recently, manufacturers are beginning to move away from Pb-Sn alloys to lead-free alloys [368]. Another common practice is to burn the PCBs to separate the recoverable metals from the plastic and solder and further subject the remaining metals to various separation methods. However, during the pyrolysis process, a significant quantity of VOCs can be released (benzene, toluene, and phenol), which may pose both carcinogenic and noncarcinogenic health hazards to the workforce and the nearby general population [369, 370].

1.6.3.3 Photovoltaic Panels

The potential impact of photovoltaic panels (PVs) as an emerging contaminant is significant. These devices have only recently been classified as E-waste by the European Commission [371]. Although, some have recognized the exponentially growing usage of PVs as a viable power source and have shown that many of the materials used in the construction of PVs can and should be recycled [372–375], albeit at a premium compared to the cost of landfill disposal [376]. After glass, the largest component of the solar cell tends to be the absorbing layer, the most common materials for which are copper indium gallium selenide (CIGS), cadmium telluride (CdTe), and amorphous silicon. Given the previously limited regulation on the disposal of these devices, there has naturally been less research on appropriate recycling technologies [377]. However, the health effects of the absorbing components have been studied. CIGS cells have been shown to leach small amounts of Cd into the environment after disposal which can lead to health effects previously described for PCBs [378]. Lung pathologies have also been shown to develop due to prolonged inhalation exposure to CIGS [379]. Another absorber component, CdTe, under landfill conditions has been shown to leach Cd as well as Te into the environment [380, 381]. In addition to the toxicity of Cd, most Te compounds are considered toxic. Studies have found that although Te concentrations in soils in sediments tend to be low, except near sources of contamination, the presence of Te is ubiquitous in soils, sediments, and waters. Te also has a high bioavailability in living organisms despite having no known biological role [382]. However, due to the sheer volume of PVs expected to be generated (500 GW capacity in 2016 to 4500 GW capacity by 2050) [376], simply disposing of these materials could create hotspots of contamination near landfills or improper recycling facilities, demonstrating the need for proper practices and regulations worldwide. The older, and therefore most common component to be discarded in the near term, is amorphous silicon. To date there is limited data on the effects of amorphous silica on the environment, but studies have shown that inhalation by the workforce can lead to chronic bronchitis, COPD, or emphysema [383].

1.6.3.4 Planned Obsolescence

In addition to the health and environmental concerns specific to a given technology, the potential contaminant-based risks are compounded due to the sheer volume of electronics that are manufactured worldwide. Naturally more devices will be produced as the technologies become cheaper and more widely available to additional consumers. However, growth is also fueled by planned obsolescence, i.e., the continuing decrease in the lifetime of E-products through both technological advances and marketing and advertising practices [384–387]. This concept is particularly noticeable for portable electronics such as cellphones and laptop computers; personal computer lifespan decreased from 6 to 4 years between 1997 and 2005 and cellphone lifespan in the USA decreased from approximately 3 years to less than 2 between 1995 and 2005 [388, 389]. These trends have only increased, particularly for companies who actively encourage upgrading to the latest devices each year! More often than not, these products are not actually defective or obsolete, rather they are perceived to be by consumers [385]. On the other hand, products are more frequently manufactured such that it is difficult or impossible to repair defects, e.g., integrated Li ion batteries that cannot be replaced [390]. Such practices exacerbate previously discussed chemical and health issues surrounding the recycling and disposal of a variety of electronic devices.

1.7 Fate and Lifecycle of Antibiotics

Antibiotics have become a common therapeutic in mammalian medicine to treat infections, inflammations, and diseases [391]. Applications of antibiotic medicine, such as penicillin, have greatly reduced the mortality and morbidity rates that result from common infections [392–394]. The success of antibiotics in treating common infections has resulted in high levels of consumption by both humans and animals [395–398]. It has been estimated that in 2010, the total global consumption by livestock was $63{,}151 \pm 1{,}560$ tons based on a global map including 228 countries [399]. It was also predicted, based upon globally observed trends, that the global antimicrobial consumption will increase to $105{,}596 \pm 3{,}605$ tons by 2030 [399]. This is a worldwide 67% increase in the consumption of antibiotics in livestock; and in a select few countries, the consumption rates will nearly double [399]. The increase in consumption of antibiotics by humans and animals thus leads to several vectors by which environmental contamination can occur in terrestrial, freshwater, and marine environments [400]. In particular, large quantities of pharmaceutical antibiotics have been found in human-made environments such as sewage and wastewater treatment plants [401].

Mammalian-generated antibiotic contamination of the environment can start directly from the human source via manufacturing locations which release them into the air or nearby water sources [402–404]. However, a major contributor to antibiotic pollution in the environment involves mammalian consumption and transport. The lifecycle of an antibiotic does not simply end when it is taken by a human or administered to an animal [404]. As much as 90% of the antibiotic is capable of being excreted in an unaltered form after passing through a mammalian digestive system [404, 405]. For example, excretion of antibiotically active compounds can exceed 50% for phenicols, trimethoprim, tetracyclines, and beta-lactams [404]. Even the portion of the antibiotic that is broken down can result in metabolites that retain their antibiotic activity and are subsequently excreted [404]. Release of these active compounds can lead to emergence of antibiotic-resistant genes (ARGs) and antibiotic-resistant bacteria (ARBs).

ARGs and ARBs are two of the main issues related to antibiotic pollution in the environment. Antibiotic resistance is defined as a reduction in susceptibility to antibiotics in a bacterial strain as compared to the wild type [406, 407]. This resistance makes antibiotics either less effective or not effective at all in treating bacterial infections. It is important to note that the environment contains a natural state of resistance to antibiotics [408–412]. ARGs have been found in permafrost samples that predate human usage of antibiotics [410, 411]. Some of these samples are dated to be 5,000–30,000 years old, indicating that antibiotic resistance could be an ancient phenomenon [410, 411]. Naturally occurring ARGs appear to originate from microorganisms that produce antibiotics to gain advantages in their environment such as growth, defense against competition, and the ability to act as messengers within the microbial community. This natural level of resistance is often not considered for environmental bacteria unless the ARGs can be transferred into pathogens [405]. This is of particular importance for superbugs such as methicillin-resistant Staphylococcus aureus (MRSA), vancomycin-resistant Enterococci (VRE), extended-spectrum beta-lactamases (ESBL)-carrying *E. coli*, and *Mycobacterium tuberculosis* which are risks to human health because their loss of vulnerability to multiple antibiotics makes them highly challenging to treat [404, 405].

Contamination of the environment by mammalian applications of antibiotics has shown to lead to an increase in antibiotic resistance when comparing ancient samples to current ones, in a phenomenon referred to as antibiotic resistance pollution [413]. As antibiotics have been used on a large scale to treat infections, a concurrent rise in ARBs has been observed and the speed at which resistance arises is extraordinary [414–418]. Some bacteria can become resistant to treatment in less than a year's time. In 1952, erythromycin was introduced to treat bacterial infections that were already resistant to penicillin [415]. One such bacterial infection included strains of *S. aureus* and 70% of those strains became resistant to erythromycin within six months [415]. Environmental samples show that human antibiotic-resistant pollution has contributed to ARG abundance for all classes of antibiotics that has only increased over the years [418].

There is a vast range of mechanisms by which antibiotic resistance arises both biologically and in the environment [419, 420]. Antibiotics can be transported outside of the cell such as when gram-positive and gram-negative bacteria become resistant to tetracycline antibiotics [421, 422]. The resistant bacteria in this case can overproduce certain membrane proteins that serve as efflux pumps which can cause the antibiotics to be pumped out at a faster rate than they can enter [421, 422]. The antibiotic is thus unable to properly diffuse to a concentration where it would be effective [419]. Staphylococci have also been shown to become resistant to erythromycin via drug efflux pumps [422, 423]. Furthermore, they can be inactivated by chemical decomposition such as derivatives of penicillin, a beta-lactam, being cleaved by beta-lactamase [424]. The target of the antibiotic can be modified by mutations that prevent the antibiotic from targeting it. Ciprofloxacin, a fluoroquinolone antibiotic, targets the DNA gyrase enzyme responsible for uncoiling double-stranded bacterial DNA at the end of each replication round [425]. The antibiotic forms a complex with the enzyme and the attached DNA which inhibits the process which eventually leads to bacterial cell death [425]. If the DNA gyrase enzyme is mutated such that the ciprofloxacin can no longer bind to it to disrupt the process of replication then the bacteria have gained resistance to the antibiotic. This can also happen in Staphylococcus aureus infections where the target enzyme is topoisomerase IV [426].

These resistances can then be transferred to pathogens though a separate range of mechanisms. Horizontal gene transfer (HGT) is the movement of genetic information between genomes by some means other than genealogical transmission [427]. This transfer can involve existing genes being replaced or inserted by incoming DNA and RNA [427]. HGT can happen through conjugation where DNA is transferred directly from a donor to acceptor via conjugated plasmids [428–430]. Conjugated plasmids are generally described as DNA units which are formed into a covalently closed circle [428]. They are problematic because they are separate from the host chromosomes, can replicate autonomously, can be highly mobile across a broad range of hosts, can survive with little to no cost to the host or at the expense of the host, are capable of transferring multiple resistance genes at once, and are resistant to virtually all antibiotics [428, 431–433]. One weakness to this mechanism is that it does require cell-to-cell contact in order to transfer the genes [434].

Two other mechanisms of HGT are transformation and transduction, which unlike conjugation do not require the donor to be present [434]. The transformation mechanism of HGT refers to the uptake of naked DNA from the environment [435–437]. This is a widespread mechanism that is present in many bacterial species in which the DNA is taken into the cell and then the genetic information is incorporated into the host bacteria [438]. The transduction mechanism involves a phage-supported infection of the recipient cell [439, 440]. Viruses that are specialized to infect bacteria, such as bacteriophages, are capable of facilitating the transmission of genomic material from the donor to the acceptor [441]. In generalized transduction, random fragments of the donor's chromosomes are encapsulated inside the phage capsid during replication and maturation. When the bacteriophage containing the genomic material infects another bacterium, the genetic material can be transferred into the new cell [442]. In specialized transduction the bacteriophage actually integrates itself in the bacterial chromosomes so that upon induction a small portion of the bacterial chromosome comes off and replaces some of the bacteriophage DNA. During replication of the phage, the bacterial portion is also replicated and can then be transferred to another bacterium when the phage infects the new cell [442].

Regardless of the particular mechanism by which antibiotic resistance arises or the method by which it is transferred between hosts, the pathogens that receive these resistances can then infect human and animal hosts which is particularly dangerous in the case of the aforementioned superbugs [404, 405]. It is also possible for ARGs and ARBs to pass through the same vectors as antibiotics to contaminate the environment; some of which are considered hotspots because the ARGs and ARBs are continually treated with high doses of antibiotics [443, 444]. These areas tend to be nutrient rich, which allows a high growth rate for ARGs and ARBs [443, 444].

Hospitals are one such hotspot where ARGs and ARBs have the potential to proliferate due to the prevalence of bacterial infections and persistent treatment of various infections with antibiotics [445–450]. Patients may enter the hospital already carrying an antibiotic-resistant pathogen that can then spread to other patients. It is possible for newly advanced ARGs to evolve and subsequently be transmitted between individual patients [451, 452]. Whether the ARGs evolve within the hospital or are introduced by a foreign body, it is possible for the genes to be transmitted into other genetic backgrounds via HGT individually or in combination with each other which can result in superbugs that are resistant to multiple antibiotics, such as those described earlier [404, 405, 451, 452]. ARGs, ARBs, and pathogens can then be transferred into the sewage that leaves the hospital alongside antibiotics and active metabolites [453, 454]. However, while hospitals can be a hotspot for ARBs and ARGs, they are a much more controlled environment because antibiotic distribution is more regulated as compared to general at-home usage that is not as easily monitored due to inappropriate or incomplete use of the antibiotic treatment [405]. This can cause the development of ARGs, ARBs, and pathogens that can leave the homes of the general community alongside antibiotics and active metabolites as sewage [407].

The sewage that leaves the community is later combined with the sewage from the hospitals in the area, which generates a nutrient rich environment perfect for the proliferation of bacteria, ARBs, and ARGs. An even bigger issue is that the combined sewage environment can also cause further HGT to create more superbugs [455, 456]. The sewage containing antibiotics, ARBs, and ARGs is eventually transferred to a wastewater treatment plant, but it is important to note these plants are not currently tasked with the specific removal of these substances [401, 453, 454, 456]. The majority of these plants utilize various three-step processes to treat sewage involving mechanical, biological, and chemical processes to biodegrade sewage. These processes attempt to remove contaminants and reduce the number of pathogens, but they are unable to remove all of the antibiotic residues. Unfortunately, this means that the antibiotic residues can then be transported into the environment via treated water being released into rivers, lakes, and drinking water. It has also been shown that while the processes used to treat sewage does result in reduction of the amount of ARGs and ARBs in the liquid portion of the sewage, they are still detected in the effluents that leave the plants and thus can be released into the environment [454, 457, 458].

Furthermore, solid waste allows for antibiotic pollutants to enter the environment. Solid waste is produced from the wastewater treatment plant in the form of sewage sludge as contaminants are removed from the water phase [457, 459, 460]. This sludge can contain antibiotics, such as fluoroquinolones, which were removed from the wastewater [404]. Sewage sludge is then used in the environment by either burying it, using it as filling material, or using it in fertilizers that can come in contact with soils, crops, and livestock [404]. This introduces more antibiotic-related contaminants to the environment. In both the case of the liquid fraction and the solid fraction of sewage that is treated, removal of these contaminants is often not a specific priority and every wastewater treatment plant uses a different methodology [461]. This means that the amounts of ARGs, ARBs, pathogens, and antibiotic residues that can potentially be released into the environment is strongly dependent on the location of treatment.

One of the more dangerous aspects when considering hospitals, sewage, and wastewater treatment plants as hotspots for antibiotic-resistant pathogens is that these species do not have to cross an animal barrier to affect humans working in these locations. Further, as there is an increase in antibiotic resistance in the environment and general population, it becomes more difficult to treat infections. We must then turn to other types of antibiotics and medicines that may only promote the development of new ARBs, ARGs, and pathogens. While these locations are the common human-generated hotspots for antibiotic resistance, there are also ARGs and ARBs that come from the antibiotic treatment of livestock that become environmental contaminants. In fact, while it is difficult to measure the exact amount of antibiotics used to treat livestock for a variety of reasons (i.e., preemptive usages, specific treatments, promoting growth) it is certainly much more than the amount used in hospitals [462–464]. One estimate suggests in the USA that 80% of the antibiotics sold are for animal consumption [465].

Manure created from livestock excretement and wastewater is therefore often contaminated with antibiotics and their active metabolites as well as ARBs. Animal wastes are usually kept in waste lagoons rather than transported through the general sewage system or treated at a wastewater treatment plant, creating a new hotspot arena [463]. Waste lagoons have been found to be highly enriched for ARBs as the manure is left to degrade before potential use as fertilizer for crops [462, 466–468]. Manure that is used to treat crops exposes the soil to ARGs and ARBs and any liquid runoff from these crops can also contaminate ground and/or surface water [463]. These vectors lead to contamination of the environment and can reintroduce those same antibiotic residues to both humans and animals via water sources and food grown using fertilizers [463].

Surface waters receive contamination by antibiotic residues, ARGs, and ARBs from nearly every vector we have discussed. This makes them an excellent transport for carrying the contaminants to environments away from the release source since most ARBs can persist in aquatic environments for weeks or months [467, 469–471].

Surface waters have also been found to have conditions which facilitate HGT as well as certain pathogens that may be capable of surviving longer or even thriving in such environments [472]. Although difficult to track, antibiotic resistance has been detected in particulate matter in the air, and smog has been found to be enriched for ARGs [473, 474]. Air also provides a method of easy transport since these contaminants can be dispersed via wind, fog, and precipitation that can lead to further contamination of surface waters and soil biomes [475–482]. Persistence of ARGs in soil is particularly variable as it often depends on a variety of factors (i.e., pH, temperature, moisture, and the microbial population) while ARBs persistence, much like in water, can range from weeks to days [483–487]. Soil bacteria and human pathogens are highly susceptible for HGT due to overlapping resistomes and ARGs have been shown to be capable of transforming some types of soil bacteria [488–490]. Regardless of the length of survival, HGT can occur in most environmental settings which can lead to increased resistances and exposure to new combinations of resistances for environmental microbes, humans, and animals.

Due to the risks associated with ARGs and ARBs it would seem imperative to begin to take measures to prevent the release of these contaminants to the environment [404]. Restricting the release of pathogens, antibiotics, and active metabolites into the environment would serve as a preventative measurement to protect human health. In order to accomplish this, more research is needed on the potentially harmful environmental impacts antibiotics can have in the generation of antibiotic resistance. It may be useful to evaluate the environmental risks as part of the authorization process for antibiotic approval for human usage so that there is available data to evaluate the potential environmental impact [404, 405]. It may be helpful to limit antibiotic use to more specific applications as a recent study revealed that 86% of the respondents used nonprescribed antibiotics to treat colds and coughs [491]. Indeed, better education about correct practices for taking and disposing of antibiotics could assist in minimizing contamination of the environment. Limiting or even banning the use of certain antibiotics to promote growth and/or use as preventative treatments for livestock could reduce the potential contamination from farms [463, 464].

While these methods may reduce potential contamination by limiting the amount of antibiotics used, it is important to provide proper remediation strategies for those antibiotics that will still be implemented in the treatment of infections. In the case of human application, the majority of antibiotic-related contaminants passes though wastewater treatment plants where specific measures could be taken to remove and dispose of antibiotic-related contaminants. Research has been performed with the focus of removing other pharmaceuticals from processed waste using additional steps in the treatment process, and further steps in the water treatment process might also remove and therefore limit environmental contamination from the treatment plants [492]. Fertilizers and manure could potentially be stored or treated using specific processes that could potentially limit the ability to allow ARGs and ARBs to persist long enough to contaminate the environments where they are used in animal farming and crop cultivation.

1.8 A Brief Comment on Microplastics

Several emerging materials and their potential for impact on the environment have been discussed thus far. While plastics are not an emerging material, the concerns about microplastics in the environment have been continuously growing over the past decade. As such, a brief overview of the origin and fate of microplastics will be discussed. Several in-depth reviews on the subject exist, which the authors would refer the reader to for further information [493–495].

Plastics are used in every day materials, from packaging, household products, electronics, fishing materials; there is an endless list of products containing plastic. Indeed, over 300 million tons of plastics were produced in 2015 [496] with a significant amount of these products designated for packaging purposes which fall under the category of single use and disposable. Due to the large quantities of plastic produced, this material inevitably finds its way into the environment, with a considerable amount of plastic material eventually ending up in the ocean each year [497]. Microplastics are synthetic solid polymeric particles that are insoluble in water and have a size range of 1 micron to 5 mm [493]. They have been found in an extensively large range of locations including polar regions, rivers, seawater, beaches, and sediment [497–500]. While microplastics have been found in air and soil, the vast majority of research has thus far focused on microplastics in marine environments [501, 502]. The composition of these microplastics varies significantly including, but not limited to, polyethylene, polystyrene, polypropylene, polyvinyl chloride, and polyethylene terephthalate [493–495, 503]. Sources of microplastics typically are defined as primary and secondary sources. Primary sources are the direct plastics from materials such as plastic pellets, industrial materials, household products, while secondary sources are degradation products of primary source materials, typically via aging/weathering [494, 500, 503, 504]. These aging/weathering mechanisms can include solar radiation, mechanical forces (hydraulic shear force, sand abrasion), oxidation, and to a lesser extent, degradation via microbial action [495]. Properties changed by the degradation may include particle size, surface morphology, and color [494, 505]. The shift in properties can affect the actions of the plastics; for instance, as the particle sizes decrease, the surface area exposed for degradation increases [506].

How these microplastics degrade and the resulting property changes are vital, as it has been found that microplastics can accumulate persistent organic pollutants and metals. The class of persistent organic pollutants that have been found to be absorbed by microplastics includes polycyclic aromatic hydrocarbons, per- and polyfluoroalkyl substances, polychlorinated biphenyls, and even antibiotics [493, 494, 507–509]. Additionally, heavy metals have been found to sorb onto plastics [510, 511]. However, the ability of the microplastics to accumulate these materials is dependent on the amount of pollution that is in the direct environment of the microplastics [494, 512]. Generally speaking, the sorption mechanism between these materials and microplastics can be varied and complicated depending on factors such as the polarity of the polymer and the pH of the environment [494, 509,

511]. For a more detailed discussion on how microplastics behave via degradation and accumulation, the work of Guo et al. [494] is enlightening.

Lastly, microplastics in the environment cannot be discussed without shedding light on their potential for toxicity. Marine species can ingest microplastics, causing physical damage to themselves from the plastics blocking their intestines, piercing the intestinal wall, or even with the microplastics moving toward their liver [513]. These physical damages cause the species to eat less, which may potentially lead to starvation and higher mortality rates. Indeed, it has been found that fish may confuse microplastics as prey and subsequently show decreased predatory performance [513]. Additionally, the pollutants adsorbed to the microplastics can accumulate and transfer to the next link of the food chain [513, 514]. As a result, plastics have been readily found in fish and other marine species [500, 503, 515]. Most research on the toxicity of microplastics have focused on marine biota, but few studies have focused on adverse effects in humans and other mammals. Microplastics have been detected in foods commonly consumed by humans including seafood, sea salt, honey, sugar, beer, and plain drinking water [493]. It has been shown that paper coffee cups can cause the ingestion of 25,000 microplastics per cup due to the degradation of the plastic lining when exposed to hot liquid [516]. Most microplastics ingested by humans have been shown to be excreted via feces, but it is possible the ingested microplastics could damage the human body along the way via adsorbed pollutants desorbing once in the body [517, 518]. The remainder of microplastics not excreted can persist and accumulate, creating health risks. In addition to microplastic exposure in food and water, it has been shown that humans could be inhaling up to 272 microplastic particles per day from the air in their own homes [519]. Depending on the size of these microplastic particles, they may persist in the lungs or be eventually cleared via various mechanisms [517]. As with ingested microplastics, it is possible that inhaled microplastics could damage the lungs and other organs if transported. It has been found that biofilm grows on the surfaces of microplastics in marine environments [493]; it is possible that infectious microbes from urban aerosols could latch onto microplastics and then be inhaled by humans or other animals, ultimately causing infection [517]. Microplastics have also been found to be a threat to mammalian male fertility [516].

1.9 Concluding Remarks

We presented a broad overview of some emerging materials either used in cutting-edge technologies or currently under investigation [520–530]. Some materials and their uses should be well known to the reader, and we hope to have addressed some of the niche technologies as well. Obviously, the advances in various fields will be driven by developments in materials science and engineering. Innovation is a continuous process and will continue to evolve as existing materials are improved upon and needs/use-cases arise. Thus, it is imperative also to continually assess the environmental impacts accompanying new advancements and innovations to safeguard the health and safety of our environment.

Three-dimensional printing (3DP) techniques have the capability to change the way products are manufactured. Three types of 3DP methods and several applications have been discussed. It is difficult to study the environmental and health impacts of these materials due to variability in the processes and materials. Each method requires a different type of printing material; and as research expands on the subject, the number and types of polymers, resins, and additives will continue to grow to match the versatility of the application. Due to the increased popularity of desktop 3DP, the FDM printer and its most common filaments, polylactic acid and acrylonitrile butadiene styrene, were discussed. While these polymer systems have different ratios which tend to be proprietary, some information is available on the degradation mechanisms. Studies detailing the particulate matter and gaseous materials produced while operating a 3DP have been presented. Direct investigations into the health effects of 3D-printed parts and emissions were detailed, and the topic of recycling filaments was presented. Overall, 3DP techniques have a wide range of possible applications that fuel the growth of fascinating research in this field. While some applications are in its infancy, the potential economic and environmental benefits will continue to spur growth in the field. Additionally, thermoplastics have long played a role in the world's industry, and continue to grow in usage. Given the high stability and prevalence of these plastics, the potential use of recycled filaments for 3DP has strong implications for transforming this technology into a more ecofriendly industry.

As long as plastics play a large role in processes, microplastics will continue to be generated and propagated through the environment. More research is necessary to elucidate the exact mechanisms through which microplastics affect biota and the associated health risks; nonetheless, the research conducted and reported thus far tell us that microplastic contaminants pose a significant threat to the Earth's environments and inhabitants.

Release of nanoparticles from nanocomposites may occur anytime throughout a nanocomposite's life cycle via multiple mechanisms. Different combinations of nanoparticles and matrices may be affected by different release mechanisms and thus determine whether the nanoparticles are released as a nanoparticle/matrix fragment, dissociated nanoparticles, or as ions. In addition, service conditions may alter the release mechanism due to both short and long-term exposure to rain, sunlight, extreme temperatures, etc. Numerous nanoparticle/matrix combinations are currently in use with more useful nanoparticle/matrix combinations constantly being discovered. Currently, only a few nanoparticles, such as the well-known carbon nanotubes (CNTs), have been explored in terms of their release mechanisms. There are numerous nanoparticles with little to no information on how they may be released into the environment. As research continues to highlight the potential toxicity of nanoparticles, it is imperative to understand the release mechanisms of these nanoparticles and to develop methods to prevent their release. If release is unavoidable, methods to detect and remove nanoparticles from environmental matrices must be improved as they are currently insufficient. As nanoparticles are the future of materials development, the need for understanding release mechanisms and developing sufficient environmental detection/mitigation techniques cannot be overstated.

Biomaterials that are used to make devices to replace a part or function of a body have to be biocompatible. Such biomaterials are required not to release any toxic chemicals throughout their application lifetime. The advances in the field will be driven by developments in materials science and engineering as new biomaterials become more widely available. Innovation will continue to progress as existing materials are improved upon.

Emerging cellulosic materials are a viable sustainable replacement for many harmful petroleum-derived chemicals and persistent materials. Cellulose can be extracted from many natural sources and waste products making it an ideal environmental building block. Furthermore, the wide array of cellulosic nanomaterials (CNMs) and tunability of the physicochemical properties make CNMs an intriguing material for various emerging applications. Production challenges for preparing CNMs remain a primary concern for future implementation as extraction and refinement of cellulose is an energy-intensive process. Increased scientific research is geared toward lowering the energy demands and environmental impacts associated with CNM preparation and product manufacture. Additionally, it is still uncertain how toxic CNMs are to human and environmental health. The few studies that have been performed indicate that several variables related to the CNM physicochemical property, cell type, exposure route, and concentration can influence the overall toxicity of the material. Therefore, there is a pressing need for further investigations prior to increased usage of cellulosic materials. Compared to other carbon and metal containing nanomaterials, CNMs have been identified to contain similar tunability and exciting properties, lower toxicity, and increased biodegradation that make them promising emerging materials for future commercialization and use.

It is critical that policy and regulation of E-waste be actively developed and maintained in order to keep up with the increasingly rapid development and use of electronic devices. It is important to study the effects that emerging materials will have on our world so as to enable the development of prudent restrictions where necessary and maintain practices that minimize or eliminate long-term hazards. For VOCs and electronics in our everyday lives, continued research and discussion will be required in order to educate consumers and policy-makers, as well as evaluate the merits of emerging materials production, use, and disposal in regard to the long-term consequences that might otherwise be neglected.

Antibiotics, active metabolites, ARGs, ARBs, and potentially dangerous pathogens such as superbugs can contaminate the environment by many vectors. Direct pathways can come from industrial release and treatment of various forms of livestock and aquaculture with antibiotics. Hospitals, the general population, and livestock operations can also contaminate the environment through sewage, wastewater, and manure that may be processed through a wastewater treatment plant or directly released into the environment. Humans, animals, and plants run the risk of exposure and even re-exposure to the contaminants, which can lead to an increase in antibiotic resistance and therefore negatively impact human health by decreasing the efficacy of antibiotics for infections. However, increased monitoring and specific remediation steps at municipal and agricultural wastewater treatment sites could play a significant role in minimizing the amount of antibiotic-related contamination and effects on the environment. Further, although not covered in the present chapter, the long term effect of the extensive use of cleaning supplies and antibacterial agents due to the COVID-19 pandemic may have impacts in the coming years and should be considered in future studies.

Acknowledgements This project was supported in part by an appointment to the Research Participation Program for the Sponsor administered by the Oak Ridge Institute for Science and Education (ORISE) through an agreement between the U.S. Department of Energy and the Agency. The use of trade, product, or firm names in this report is for descriptive purposes only and does not imply endorsement by the U.S. Government. The tests described and the resulting data presented herein, unless otherwise noted, were obtained from research conducted under the Installations and Operational Environments (IOE) Program of the United States Army Corps of Engineers. Permission was granted by the Chief of Engineers to publish this information. The findings of this report are not to be considered as an official Department of the Army position unless so designated by other authorized documents. Authors acknowledge the HPC resources at ERDC for computational time to perform this research. This document has been approved for public release (Distribution Statement A).

References

1. Kazan-Allen L (2000) Asbestos: properties, uses and problems. http://ibasecretariat.org/lka_prop.php. Accessed: 30 March 2022
2. King A (2017) Asbestos, explained. https://www.chemistryworld.com/news/why-asbestos-is-still-used-around-the-world/3007504.article. Accessed: 30 March 2022
3. EWG Action Fund (2021) Asbestos bans around the world. http://www.asbestosnation.org/facts/asbestos-bans-around-the-world/. Accessed: 30 March 2022
4. Schmaltz E, Melvin EC, Diana Z, Gunady EF, Rittschof D, Somarelli JA, Virdin J, Dunphy-Daly MM (2020) Plastic pollution solutions: emerging technologies to prevent and collect marine plastic pollution. Environ Int 144:106067
5. Lebreton L, Slat B, Ferrari F, Sainte-Rose B, Aitken J, Marthouse R, Hajbane S, Cunsolo S, Schwarz A, Levivier A, Noble K, Debeljak P, Maral H, Schoeneich-Argent R, Brambini R, Reisser J (2018) Evidence that the Great Pacific Garbage Patch is rapidly accumulating plastic. Sci Rep 8(1):1–15
6. Cox KD, Covernton GA, Davies HL, Dower JF, Juanes F, Dudas SE (2019) Human consumption of microplastics. Environ Sci Technol 53(12):7068–7074
7. Barboza LGA, Dick Vethaak A, Lavorante BRBO, Lundebye AK, Guilhermino L (2018) Marine microplastic debris: an emerging issue for food security, food safety and human health. Mar Pollut Bull 133(May):336–348
8. Ragusa A, Svelato A, Santacroce C, Catalano P, Notarstefano V, Oliana Carnevali, Fabrizio Papa, Rongioletti MCA, Baiocco F, Draghi S, D'Amore E, Denise Rinaldo, Matta M, Giorgini E (2021) Plasticenta: first evidence of microplastics in human placenta. Environ Int 146:106274
9. Ahrens L, Benskin JP, Cousins IT, Crimi M, Higgins CP (2019) Themed issues on per- and polyfluoroalkyl substances. Environ Sci Processes Impacts 21:1797–1802
10. EWG updates PFAS map: Toxic 'forever chemicals' contaminate 3,186 locations in U.S. https://www.ewg.org/news-insights/news-release/2023/08/ewg-updates-pfas-map-toxic-forever-chemicals-contaminate-3186/. Accessed: 31 March 2023

11. Barbo N, Stoiber T, Naidenko OV, Andrews DQ (2023) Locally caught freshwater fish across the United States are likely a significant source of exposure to PFOS and other perfluorinated compounds. Environ Res 220:115165
12. Per- and Polyfluoroalkyl Substances (PFAS)—Proposed PFAS National Primary Drinking Water Regulation. https://www.epa.gov/sdwa/and-polyfluoroalkyl-substances-pfas/. Accessed: 31 August 2023
13. Etz BD, Shukla MK, PFAS chemical degradation strategies: insights into the underlying reaction mechanism (2023) Curr Opin Chem Eng 42:100956
14. Cousins IT, Goldenman G, Herzke D, Lohmann R, Miller M, Ng CA, Patton S, Scheringer M, Trier X, Vierke L, Wang Z, Dewitt JC (2019) The concept of essential use for determining when uses of PFASs can be phased out. Environ Sci Process Impacts 21(11):1803–1815
15. Mahinroosta R, Senevirathna L (2020) A review of the emerging treatment technologies for PFAS contaminated soils. J Environ Manage 255:109896
16. Lu D, Sha S, Luo J, Huang Z, Jackie XZ (2020) Treatment train approaches for the remediation of per- and polyfluoroalkyl substances (PFAS): a critical review. J Hazard Mater 386:121963
17. Ahmed MB, Alam MM, Zhou JL, Xu B, Johir MAH, Karmakar AK, Rahman MS, Hossen J, Hasan AK, Moni MA (2020) Advanced treatment technologies efficacies and mechanism of per- and poly-fluoroalkyl substances removal from water. Process Saf Environ Prot 136:1014
18. Kreiger EL, Kreiger MA, Case MP (2019) Development of the construction processes for reinforced additively constructed concrete. Addit Manuf 28:39–49
19. Shahrubudin N, Lee TC, Ramlan R (2019) An overview on 3D printing technology: technological, materials, and applications. Procedia Manuf 35:1286–1296
20. Campbell T, Williams C, Ivanova O, Garrett B (2011) Could 3D printing change the world? http://www.atlanticcouncil.org/publications/reports/could-3d-printing-change-the-world. Accessed: 30 March 2022
21. Dawood A, Marti Marti B, Sauret-Jackson V, Darwood A (2015) 3D printing in dentistry. Br Dent J 219(11):521–529
22. Kurenov SN, Ionita C, Sammons D, Demmy TL (2015) Three-dimensional printing to facilitate anatomic study, device development, simulation, and planning in thoracic surgery. J Thorac Cardiovasc Surg 149(4):973-979.e1
23. Chen J, Zhang Z, Chen X, Zhang C, Zhang G, Zhewu X (2014) Design and manufacture of customized dental implants by using reverse engineering and selective laser melting technology. J Prosthetic Dentistry 112(5):1088-1095.e1
24. Popov VK, Evseev AV, Ivanov AL, Roginski VV, Volozhin AI, Howdle SM (2004) Laser stereolithography and supercritical fluid processing for custom-designed implant fabrication. J Mater Sci Mater Med 15(2):123–128
25. Wenyang X, Zhang X, Yang P, Långvik O, Wang X, Zhang Y, Cheng F, Österberg M, Willför S, Chunlin X (2019) Surface engineered biomimetic inks based on UV cross-linkable wood biopolymers for 3D printing. ACS Appl Mater Interfaces 11(13):12389–12400
26. Gupta MK, Meng F, Johnson BN, Kong YL, Tian L, Yeh YW, Masters N, Singamaneni S, McAlpine MC (2015) 3D printed programmable release capsules. Nano Lett 15(8):5321–5329
27. Yan Q, Dong H, Jin S, Han J, Song B, Wei Q, Shi Y (2018) A review of 3D printing technology for medical applications. Engineering 4(5):729–742
28. Lee Ventola C (2014) Medical applications for 3D printing: current and projected uses. Pharm Therapeut 39(10):704–711
29. Melchels FPW, Feijen J, Grijpma DW (2010) A review on stereolithography and its applications in biomedical engineering. Biomaterials 31(24):6121–6130
30. Schubert C, Van Langeveld MC, Donoso LA (2014) Innovations in 3D printing: a 3D overview from optics to organs. Br J Ophthalmol 98(2):159–161
31. Gross BC, Erkal JL, Lockwood SY, Chen C, Spence DM (2014) Evaluation of 3D printing and its potential impact on biotechnology and the chemical sciences. Anal Chem 86(7):3240–3253
32. Ursan ID, Chiu L, Pierce A (2013) Three-dimensional drug printing: a structured review. J Am Pharm Assoc 53(2):136–144
33. Bartlett S (2013) Printing organs on demand. Lancet Respir Med 1(9):684

34. Espalin D, Muse DW, MacDonald E, Wicker RB (2014) 3D Printing multifunctionality: structures with electronics. Int J Adv Manuf Technol 72(5–8):963–978
35. Joshi SC, Sheikh AA (2015) 3D printing in aerospace and its long-term sustainability. Virtual Phys Prototyp 10(4):175–185
36. Wang YC, Chen T, Yeh YL (2019) Advanced 3D printing technologies for the aircraft industry: a fuzzy systematic approach for assessing the critical factors. Int J Adv Manuf Technol 105(10):4059–4069
37. Dankar I, Pujolà M, El Omar F, Sepulcre F, Haddarah A (2018) Impact of mechanical and microstructural properties of potato puree-food additive complexes on extrusion-based 3D printing. Food Bioprocess Technol 11(11):2021–2031
38. Liu Z, Zhang M, Bhandari B, Wang Y (2017) 3D printing: printing precision and application in food sector. Trends Food Sci Technol 69:83–94
39. Vanderploeg A, Lee S-E, Mamp M (2017) The application of 3D printing technology in the fashion industry. Int J Fash Des Technol Educ 10(2):170–179
40. Jammalamadaka U, Tappa K (2018) Recent advances in biomaterials for 3D printing and tissue engineering. J Funct Biomater 9(1):22
41. Liska R, Schuster M, Inführ R, Turecek C, Fritscher C, Seidl B, Schmidt V, Kuna L, Haase A, Varga F, Lichtenegger H, Stampfl J (2007) Photopolymers for rapid prototyping. J Coatings Technol Res 4(4):505–510
42. Schuster M, Turecek C, Kaiser B, Stampfl J, Liska R, Varga F (2007) Evaluation of biocompatible photopolymers I: photoreactivity and mechanical properties of reactive diluents. J Macromol Sci Part A 44(5):547–557
43. Schuster M, Turecek C, Mateos A, Stampfl J, Liska R, Varga F (2007) Evaluation of biocompatible photopolymers II: further reactive diluents. Monatshefte fur Chemie 138(4):261–268
44. Kumar S (2003) Selective laser sintering: a qualitative and objective approach. JOM 55(10):43–47
45. Cabrera A, Pellejero I, Oroz-Mateo T, Salazar C, Navajas A, Fernández-Acevedo C, Gandía LM (2020) Three-dimensional printing of acrylonitrile butadiene styrene microreactors for photocatalytic applications. Ind Eng Chem Res 59(47):20686–20692
46. Zhao XG, Hwang KJ, Lee D, Kim T, Kim N (2018) Enhanced mechanical properties of self-polymerized polydopamine-coated recycled PLA filament used in 3D printing. Appl Surf Sci 441(March):381–387
47. Potter PM, Al-Abed SR, Lay D, Lomnicki SM (2019) VOC emissions and formation mechanisms from carbon nanotube composites during 3D printing. Environ Sci Technol 53(8):4364–4370
48. Alex A, Ilango NK, Ghosh P (2018) Comparative role of chain scission and solvation in the biodegradation of polylactic acid (PLA). J Phys Chem B 122(41):9516–9526
49. Mendes L, Kangas A, Kukko K, Mølgaard B, Säämänen A, Kanerva T, Flores Ituarte I, Huhtiniemi M, Stockmann-Juvala H, Partanen J, Hämeri K, Eleftheriadis K, Viitanen AK (2017) Characterization of emissions from a desktop 3D printer. J Ind Ecol 21:S94–S106
50. Chamas A, Moon H, Zheng J, Qiu Y, Tabassum T, Jang JH, Abu-Omar M, Scott SL, Suh S (2020) Degradation rates of plastics in the environment. ACS Sustain Chem Eng 8(9):3494–3511
51. Navarro M, Ginebra MP, Planell JA, Barrias CC, Barbosa MA (2005) In vitro degradation behavior of a novel bioresorbable composite material based on PLA and a soluble CaP glass. Acta Biomater 1(4):411–419
52. Russias J, Saiz E, Nalla RK, Gryn K, Ritchie RO, Tomsia AP (2006) Fabrication and mechanical properties of PLA/HA composites: a study of in vitro degradation. Mater Sci Eng C, Biomim Supramol Syst 26(8):1289–1295
53. Gupta MC, Deshmukh VG (1982) Thermal oxidative degradation of poly-lactic acid–Part II: molecular weight and electronic spectra during isothermal heating. Colloid Polym Sci 260(5):514–517
54. Rutkowski JV, Levin BC (1986) Acrylonitrile-butadiene-styrene copolymers (ABS): pyrolysis and combustion products and their toxicity–a review of the literature. Fire Mater 10(3–4):93–105

55. Tiganis BE, Burn LS, Davis P, Hill AJ (2002) Thermal degradation of acrylonitrile-butadiene-styrene (ABS) blends. Polym Degrad Stab 76(3):425–434
56. Lithner D, Larsson A, Dave G (2011) Environmental and health hazard ranking and assessment of plastic polymers based on chemical composition. Sci Total Environ 409(18):3309–3324
57. Unwin J, Coldwell MR, Keen C, McAlinden JJ (2013) Airborne emissions of carcinogens and respiratory sensitizers during thermal processing of plastics. Ann Occup Hyg 57(3):399–406
58. Stephens B, Azimi P, El Orch Z, Ramos T (2013) Ultrafine particle emissions from desktop 3D printers. Atmos Environ 79:334–339
59. Kim Y, Yoon C, Ham S, Park J, Kim S, Kwon O, Tsai P (2015) Emissions of nanoparticles and gaseous material from 3D printer operation. Environ Sci Technol 49(20):12044–12053
60. Zontek TL, Ogle BR, Jankovic JT, Hollenbeck SM (2017) An exposure assessment of desktop 3D printing. J Chem Heal Saf 24(2):15–25
61. Kwon O, Yoon C, Ham S, Park J, Lee J, Yoo D, Kim Y (2017) Characterization and control of nanoparticle emission during 3D printing. Environ Sci Technol 51(18):10357–10368
62. Vance ME, Pegues V, Van Montfrans S, Leng W, Marr LC (2017) Aerosol emissions from fuse-deposition modeling 3D printers in a chamber and in real indoor environments. Environ Sci Technol 51(17):9516–9523
63. Zontek TL, Hollenbeck S, Jankovic J, Ogle BR (2019) Modeling particle emissions from three-dimensional printing with acrylonitrile-butadiene-styrene polymer filament. Environ Sci Technol 53(16):9656–9663
64. Azimi P, Zhao S, Pouzet C, Crain NE, Stephens B (2016) Emissions of ultrafine particles and volatile organic compounds from commercially available desktop three-dimensional printers with multiple filaments. Environ Sci Technol 50(3):1260–1268
65. Wojtyła S, Klama P, Baran T (2017) Is 3D printing safe? Analysis of the thermal treatment of thermoplastics: ABS, PLA, PET, and nylon. J Occup Environ Hyg 14(6):D80–D85
66. Stefaniak AB, Lebouf RF, Yi J, Ham J, Nurkewicz T, Schwegler-Berry DE, Chen BT, Raymond Wells J, Duling MG, Lawrence RB, Martin SB, Johnson AR, Abbas Virji M (2017) Characterization of chemical contaminants generated by a desktop fused deposition modeling 3-dimensional printer. J Occup Environ Hyg. 14(7):540–550
67. House R, Rajaram N, Tarlo SM (2017) Case report of asthma associated with 3D printing. Occup Med (Chic Ill)
68. Stefaniak AB, LeBouf RF, Duling MG, Yi J, Abukabda AB, McBride CR, Nurkiewicz TR (2017) Inhalation exposure to three-dimensional printer emissions stimulates acute hypertension and microvascular dysfunction. Toxicol Appl Pharmacol 335:1–5
69. Oskui SM, Diamante G, Liao C, Shi W, Gan J, Schlenk D, Grover WH (2016) Assessing and reducing the toxicity of 3D-printed parts. Environ Sci Technol Lett 3(1):1–6
70. Inoue Y, Ikuta K (2013) Detoxification of the photoeurable polymer by heat treatment for microstereolithography. Procedia CIRP 5:115–118
71. Ong TK, Choo HL, Choo WJ, Koay SC, Pang MM (2020) Recycling of polylactic acid (PLA) wastes from 3D printing laboratory. Advances in manufacturing engineering. Springer Singapore, Singapore, pp 725–732
72. Woern AL, Byard DJ, Oakley RB, Fiedler MJ, Snabes SL, Pearce JM (2018) Fused particle fabrication 3-D printing: recycled materials' optimization and mechanical properties. Materials 11(8):1413
73. Mohammed MI, Wilson D, Gomez-Kervin E, Tang B, Wang J (2019) Investigation of closed-loop manufacturing with acrylonitrile butadiene styrene over multiple generations using additive manufacturing. ACS Sustain Chem Eng 7(16):13955–13969
74. Mohammed MI, Wilson D, Gomez-Kervin E, Vidler C, Rosson L, Long J (2018) The recycling of E-Waste ABS plastics by melt extrusion and 3D printing using solar powered devices as a transformative tool for humanitarian aid. In: Solid freeform fabrication 2018: proceedings of the 29th annual international solid freeform fabrication symposium—an additive manufacturing conference, pp 80–92
75. Gaikwad V, Ghose A, Cholake S, Rawal A, Iwato M, Sahajwalla V (2018) Transformation of E-waste plastics into sustainable filaments for 3D printing. ACS Sustain Chem Eng 6(11):14432–14440

76. Nowack B, Ranville JF, Diamond S, Gallego-Urrea JA, Metcalfe C, Rose J, Horne N, Koelmans AA, Klaine SJ (2012) Potential scenarios for nanomaterial release and subsequent alteration in the environment. Environ Toxicol Chem 31(1):50–59
77. Kingston C, Zepp R, Andrady A, Boverhof D, Fehir R, Hawkins D, Roberts J, Sayre P, Betsy Shelton, Sultan Y, Vejins V, Wohlleben W (2014) Release characteristics of selected carbon nanotube polymer composites. Carbon N Y 68:33–57
78. Bello D, Wardle BL, Yamamoto N, deVilloria RG, Garcia EG, Hart AJ, Ahn K, Ellenbecker MJ, Hallock M (2009) Exposure to nanoscale particles and fibers during machining of hybrid advanced composites containing carbon nanotubes. J Nanoparticle Res 11(1):231–249
79. Bello B, Wardle BL, Zhang J, Yamamoto N, Santeufemio C, Hallock M, Abbas Virji M (2010) Characterization of exposures to nanoscale particles and fibers during solid core drilling of hybrid carbon nanotube advanced composites. Int J Occup Environ Health 16(4):434–450
80. Ogura I, Shigeta M, Kotake M, Uejima M, Honda K (2015) Particle release from single-wall and multiwall carbon nanotubes in polystyrene-based composites during grinding. J Phys Conf Ser 617:12028
81. Wohlleben W, Brill S, Meier MW, Mertler Cox G, Hirth S, von Vacano B, Strauss V, Treumann S, Wiench K, Ma-Hock L, Landsiedel R (2011) On the lifecycle of nanocomposites: comparing released fragments and their in-vivo hazards from three release mechanisms and four nanocomposites. Small 7(16):2384–2395
82. Wohlleben W, Meier MW, Vogel S, Landsiedel R, Cox G, Hirth S, Tomović Ž (2013) Elastic CNT-polyurethane nanocomposite: synthesis, performance and assessment of fragments released during use. Nanoscale 5(1):369–380
83. Cena LG, Peters TM (2011) Characterization and control of airborne particles emitted during production of epoxy/carbon nanotube nanocomposites. J Occup Environ Hyg 8(2):86–92
84. Huang G, Park JH, Cena LG, Shelton BL, Peters TM (2012) Evaluation of airborne particle emissions from commercial products containing carbon nanotubes. J Nanoparticle Res 14(11):1231
85. Schlagenhauf L, Chu BTT, Buha J, Nüesch F, Wang J (2012) Release of carbon nanotubes from an epoxy-based nanocomposite during an abrasion process. Environ Sci Technol 46(13):7366–7372
86. Golanski L, Gaborieau A, Guiot A, Uzu G, Chatenet J, Tardif F (2011) Characterization of abrasion-induced nanoparticle release from paints into liquids and air. J Phys Conf Ser 304:12062
87. Wohlleben W, Neubauer N (2016) Quantitative rates of release from weathered nanocomposites are determined across 5 orders of magnitude by the matrix, modulated by the embedded nanomaterial. NanoImpact 1:39–45
88. Nguyen T, Pellegrin B, Bernard C, Gu X, Gorham JM, Stutzman P, Stanley D, Shapiro A, Byrd E, Hettenhouser R, Chin J (2011) Fate of nanoparticles during life cycle of polymer nanocomposites. J Phys Conf Ser 304:012060
89. Hirth S, Cena L, Cox G, Tomović Ž, Peters T, Wohlleben W (2013) Scenarios and methods that induce protruding or released CNTs after degradation of nanocomposite materials. J Nanoparticle Res 15(4):1504
90. Bernard C, Nguyen T, Pellegrin B, Holbrook RD, Zhao M, Chin J (2011) Fate of graphene in polymer nanocomposite exposed to UV radiation. J Phys Conf Ser 304:012063
91. George S, Gardner H, Seng EK, Chang H, Wang C, Fang CHY, Richards M, Valiyaveettil S, Chan WK (2014) Differential effect of solar light in increasing the toxicity of silver and titanium dioxide nanoparticles to a fish cell line and zebrafish embryos. Environ Sci Technol 48(11):6374–6382
92. Park H-G, Kim JI, Kang M, Yeo M-K (2014) The effect of metal-doped TiO_2 nanoparticles on zebrafish embryogenesis. Mol Cell Toxicol 10(3):293–301
93. Shin Y-J, Nam S-H, An Y-J (2016) Continuous ultraviolet irradiation increases the adverse effects of photoreactive nanoparticles on the early development of Oryzias latipes. Environ Toxicol Chem 35(5):1195–1200

94. Laux P, Riebeling C, Booth AM, Brain JD, Brunner J, Cerrillo C, Creutzenberg O, Estrela-Lopis I, Gebel T, Johanson G, Jungnickel H, Kock H, Tentschert J, Tlili A, Schäffer A, Sips AJAM, Yokel RA, Luch A (2018) Challenges in characterizing the environmental fate and effects of carbon nanotubes and inorganic nanomaterials in aquatic systems. Environ Sci Nano 5(1):48–63
95. Hennig MP, Maes HM, Ottermanns R, Schäffer A, Siebers N (2019) Release of radiolabeled multi-walled carbon nanotubes (14C-MWCNT) from epoxy nanocomposites into quartz sand-water systems and their uptake by Lumbriculus variegatus. NanoImpact 14:100159
96. Ging J, Tejerina-Anton R, Ramakrishnan G, Nielsen M, Murphy K, Gorham JM, Nguyen T, Orlov A (2014) Development of a conceptual framework for evaluation of nanomaterials release from nanocomposites: environmental and toxicological implications. Sci Total Environ 473–474:9–19
97. Petersen EJ, Lam T, Gorham JM, Scott KC, Long CJ, Stanley D, Sharma R, Alexander Liddle J, Pellegrin B, Nguyen T (2014) Methods to assess the impact of UV irradiation on the surface chemistry and structure of multiwall carbon nanotube epoxy nanocomposites. Carbon 69:194–205
98. Schlagenhauf L, Kianfar B, Buerki-Thurnherr T, Kuo Y-Y, Wichser A, Nüesch F, Wick P, Wang J (2015) Weathering of a carbon nanotube/epoxy nanocomposite under UV light and in water bath: impact on abraded particles. Nanoscale 7:18524–18536
99. Wohlleben W, Meyer J, Muller J, Müller P, Vilsmeier K, Stahlmecke B, Kuhlbusch TAJ (2016) Release from nanomaterials during their use phase: combined mechanical and chemical stresses applied to simple and multi-filler nanocomposites mimicking wear of nano-reinforced tires. Environ Sci Nano 3:1036–1051
100. Koivisto AJ, Jensen ACØ, Kling KI, Nørgaard A, Brinch A, Christensen F, Jensen KA (2017) Quantitative material releases from products and articles containing manufactured nanomaterials: towards a release library. NanoImpact 5:119–132
101. Nguyen T, Petersen EJ, Pellegrin B, Gorham JM, Lam T, Zhao M, Sung L (2017) Impact of UV irradiation on multiwall carbon nanotubes in nanocomposites: formation of entangled surface layer and mechanisms of release resistance. Carbon 116:191–200
102. Zulin H, Zhang J, Bai X, Ye Z, Zhiqiang T, Lu L, Yuqi L (2016) Aggregation of TiO_2-graphene nanocomposites in aqueous environment: influence of environmental factors and UV irradiation. Sci Total Environ 539:196–205
103. Donia DT, Carbone M (2019) Fate of the nanoparticles in environmental cycles. Int J Environ Sci Technol 16(1):583–600
104. Troester M, Brauch H-J, Hofmann T (2016) Vulnerability of drinking water supplies to engineered nanoparticles. Water Res 96:255–279
105. Zanna S, Saulou C, Mercier-Bonin M, Despax B, Raynaud P, Seyeux A, Marcus P (2010) Ageing of plasma-mediated coatings with embedded silver nanoparticles on stainless steel: an XPS and ToF-SIMS investigation. Appl Surf Sci 256(22):6499–6505
106. Kaegi R, Sinnet B, Zuleeg S, Hagendorfer H, Mueller E, Vonbank R, Boller M, Burkhardt M (2010) Release of silver nanoparticles from outdoor facades. Environ Pollut 158(9):2900–2905
107. Froggett SJ, Clancy SF, Boverhof DR, Canady RA (2014) A review and perspective of existing research on the release of nanomaterials from solid nanocomposites. Part Fibre Toxicol 11(1):1–28
108. Kulthong K, Srisung S, Boonpavanitchakul K, Kangwansupamonkon W, Maniratanachote R (2010) Determination of silver nanoparticle release from antibacterial fabrics into artificial sweat. Part Fibre Toxicol 7(1):8
109. von Goetz N, Lorenz C, Windler L, Nowack B, Heuberger M, Hungerbühler K (2013) Migration of Ag- and TiO_2-(nano)particles from textiles into artificial sweat under physical stress: experiments and exposure modeling. Environ Sci Technol 47(17):9979–9987
110. Benn TM, Westerhoff P (2008) Nanoparticle silver released into water from commercially available sock fabrics. Environ Sci Technol 42(18):7025–7026

111. Geranio L, Heuberger M, Nowack B (2009) The behavior of silver nanotextiles during washing. Environ Sci Technol 43(21):8113–8118
112. Pasricha A, Jangra SL, Singh N, Dilbaghi N, Sood KN, Arora K, Pasricha R (2012) Comparative study of leaching of silver nanoparticles from fabric and effective effluent treatment. J Environ Sci 24(5):852–859
113. Benn T, Cavanagh B, Hristovski K, Posner JD, Westerhoff P (2010) The release of nanosilver from consumer products used in the home. J Environ Qual 39(6):1875–1882
114. Lorenz C, Windler L, von Goetz N, Lehmann RP, Schuppler M, Hungerbühler K, Heuberger M, Nowack B (2012) Characterization of silver release from commercially available functional (nano)textiles. Chemosphere 89(7):817–824
115. Windler L, Lorenz C, von Goetz N, Hungerbühler K, Amberg M, Heuberger M, Nowack B (2012) Release of titanium dioxide from textiles during washing. Environ Sci Technol 46(15):8181–8188
116. Stebounova LV, Guio E, Grassian VH (2011) Silver nanoparticles in simulated biological media: a study of aggregation, sedimentation, and dissolution. J Nanoparticle Res 13(1):233–244
117. Beddow J, Stolpe B, Cole PA, Lead JR, Sapp M, Lyons BP, Colbeck I, Whitby C (2017) Nanosilver inhibits nitrification and reduces ammonia-oxidising bacterial but not archaeal amoA gene abundance in estuarine sediments. Environ Microbiol 19(2):500–510
118. Dulger M, Sakallioglu T, Temizel I, Demirel B, Copty NK, Onay TT, Uyguner-Demirel CS, Karanfil T (2016) Leaching potential of nano-scale titanium dioxide in fresh municipal solid waste. Chemosphere 144:1567–1572
119. Sakallioglu T, Bakirdoven M, Temizel I, Demirel B, Copty NK, Onay TT, Uyguner Demirel CS, Karanfil T (2016) Leaching of nano-ZnO in municipal solid waste. J Hazard Mater 317:319–326
120. Sun J, Guo L-H, Zhang H, Zhao L (2014) UV irradiation induced transformation of TiO_2 nanoparticles in water: aggregation and photoreactivity. Environ Sci Technol 48(20):11962–11968
121. Romanello MB, de Cortalezzi MMF (2013) An experimental study on the aggregation of TiO_2 nanoparticles under environmentally relevant conditions. Water Res 47(12):3887–3898
122. Lei W, Liu L, Gao B, Muñoz-Carpena R, Zhang M, Chen H, Zhou Z, Wang H (2013) Aggregation kinetics of graphene oxides in aqueous solutions: experiments, mechanisms, and modeling. Langmuir 29(49):15174–15181
123. Merrifield RC, Stephan C, Lead J (2017) Determining the concentration dependent transformations of Ag nanoparticles in complex media: using SP-ICP-MS and Au@Ag core-shell nanoparticles as tracers. Environ Sci Technol 51(6):3206–3213
124. Thio BJR, Zhou D, Keller AA (2011) Influence of natural organic matter on the aggregation and deposition of titanium dioxide nanoparticles. J Hazard Mater 189(1–2):556–563
125. Bouillard JX, R'Mili B, Moranviller D, Vignes A, Le Bihan O, Ustache A, Bomfim JAS, Frejafon E, Fleury D (2013) Nanosafety by design: risks from nanocomposite/nanowaste combustion. J Nanoparticle Res 15(4):1519
126. Sotiriou GA, Singh D, Zhang F, Chalbot MG, Spielman-Sun E, Hoering L, Kavouras IG, Lowry GV, Wohlleben W, Demokritou P (2016) Possible environmental health and safety implications. Thermal decomposition of nano-enabled thermoplastics. J Hazard Mater 305:87–95
127. Vejerano EP, Ma Y, Holder AL, Pruden A, Elankumaran S, Marr LC (2015) Toxicity of particulate matter from incineration of nanowaste. Environ Sci Nano 2(2):143–154
128. Meier C, Voegelin A, del Real AP, Sarret G, Mueller CR, Kaegi R (2016) Transformation of silver nanoparticles in sewage sludge during incineration. Environ Sci Technol 50(7):3503–3510
129. Li S, Ma H, Wallis LK, Etterson MA, Riley B, Hoff DJ, Diamond SA (2016) Impact of natural organic matter on particle behavior and phototoxicity of titanium dioxide nanoparticles. Sci Total Environ 542:324–333

130. Zhi CY, Bando Y, Tang CC, Huang Q, Golberg D (2008) Boron nitride nanotubes: functionalization and composites. J Mater Chem 18(33):3900
131. Augustine J, Cheung T, Gies V, Boughton J, Chen M, Jakubek ZJ, Walker S, Martinez-Rubi Y, Simard B, Zou S (2019) Assessing size-dependent cytotoxicity of boron nitride nanotubes using a novel cardiomyocyte AFM assay. Nanoscale Adv 1(5):1914–1923
132. Wang H, Dong Y-n, Zhu M, Li Z, Keller AA, Wang T, Li F (2015) Heteroaggregation of engineered nanoparticles and kaolin clays in aqueous environments. Water Res 80:130–138
133. Park J, Lakes RS (2007) Biomaterials: an introduction, 3rd edn. Springer Science & Business Media, New York, NY
134. Hench LL (1991) Bioceramics: from concept to clinic. J Am Ceram Soc 74(7):1487–1510
135. Sáenz A, Rivera E, Brostow W, Castaño VM (1999) Ceramic biomaterials: an introductory overview. J Mater Educ 21(5/6):267–276
136. Prabha RD, Ding M, Bollen P, Ditzel N, Varma HK, Nair PD, Kassem M (2020) Strontium ion reinforced bioceramic scaffold for load bearing bone regeneration. Mater Sci Eng C 109:110427
137. Zamet JS, Darbar UR, Griffiths GS, Bulman JS, Brägger U, Bürgin W, Newman HN (1997) Particulate bioglass® as a grafting material in the treatment of periodontal intrabony defects. J Clin Periodontol 24(6):410–418
138. Bellucci D, Salvatori R, Cannio M, Luginina M, Orrù R, Montinaro S, Anesi A, Chiarini L, Cao G, Cannillo V (2018) Bioglass and bioceramic composites processed by spark plasma sintering (SPS): biological evaluation versus SBF test. Biomed Glasses 4(1):21–31
139. Hench LL, Splinter RJ, Allen WC, Greenlee TK (1971) Bonding mechanisms at the interface of ceramic prosthetic materials. J Biomed Mater Res 5(6):117–141
140. Oonishi H, Kushitani S, Iwaki H (1995) Comparative bone formation in several kinds of bioceramic granules. In: 8th international symposium on ceramics in medicine, vol 8, pp 137–144
141. Li RWK, Chow TW, Matinlinna JP (2014) Ceramic dental biomaterials and CAD/CAM technology: state of the art. J Prosthodont Res 58(4):208–216
142. Arcos D, Vallet-Regí M (2013) Bioceramics for drug delivery. Acta Mater 61(3):890–911
143. Hossen S, Hossain MK, Basher MK, Mia MNH, Rahman MT, Jalal Uddin M (2018) Smart nanocarrier-based drug delivery systems for cancer therapy and toxicity studies: a review. J Adv Res 15:1–18
144. Zhu C-L, Song X-Y, Zhou W-H, Yang H-H, Wen Y-H, Wang X-R (2009) An efficient cell-targeting and intracellular controlled-release drug delivery system based on MSN-PEM-aptamer conjugates. J Mater Chem 19(41):7765–7770
145. Tenland E, Pochert A, Krishnan N, Rao KU, Kalsum S, Braun K, Glegola-Madejska I, Lerm M, Robertson BD, Lindén M, Godaly G (2019) Effective delivery of the anti-mycobacterial peptide nzx in mesoporous silica nanoparticles. PLoS One 14(2):1–16
146. Huang J, Li X, Guo ZX (2020) Biomechanical and biochemical compatibility in innovative biomaterials. In: Boutrand JP (ed) Biocompatibility performance of medical devices, 2nd edn. Woodhead Publishing series in biomaterials, Woodhead Publishing, pp 23–46
147. Cheng GJ, Pirzada D, Cai M, Mohanty P, Bandyopadhyay A (2005) Bioceramic coating of hydroxyapatite on titanium substrate with Nd-YAG laser. Mater Sci Eng C 25(4):541–547
148. Ong JL, Lucas LC, Lacefield WR, Rigney ED (1992) Structure, solubility and bond strength of thin calcium phosphate coatings produced by ion beam sputter deposition. Biomaterials 13(4):249–254
149. Heimann RB (2006) Thermal spraying of biomaterials. Surf Coatings Technol 201(5):2012–2019
150. Hasirci V, Hasirci N (2018) Fundamentals of biomaterials. Springer, New York, NY
151. Noam Eliaz (2019) Corrosion of metallic biomaterials: a review. Materials 12(3):407
152. Li Y, Yang C, Zhao H, Qu S, Li X, Li Y (2014) New developments of Ti-based alloys for biomedical applications. Materials (Basel) 7:1709–1800
153. Wang N, Li H, Wang J, Su C, Ma Y, Zhang Z (2012) Study on the anticorrosion, biocompatibility, and osteoinductivity of tantalum decorated with tantalum oxide nanotube array films. ACS Appl Mater Interfaces 4(9):4516–4523

154. Budge MD, Kurdziel MD, Baker KC, Wiater JM (2013) A biomechanical analysis of initial fixation options for porous-tantalum-backed glenoid components. J Shoulder Elb Surg 22(5):709–715
155. Chen Q, Thouas GA (2015) Metallic implant biomaterials. Mater Sci Eng R Rep 87:1–57
156. Elahinia MH, Hashemi M, Tabesh M, Bhaduri SB (2012) Manufacturing and processing of NiTi implants: a review. Prog Mater Sci 57(5):911–946
157. Tsai M-H, Yeh J-W (2014) High-entropy alloys: a critical review. Mater Res Lett 2(3):107–123
158. Todai M, Nagase T, Hori T, Matsugaki A, Sekita A, Nakano T (2017) Novel TiNbTaZrMo high-entropy alloys for metallic biomaterials. Scr Mater 129:65–68
159. Yuan Y, Yuan W, Yang Z, Liang X, Lei Z, Huang H, Wang H, Liu X, An K, Wei Wu, Lu Z (2019) Formation, structure and properties of biocompatible TiZrHfNbTa high-entropy alloys. Mater Res Lett 7(6):225–231
160. Chen M (2011) A brief overview of bulk metallic glasses. NPG Asia Mater 3(9):82–90
161. Szyba D, Rynkiewicz EC, Babilas R (2019) Crystallization and corrosion study of $Ca_{57.5}Mg_{15}Zn_{27.5}$ bulk metallic glasses. Acta Phys Pol A 135:129–132
162. Sawyer V, Tao X, Dong H, Dashtbozorg B, Li X, Sammons R, Dong H-S (2020) Improving the tribological properties and biocompatibility of Zr-based bulk metallic glass for potential biomedical applications. Mater (Basel) 13:1960
163. Jadhav BS (2001) Bioresorbable stent. Patent Application, 20010029398A1, United States
164. Kirkland NT, Birbilis N, Walker J, Woodfield T, Dias GJ, Staiger MP (2010) In-vitro dissolution of magnesium-calcium binary alloys: clarifying the unique role of calcium additions in bioresorbable magnesium implant alloys. J Biomed Mater Res Part B Appl Biomater 95B(1):91–100
165. Shoichet MS (2010) Polymer scaffolds for biomaterials applications. Macromolecules 43(2):581–591
166. Paxton NC, Allenby MC, Lewis PM, Woodruff MA (2019) Biomedical applications of polyethylene. Eur Polym J 118:412–428
167. McKeen LW (2014) 3–Plastics used in medical devices. In: Modjarrad K, Ebnesajjad S (eds) Handbook of polymer applications in medicine and medical devices. William Andrew Publishing, Oxford, Plastics Design Library, pp 21–53
168. Scheidbach H, Tamme C, Tannapfel A, Lippert H, Köckerling F (2004) In vivo studies comparing the biocompatibility of various polypropylene meshes and their handling properties during endoscopic total extraperitoneal (TEP) patchplasty: an experimental study in pigs. Surg Endosc Other Interv Tech 18(2):211–220
169. Zheng F, Luo X, Verbiest L, Verbeken E, De Ridder D, Deprest J (2007) Cytokine production following experimental implantation of xenogenic dermal collagen and polypropylene grafts in mice. Neurourol Urodyn 26(2):280–289
170. Dinesh Kumar S, Anandan R, Ekanthamoorthy J (2015) Advances in dental-PMMA based artificial teeth through rapid prototyping technology. Biomed Res 26:6–8
171. Matellan C, del Río Hernández AE (2018) Cost-effective rapid prototyping and assembly of poly(methyl methacrylate) microfluidic devices. Sci Rep 8(1):6971
172. Heran MKS, Pham TH, Butterworth S, Robinson A (2011) Use of a microporous polytetrafluoroethylene catheter balloon to treat refractory esophageal stricture: a novel technique for delivery of mitomycin C. J Pediatr Surg 46(4):776–779
173. Raju S (1987) PTFE grafts for hemodialysis access. Techniques for insertion and management of complications. Ann Surg 206(5):666–673
174. Houis S, Engelhardt EM, Wurm F, Gries T (2010) Application of polyvinylidene fluoride (PVDF) as a biomaterial in medical textiles. In: Anand SC, Kennedy JF, Miraftab M, Rajendran S (eds) Medical and healthcare textiles. Woodhead Publishing series in textiles, Woodhead Publishing, pp 342–352
175. Cardoso VF, Correia DM, Ribeiro C, Fernandes MM, Lanceros-Méndez S (2018) Fluorinated polymers as smart materials for advanced biomedical applications. Polymers (Basel) 10(2):161

176. Altman GH, Diaz F, Jakuba C, Calabro T, Horan RL, Chen J, Lu H, Richmond J, Kaplan DL (2003) Silk-based biomaterials. Biomaterials 24(3):401–416
177. Pruitt L, Furmanski J (2009) Polymeric biomaterials for load-bearing medical devices. JOM 61(9):14–20
178. Menchaca-Campos C, Sarmiento-Bustos E, Uruchurtu J (2013) Synthesis and characterization of a nylon/titanium as hybrid biomaterial. Recent Patents Corros Sci (Discontinued) 3(1):12–21
179. Lin S-J, Hsiao W-C, Jee S-H, Yu H-S, Tsai T-F, Lai J-Y, Young T-H (2006) Study on the effects of nylon-chitosan-blended membranes on the spheroid-forming activity of human melanocytes. Biomaterials 27(29):5079–5088
180. Chen H, Xiao H, Cebe P (2008) Thermal properties and phase transitions in nylon-6 and silk fibroin blends. J Therm Anal Calorim 93:201–206
181. Maitz MF (2015) Applications of synthetic polymers in clinical medicine. Biosurface Biotribol 1(3):161–176
182. Nair LS, Laurencin CT (2007) Biodegradable polymers as biomaterials. Prog Polym Sci 32(8):762–798
183. Sionkowska A (2011) Current research on the blends of natural and synthetic polymers as new biomaterials: review. Prog Polym Sci 36(9):1254–1276
184. Wichterle O, Lím D (1960) Hydrophilic gels for biological use. Nature 185(4706):117–118
185. Yannas IV, Lee E, Orgill DP, Skrabut EM, Murphy GF (1989) Synthesis and characterization of a model extracellular matrix that induces partial regeneration of adult mammalian skin. Proc Natl Acad Sci U S A 86(3):933–937
186. Nguyen KT, West JL (2002) Photopolymerizable hydrogels for tissue engineering applications. Biomaterials 23(22):4307–4314
187. Choe G, Park J, Park H, Lee JY (2018) Hydrogel biomaterials for stem cell microencapsulation. Polymers (Basel) 10(9):997
188. Saludas L, Pascual-Gil S, Prósper F, Garbayo E, Blanco-Prieto M (2017) Hydrogel based approaches for cardiac tissue engineering. Int J Pharm 523(2):454–475
189. Peña B, Laughter M, Jett S, Rowland TJ, Taylor MRG, Mestroni L, Park D (2018) Injectable hydrogels for cardiac tissue engineering. Macromol Biosci 18(6):1800079
190. El-Sherbiny IM (2013) Yacoub MH (2013) Hydrogel scaffolds for tissue engineering: progress and challenges. Glob Cardiol Sci Pract 3:316–342
191. Otani Y, Tabata Y, Ikada Y (1999) Sealing effect of rapidly curable gelatin-poly (l-glutamic acid) hydrogel glue on lung air leak. Ann Thorac Surg 67(4):922–926
192. Stapleton F, Stretton S, Papas E, Skotnitsky C, Sweeney DF (2006) Silicone hydrogel contact lenses and the ocular surface. Ocul Surf 4(1):24–43
193. Kamoun EA, Kenawy ES, Chen X (2017) A review on polymeric hydrogel membranes for wound dressing applications: PVA-based hydrogel dressings. J Adv Res 8(3):217–233
194. Buggy M (2006) Review of: Natural fibers, biopolymers, and biocomposites edited by Amar K Mohanty, Manjusri Misra and Lawrence T Drzal. Polym Int 55(12):1462
195. Kaplan DL (1998) Introduction to biopolymers from renewable resources. Springer, Berlin Heidelberg, Berlin, Heidelberg, pp 1–29
196. Khan R, Khan MH (2013) Use of collagen as a biomaterial: an update. J Indian Soc Periodontol 17(4):539–542
197. Patino MG, Neiders ME, Andreana S, Noble B, Cohen RE (2002) Collagen as an implantable material in medicine and dentistry. J Oral Implant 28(5):220–225
198. Lynch CR, Kondiah PPD, Choonara YE, du Toit LC, Ally N, Pillay V (2020) Hydrogel biomaterials for application in ocular drug delivery. Front Bioeng Biotechnol 8:228
199. Pettian MS, de Guzzi Plepis AM, da Conceição Amaro Martins V, Dos Santos GR, Pinto CAL, Galdeano EA, Calegari ARA, de Moraes CA, da Cunha MR, (2018) Use of an anionic collagen matrix made from bovine intestinal serosa for in vivo repair of cranial defects. PLOS ONE 13(7):1–17
200. Ruoß M, Häussling V, Schügner F, Olde Damink LHH, Lee SM, Ge L, Ehnert S, Nussler AK (2018) A standardized collagen-based scaffold improves human hepatocyte shipment and allows metabolic studies over 10 days. Bioengineering 5(4):86

201. Nikkhah M, Akbari M, Paul A, Memic A, Dolatshahi-Pirouz A, Khademhosseini A (2016) Gelatin-based biomaterials for tissue engineering and stem cell bioengineering. In: Biomaterials from nature advanced devices and therapies, chap 3. Wiley, pp 37–62
202. Daamen WF, Veerkamp JH, van Hest JCM, van Kuppevelt TH (2007) Elastin as a biomaterial for tissue engineering. Biomaterials 28(30):4378–4398
203. Soucy JR, Sani ES, Lara RP, Diaz D, Dias F, Weiss AS, Koppes AN, Koppes RA, Annabi N (2018) Photocrosslinkable gelatin/tropoelastin hydrogel adhesives for peripheral nerve repair. Tissue Eng Part A 24(17–18):1393–1405
204. Su RS-C, Kim Y, Liu JC (2014) Resilin: protein-based elastomeric biomaterials. Acta Biomater 10(4):1601–1611
205. Dong J, Sun Q, Jin-Ye Wang (2004) Basic study of corn protein, zein, as a biomaterial in tissue engineering, surface morphology and biocompatibility. Biomaterials 25(19):4691–4697
206. Marcone M (2007) Review of: Soy applications in food by Mian N. Riaz. Trends Food Sci Technol 18(2):110
207. Vepari C, Kaplan DL (2007) Silk as a biomaterial. Prog Polym Sci 32(8–9):991–1007
208. Bandyopadhyay A, Chowdhury SK, Dey S, Moses JC, Mandal BB (2019) Silk: a promising biomaterial opening new vistas towards affordable healthcare solutions. J Indian Inst Sci 99(3):445–487
209. Wasupalli GK, Verma D (2018) 3—Polysaccharides as biomaterials. In: Thomas , Preetha B, Sreekala MS (eds) Fundamental biomaterials: polymers. Woodhead Publishing series in biomaterials. Woodhead Publishing, pp 37–70
210. Jani G, Shah D, Prajapati V (2009) Gums and mucilages: versatile excipients for pharmaceutical formulations. Asian J Pharm Sci 4
211. Singh D, Thomas D (2019) Advances in medical polymer technology towards the panacea of complex 3D tissue and organ manufacture. Am J Surg 217(4):807–808
212. Jovic TH, Kungwengwe G, Mills AC, Whitaker IS (2019) Plant-derived biomaterials: a review of 3D bioprinting and biomedical applications. Front Mech Eng 5:19
213. Lee KY, Mooney DJ (2012) Alginate: properties and biomedical applications. Prog Polym Sci 37(1):106–126
214. Chen H, Xie F, Chen L, Zheng B (2019) Effect of rheological properties of potato, rice and corn starches on their hot-extrusion 3D printing behaviors. J Food Eng 244:150–158
215. Coimbra P, Ferreira P, de Sousa HC, Batista P, Rodrigues MA, Correia IJ, Gil MH (2011) Preparation and chemical and biological characterization of a pectin/chitosan polyelectrolyte complex scaffold for possible bone tissue engineering applications. Int J Biol Macromol 48(1):112–118
216. Chimene D, Peak CW, Gentry JL, Carrow JK, Cross LM, Mondragon E, Cardoso GB, Kaunas R, Gaharwar AK (2018) Nanoengineered ionic-covalent entanglement (NICE) bioinks for 3D bioprinting. ACS Appl Mater Interfaces 10(12):9957–9968
217. Gomes ME, Godinho JS, Tchalamov D, Cunha AM, Reis RL (2002) Alternative tissue engineering scaffolds based on starch: processing methodologies, morphology, degradation and mechanical properties. Mater Sci Eng C 20(1):19–26
218. Yegappan R, Selvaprithiviraj V, Amirthalingam S, Jayakumar R (2018) Carrageenan based hydrogels for drug delivery, tissue engineering and wound healing. Carbohydr Polym 198:385–400
219. Akhlaghi SP, Tiong D, Berry RM, Tam KC (2014) Comparative release studies of two cationic model drugs from different cellulose nanocrystal derivatives. Eur J Pharm Biopharm 88(1):207–215
220. Li Q, Niu Y, Xing P, Wang C (2018) Bioactive polysaccharides from natural resources including Chinese medicinal herbs on tissue repair. Chin Med 13(1):7
221. Papakonstantinou E, Roth M, Karakiulakis G (2012) Hyaluronic acid: a key molecule in skin aging. Dermatoendocrinology 4(3):253–258
222. Shuborna NS, Chaiyasamut T, Sakdajeyont W, Vorakulpipat C, Rojvanakarn M, Wongsirichat N (2019) Generation of novel hyaluronic acid biomaterials for study of pain in third molar intervention: a review. J Dent Anesth Pain Med 19(1):11–19

223. Gupta RC, Lall R, Srivastava A, Sinha A (2019) Hyaluronic acid: molecular mechanisms and therapeutic trajectory. Front Vet Sci 6:192
224. Park BK, Kim M-M (2010) Applications of chitin and its derivatives in biological medicine. Int J Mol Sci 11(12):5152–5164
225. Trache D, Tarchoun AF, Derradji M, Hamidon TS, Masruchin N, Brosse N, Hazwan Hussin N (2020) Nanocellulose: from fundamentals to advanced applications. Front Chem 8:392
226. Mokhena TC, John MJ (2020) Cellulose nanomaterials: new generation materials for solving global issues, vol 27. Springer, Netherlands
227. Moon RJ, Schueneman GT, Simonsen J (2016) Overview of cellulose nanomaterials, their capabilities and applications. JOM 68(9):2383–2394
228. Kontturi E, Laaksonen P, Linder MB, Nonappa Gröschel AH, Rojas OJ, Ikkala O (2018) Advanced materials through assembly of nanocelluloses. Adv Mater 30(24):1703779
229. Galbe M, Wallberg O (2019) Pretreatment for biorefineries: a review of common methods for efficient utilisation of lignocellulosic materials. Biotechnol Biofuels 12(1):1–26
230. Catalán J, Norppa H (2017) Safety aspects of bio-based nanomaterials. Bioengineering 4(4):2–9
231. Mirshafiee V, Osborne OJ, Sun B, Xia T (2018) Safety concerns of industrial engineered nanomaterials. Elsevier
232. Kargarzadeh H, Mariano M, Gopakumar D, Ahmad I, Thomas S, Dufresne A, Huang J, Lin N (2018) Advances in cellulose nanomaterials, vol 25. Springer, Netherlands
233. Sheikhi A (2018) Emerging cellulose-based nanomaterials and nanocomposites. Elsevier
234. Li F, Mascheroni E, Piergiovanni L (2015) The potential of nanocellulose in the packaging field?: a review. Packag Technol Sci 28:475–508
235. Ahankari SS, Subhedar AR, Bhadauria SS, Dufresne A (2021) Nanocellulose in food packaging: a review. Carbohydr Polym 255:117479
236. Wang X, Yao C, Wang F, Li Z (2017) Cellulose-based nanomaterials for energy applications. Small 13(42):1–19
237. Dutta S, Kim J, Ide Y, Kim JH, Hossain MSA, Bando Y, Yamauchi Y, Wu KCW (2017) 3D network of cellulose-based energy storage devices and related emerging applications. Mater Horizons 4(4):522–545
238. Zhao D, Zhu Y, Cheng W, Chen W, Wu Y, Yu H (2021) Cellulose-based flexible functional materials for emerging intelligent electronics. Adv Mater 33(28):2000619
239. Hickey RJ, Pelling RE (2019) Cellulose biomaterials for tissue engineering. Front Bioeng Biotechnol 7
240. Yan G, Chen B, Zeng X, Sun Y, Tang X, Lin L (2020) Recent advances on sustainable cellulosic materials for pharmaceutical carrier applications. Carbohydr Polym 244:116492
241. Abouzeid RE, Khiari R, El-Wakil N, Dufresne A (2019) Current state and new trends in the use of cellulose nanomaterials for wastewater treatment. Biomacromolecules 20(2):573–597
242. Rajinipriya M, Nagalakshmaiah M, Robert M, Elkoun S (2018) Importance of agricultural and industrial waste in the field of nanocellulose and recent industrial developments of wood based nanocellulose: a review. ACS Sustain Chem Eng 6(3):2807–2828
243. Oksman K, Aitomäki Y, Mathew AP, Siqueira G, Zhou Q, Butylina S, Tanpichai S, Zhou X, Hooshmand S (2016) Review of the recent developments in cellulose nanocomposite processing. Compos Part A Appl Sci Manuf 83:2–18
244. Wang Q, Yao Q, Liu J, Sun J, Zhu Q, Chen H (2019) Processing nanocellulose to bulk materials: a review, vol 26. Springer, Netherlands
245. Vanderfleet OM, Cranston ED (2021) Production routes to tailor the performance of cellulose nanocrystals. Nat Rev Mater 6(2):124–144
246. Hubbe MA, Rojas OJ, Lucia LA (2015) Surface modification: Review. BioResources 10(3):6095–6206
247. Onwukamike KN, Grelier S, Grau E, Cramail H, Meier MAR (2019) Critical review on sustainable homogeneous cellulose modification: why renewability is not enough. ACS Sustain Chem Eng 7(2):1826–1840

248. Salama A, Hesemann P (2020) Recent trends in elaboration, processing, and derivatization of cellulosic materials using ionic liquids. ACS Sustain Chem Eng 8(49):17893–17907
249. Arvidsson R, Tillman AM, Sandén BA, Janssen M, Nordelöf A, Kushnir D, Molander S (2018) Environmental assessment of emerging technologies: recommendations for prospective LCA. J Ind Ecol 22(6):1286–1294
250. Hervy M, Evangelisti S, Lettieri P, Lee KY (2015) Life cycle assessment of nanocellulose-reinforced advanced fibre composites. Compos Sci Technol 118:154–162
251. Uihlein A, Schebek L (2009) Environmental impacts of a lignocellulose feedstock biorefinery system: an assessment. Biomass and Bioenergy 33(5):793–802
252. Paunonen S, Kamppuri T, Katajainen L, Hohenthal C, Heikkilä P, Harlin A (2019) Environmental impact of cellulose carbamate fibers from chemically recycled cotton. J Clean Prod 222:871–881
253. Shen L, Worrell E, Patel MK (2012) Comparing life cycle energy and GHG emissions of bio- based PET, recycled PET, PLA, and man-made cellulosics. Biofuels Bioprod Biorefining 6:625–639
254. Murphy CW, Kendall A (2015) Life cycle analysis of biochemical cellulosic ethanol under multiple scenarios. GCB Bioenergy 7:1019–1033
255. Li Q, McGinnis S, Sydnor C, Wong A, Renneckar S (2013) Nanocellulose life cycle assessment. ACS Sustain Chem Eng 1(8):919–928
256. Arvidsson R, Nguyen D, Svanström M (2015) Life cycle assessment of cellulose nanofibrils production by mechanical treatment and two different pretreatment processes. Environ Sci Technol 49(11):6881–6890
257. de Figueirêdo MCB, Rosa MDF, Ugaya CML, Souza Filho MDSMD, Silva Braid ACCD, Melo LFLD (2012) Life cycle assessment of cellulose nanowhiskers. J Clean Prod 35:130–139
258. do Nascimento DM, Dias AF, de Araújo Junior CP, de Freitas Rosa M, Morais JPS, de Figueirêdo MCB, (2016) A comprehensive approach for obtaining cellulose nanocrystal from coconut fiber. Part II. Environmental assessment of technological pathways. Ind Crop Prod 93:58–65
259. Piccinno F, Hischier R, Seeger S, Som C (2015) Life cycle assessment of a new technology to extract, functionalize and orient cellulose nanofibers from food waste. ACS Sustain Chem Eng 3(6):1047–1055
260. Shatkin JA, Kim B (2015) Cellulose nanomaterials: life cycle risk assessment, and environmental health and safety roadmap. Environ Sci Nano 2(5):477–499
261. Shatkin JA, Kim B (2017) Environmental health and safety of cellulose nanomaterials and composites, 1st edn. Wiley Online Library
262. Wigger H, Kägi R, Wiesner M, Nowack B (2020) Exposure and possible risks of engineered nanomaterials in the environment-current knowledge and directions for the future. Rev Geophys 58(4):1–25
263. Endes C, Camarero-Espinosa S, Mueller S, Foster EJ, Petri-Fink A, Rothen-Rutishauser B, Weder C, Clift MJD (2016) A critical review of the current knowledge regarding the biological impact of nanocellulose. J Nanobiotechnol 14(1):1–14
264. Ventura C, Pinto F, Lourenço AF, Ferreira PJT, Louro H, Silva MJ (2020) On the toxicity of cellulose nanocrystals and nanofibrils in animal and cellular models, vol 27
265. Boros BV, Ostafe V (2020) Evaluation of ecotoxicology assessment methods of nanomaterials and their effects. Nanomater 10(4)
266. Hohenthal C, Ovaskainen M, Bussini D, Sadocco P, Pajula T, Lehtinen H, Kautto J, Salmenkivi K (2012) Final assessment of nano-enhanced new products. Technical report, VTT Technical Research Centre of Finland, InnovHub—SSCCP, Pöyry Management Consulting Oy
267. Vartiainen J, Pöhler T, Sirola K, Pylkkänen L, Alenius H, Hokkinen J, Tapper U, Lahtinen P, Kapanen A, Putkisto K, Hiekkataipale P, Eronen P, Ruokolainen J, Laukkanen A (2011) Health and environmental safety aspects of friction grinding and spray drying of microfibrillated cellulose. Cellulose 18(3):775–786
268. Stoudmann N, Nowack B, Som C (2019) Prospective environmental risk assessment of nanocellulose for Europe. Environ Sci Nano 6(8):2520–2531

269. Kovacs T, Naish V, O'Connor B, Blaise C, Gagne F, Hall L, Trudeau V, Martel P (2010) An ecotoxicological characterization of nanocrystalline cellulose (NCC). Nanotoxicology 4(3):255–270
270. Harper BJ, Clendaniel A, Sinche F, Way D, Hughes M, Schardt J, Simonsen J, Stefaniak AB, Harper SL (2016) Impacts of chemical modification on the toxicity of diverse nanocellulose materials to developing zebrafish. Cellulose 23(3):1763–1775
271. Pereira MM, Mouton L, Yéprémian C, Couté A, Lo J, Marconcini JM, Ladeira LO, Raposo NRB, Brandão HM, Brayner R (2014) Ecotoxicological effects of carbon nanotubes and cellulose nanofibers in Chlorella vulgaris. J Nanobiotechnology 12(1):1–13
272. Kümmerer K, Menz J, Schubert T, Thielemans W (2011) Biodegradability of organic nanoparticles in the aqueous environment. Chemosphere 82(10):1387–1392
273. Li K, Huang J, Gao H, Zhong Y, Cao X, Chen Y, Zhang L, Cai J (2016) Reinforced mechanical properties and tunable biodegradability in nanoporous cellulose gels: poly(l-lactide-co-caprolactone) nanocomposites. Biomacromolecules 17(4):1506–1515
274. Morillo J, Usero J, Gracia I (2004) Heavy metal distribution in marine sediments from the southwest coast of Spain. Chemosphere 55(3):431–442
275. Jewel M, Abu Sayed M, Haque RA, Hasan J, Mondal S, Sharif Ahmed (2020) Heavy metal contamination and human health risk associated with sediment of Ganges River (Northwestern Bangladesh). Nat Environ Pollut Technol 19:783–790
276. Schwarzenbach R, Egli T, Hofstetter TB, Von Gunten U, Wehrli B (2010) Global water pollution and human health. Annu Rev Environ Resour 35:109–136
277. Lin N, Kopp RE, Horton BP, Donnelly JP (2016) Hurricane Sandy's flood frequency increasing from year 1800 to 2100. Proc Natl Acad Sci 113(43):12071–12075
278. Vasconcelos DC, Oliveira AH, Silva MRS, Penna R, Santos TO, Pereira C, Rocha Z, Menezes MÂBC (2009) Natural radioactivity in extreme south of Bahia, Brazil, using gamma-ray spectrometry. In: International nuclear Atlantic conference—INAC 2009, RJ, Brazil, Rio de Janeiro, p 2009
279. Thorne MC (2003) Background radiation: natural and man-made. J Radiol Prot 23(1):29
280. Kennedy HW (1963) Legislative and regulatory action in air pollution control. Public Health Rep 78(9):799
281. Chowkwanyun M (2019) Two cheers for air pollution control: triumphs and limits of the mid-century fight for air quality. Public Health Rep 134(3):307–312
282. Sendesi SMT, Noh Y, Nuruddin M, Boor BE, Howarter JA, Youngblood JP, Jafvert CT, Whelton AJ (2020) An emerging mobile air pollution source: outdoor plastic liner manufacturing sites discharge VOCs into urban and rural areas. Environ Sci Process Impacts 22(9):1828–1841
283. Merget R, Rosner G (2001) Evaluation of the health risk of platinum group metals emitted from automotive catalytic converters. Sci Total Environ 270(1–3):165–173
284. Kukutschová J, Moravec P, Tomášek V, Matějka V, Smolík J, Schwarz J, Seidlerová J, Šafářová K, Filip P (2011) On airborne nano/micro-sized wear particles released from low-metallic automotive brakes. Environ Pollut 159(4):998–1006
285. Garlapati VK (2016) E-waste in India and developed countries: management, recycling, business and biotechnological initiatives. Renew Sustain Energy Rev 54:874–881
286. Little JC, Hodgson AT, Gadgil AJ (1994) Modeling emissions of volatile organic compounds from new carpets. Atmos Environ 28(2):227–234
287. Cao X-L, Sparling M, Dabeka R (2016) Occurrence of 13 volatile organic compounds in foods from the Canadian total diet study. Food Addit Contam Part A 33(2):373–382
288. He C, Cheng J, Zhang X, Douthwaite M, Pattisson S, Hao Z (2019) Recent advances in the catalytic oxidation of volatile organic compounds: a review based on pollutant sorts and sources. Chem Rev 119(7):4471–4568
289. Farrow A, Taylor H, Northstone K, Golding J (2003) Symptoms of mothers and infants related to total volatile organic compounds in household products. Arch Environ Heal Int J 58(10):633–641

290. Van Lente R, Herman RJ (2001) The smell of success—exploiting the leather aroma. Technical Report 2001-01-0047, SAE Technical Paper
291. Cometto-Muñiz JE, Cain WS, Abraham MH (2004) Detection of single and mixed VOCs by smell and by sensory irritation. Indoor Air 14:108–117
292. Kandyala R, Raghavendra SPC, Rajasekharan ST (2010) Xylene: an overview of its health hazards and preventive measures. J Oral Maxillofac Pathol 14(1):1–5
293. Zhang Z, Chen J, Gao Y, Ao Z, Li G, An T, Hu Y, Li Y (2018) A coupled technique to eliminate overall nonpolar and polar volatile organic compounds from paint production industry. J Clean Prod 185:266–274
294. Mølgaard B, Viitanen A-K, Kangas A, Huhtiniemi M, Larsen ST, Vanhala E, Hussein T, Boor BE, Hämeri K, Koivisto AJ (2015) Exposure to airborne particles and volatile organic compounds from polyurethane molding, spray painting, lacquering, and gluing in a workshop. Int J Environ Res Public Health 12(4):3756–3773
295. Adeniran JA, Yusuf RO, Mustapha SI, Sonibare JA (2017) Exposure to total volatile organic compounds (TVOCs) from household spray products. Environ Res Eng Manag 73(4):21–30
296. Steinemann AC, MacGregor IC, Gordon SM, Gallagher LG, Davis AL, Ribeiro DS, Wallace LA (2011) Fragranced consumer products: chemicals emitted, ingredients unlisted. Environ Impact Assess Rev 31(3):328–333
297. Langolf BM, Kleinheinz GT (2006) A lava rock-based biofilter for the treatment of alpha-pinene. Bioresour Technol 97(15):1951–1958
298. Singal M, Vitale D, Smith L (2011) Fragranced products and VOCs. Environ Health Perspect 119(5):A200–A200
299. Moraes TM, Kushima H, Moleiro FC, Santos RC, Machado Rocha LR, Marques MO, Vilegas W, Hiruma-Lima CA (2009) Effects of limonene and essential oil from Citrus aurantium on gastric mucosa: role of prostaglandins and gastric mucus secretion. Chem Biol Interact 180(3):499–505
300. Elansary HO, Ashmawy NA (2013) Essential oils of mint between benefits and hazards. J Essent Oil Bear Plants 16(4):429–438
301. Smith N (2019) The seven ages of materials. E&T Eng Technol 14(9):22
302. Lv M, Huang W, Rong X, He J, Yang X (2020) Source apportionment of volatile organic compounds (VOCs) in vehicle cabins diffusing from interior materials. Part I: Measurements of VOCs in new cars in China. Build Environ 175:106796
303. Bin X, Chen X, Xiong J (2018) Air quality inside motor vehicles' cabins: a review. Indoor Built Environ 27(4):452–465
304. Tong Z, Liu H (2020) Modeling in-vehicle VOCs distribution from cabin interior surfaces under solar radiation. Sustainability 12(14):5526
305. Lim MCH, Ayoko GA, Morawska L, Ristovski ZD, Jayaratne ER (2007) The effects of fuel characteristics and engine operating conditions on the elemental composition of emissions from heavy duty diesel buses. Fuel 86(12–13):1831–1839
306. Nader R (2011) Unsafe at any speed: the designed-in dangers of the American automobile. 1965. Am J Public Health 101(2):254–256
307. Nixon R (1970) 42 U.S.C. §§7401 et seq. Clean Air Act
308. Nriagu JO (1990) The rise and fall of leaded gasoline. Sci Total Environ 92:13–28
309. Kiener K (1977) Internal combustion process and engine. Patent, US4040400A, United States
310. Baren RE, Hershberger JF (2002) Kinetics of the NCN radical. J Phys Chem A 106(46):11093–11097
311. Meyer JP, Hershberger JF (2006) Ab initio study of the HCCO+ NO_2 reaction. Chem Phys 325(2–3):545–553
312. Miller JA, Bowman CT Mechanism and modeling of nitrogen chemistry in combustion. Prog Energy Combust Sci 15(4):287–338
313. Lu Z, Streets DG, de Foy B, Lamsal LN, Duncan BN, Xing J (2015) Emissions of nitrogen oxides from US urban areas: estimation from Ozone monitoring instrument retrievals for 2005–2014. Atmos Chem Phys 15(18):10367–10383

314. Munro AM, Chandler C, Garling M, Chai D, Popovich V, Lystrom L, Kilina S (2016) Phenyldithiocarbamate ligands decompose during nanocrystal ligand exchange. J Phys Chem C 120(51):29455–29462
315. Kilina S, Kilin D, Tretiak S (2015) Light-driven and phonon-assisted dynamics in organic and semiconductor nanostructures. Chem Rev 115(12):5929–5978
316. Puzyn T, Rasulev B, Gajewicz A, Hu X, Dasari TP, Michalkova A, Hwang H-M, Toropov A, Leszczynska D, Leszczynski J (2011) Using nano-QSAR to predict the cytotoxicity of metal oxide nanoparticles. Nat Nanotechnol 6(3):175–178
317. Gajewicz A, Schaeublin N, Rasulev B, Hussain S, Leszczynska D, Puzyn T, Leszczynski J (2015) Towards understanding mechanisms governing cytotoxicity of metal oxides nanoparticles: hints from nano-QSAR studies. Nanotoxicology 9(3):313–325
318. Kašpar J, Fornasiero P, Hickey N (2003) Automotive catalytic converters: current status and some perspectives. Catal Today 77(4):419–449
319. Sørensen SN, Engelbrekt C, Lützhøft H-CH, Jiménez-Lamana J, Noori JS, Alatraktchi FA, Delgado G, Slaveykova VI, Baun A (2016) A multimethod approach for investigating algal toxicity of platinum nanoparticles. Environ Sci Technol 50(19):10635–10643
320. Dietz K-J, Herth S (2011) Plant nanotoxicology. Trends Plant Sci 16(11):582–589
321. Wiseman CLS, Zereini F (2009) Airborne particulate matter, platinum group elements and human health: a review of recent evidence. Sci Total Environ 407(8):2493–2500
322. Singh N, Duan H, Tang Y (2020) Toxicity evaluation of E-waste plastics and potential repercussions for human health. Environ Int 137:105559
323. Forti V, Baldé CP, Kuehr R, Bel G (2020) The global E-waste monitor 2020: quantities, flows and the circular economy potential. United Nations University/United Nations Institute for Training and Research, International Telecommunication Union, and International Solid Waste Association, Bonn, Geneva and Rotterdam
324. Peiry KK (2010) Basel convention on the control of transboundary movements of hazardous wastes and their disposal. https://legal.un.org/avl/pdf/ha/bcctmhwd/bcctmhwd_e.pdf. United Nations Audiovisual Library of International Law. Accessed: 30 March 2022
325. Massari M, Monzini P (2004) Dirty businesses in Italy: a case-study of illegal trafficking in hazardous waste. Glob Crime 6(3–4):285–304
326. Andrews A (2009) Beyond the ban–can the Basel convention adequately safeguard the interests of the world's poor in the international trade of hazardous waste? Law Environ Dev J 5(2):167–184
327. Grant K, Goldizen FC, Sly PD, Brune M-N, Neira M, van den Berg M, Norman RE (2013) Health consequences of exposure to e-waste: a systematic review. Lancet Glob Health 1(6):e350–e361
328. Chen A, Dietrich KN, Huo X, Ho SM (2011) Developmental neurotoxicants in e-waste: an emerging health concern. Environ Health Perspect 119(4):431–438
329. Stanley Whittingham M (2004) Lithium batteries and cathode materials. Chem Rev 104(10):4271–4302
330. Padhi AK, Nanjundaswamy KS, Goodenough JB (1997) Phospho-olivines as positive-electrode materials for rechargeable lithium batteries. J Electrochem Soc 144(4):1188–1194
331. Delacourt C, Poizot P, Morcrette M, Tarascon JM, Masquelier C (2004) One-step low-temperature route for the preparation of electrochemically active $LiMnPO_4$ powders. Chem Mater 16(1):93–99
332. Smil V (2000) Phosphorus in the environment: natural flows and human interferences. Annu Rev Energy Environ 25(1):53–88
333. Léonard A, Lauwerys R (1990) Mutagenicity, carcinogenicity and teratogenicity of cobalt metal and cobalt compounds. Mutat Res Genet Toxicol 239(1):17–27
334. Domingo JL (1989) Cobalt in the environment and its toxicological implications. Reviews of environmental contamination and toxicology. Springer, New York, New York, NY, pp 105–132
335. Ragsdale SW (2009) Nickel-based enzyme systems. J Biol Chem 284(28):18571–18575

336. Boer JL, Mulrooney SB, Hausinger RP (2014) Nickel-dependent metalloenzymes. Arch Biochem Biophys 544:142–152
337. Maroney MJ, Ciurli S (2014) Nonredox nickel enzymes. Chem Rev 114(8):4206–4228
338. Desguin B, Fellner M, Riant O, Hu J, Hausinger RP, Hols P, Soumillion P (2018) Biosynthesis of the nickel-pincer nucleotide cofactor of lactate racemase requires a CTP-dependent cyclometallase. J Biol Chem 293(32):12303–12317
339. Sreekanth TVM, Nagajyothi PC, Lee KD, Prasad TNVKV (2013) Occurrence, physiological responses and toxicity of nickel in plants. Int J Environ Sci Technol 10(5):1129–1140
340. Zhao J, Shi X, Castranova V, Ding M (2009) Occupational toxicology of nickel and nickel compounds. J Environ Pathol Toxicol Oncol 28(3):177–208
341. McGregor DB, Baan RA, Partensky C, Rice JM, Wilbourn JD (2000) Evaluation of the carcinogenic risks to humans associated with surgical implants and other foreign bodies—a report of an IARC Monographs Programme Meeting. International Agency for Research on Cancer. Eur J Cancer 36(3):307–313
342. Seilkop SK, Oller AR (2003) Respiratory cancer risks associated with low-level nickel exposure: an integrated assessment based on animal, epidemiological, and mechanistic data. Regul Toxicol Pharmacol 37(2):173–190
343. Sivulka DJ (2005) Assessment of respiratory carcinogenicity associated with exposure to metallic nickel: a review. Regul Toxicol Pharmacol 43(2):117–133
344. Chiou YH, Wong RH, Chao MR, Chen CY, Liou SH, Lee H (2014) Nickel accumulation in lung tissues is associated with increased risk of p53 mutation in lung cancer patients. Environ Mol Mutagen 55(8):624–632
345. Zambelli B, Uversky VN (1864) Ciurli S (2016) Nickel impact on human health: an intrinsic disorder perspective. Biochim Biophys Acta 12:1714–1731
346. Sivulka DJ, Seilkop SK, Lascelles K, Conard BR, Jones SF, Collinson EC (2014) Reconstruction of historical exposures at a Welsh nickel refinery (1953–2000). Ann Occup Hyg 58(6):739–760
347. Jose CC, Jagannathan L, Tanwar VS, Zhang X, Zang C, Cuddapah S (2018) Nickel exposure induces persistent mesenchymal phenotype in human lung epithelial cells through epigenetic activation of ZEB1. Mol Carcinog 57(6):794–806
348. Nasir AFAA, Cameron SF, von Hippel FA, Postlethwait J, Niehaus AC, Blomberg S, Wilson RS (2018) Manganese accumulates in the brain of northern quolls (Dasyurus hallucatus) living near an active mine. Environ Pollut 233:377–386
349. Josephs KA, Ahlskog JE, Klos KJ, Kumar N, Fealey RD, Trenerry MR, Cowl CT (2005) Neurologic manifestations in welders with pallidal MRI T1 hyperintensity. Neurology 64(12):2033–2039
350. Antonini JM (2003) Health effects of welding. Crit Rev Toxicol 33(1):61–103
351. Bailey LA, Kerper LE, Goodman JE (2018) Derivation of an occupational exposure level for manganese in welding fumes. Neurotoxicology 64:166–176
352. Ma RE, Ward EJ, Yeh C-L, Snyder S, Long Z, Yavuz FG, Zauber SE, Dydak U (2018) Thalamic GABA levels and occupational manganese neurotoxicity. Association with exposure levels and brain MRI. Neurotoxicology 64:30–42
353. Seo J, Chang Y, Jang KE, Park JW, Kim Y-T, Park S-J, Jeong KS, Kim A, Kim SH, Kim Y (2016) Altered executive function in the welders: a functional magnetic resonance imaging study. Neurotoxicol Teratol 56:26–34
354. Criswell SR, Nielsen SS, Warden MN, Flores HP, Lenox-Krug J, Racette S, Sheppard L, Checkoway H, Racette B (2019) MRI signal intensity and parkinsonism in manganese-exposed workers. J Occup Environ Med 61(8):641–645
355. Criswell SR, Perlmutter JS, Huang JL, Golchin N, Flores HP, Hobson A, Aschner M, Erikson KM, Checkoway H, Racette BA (2012) Basal ganglia intensity indices and diffusion weighted imaging in manganese-exposed welders. Occup Environ Med 69(6):437–443
356. Lee E-Y, Flynn MR, Du G, Lewis MM, Herring AH, Van Buren E, Van Buren S, Kong L, Mailman RB, Huang X (2016) Editor's Highlight: Lower fractional anisotropy in the globus pallidus of asymptomatic welders, a marker for long-term welding exposure. Toxicol Sci 153(1):165–173

357. Pesch B, Dydak U, Lotz A, Casjens S, Quetscher C, Lehnert M, Abramowski J, Stewig C, Yeh C-L, Weiss T, van T C, Herrmann L, Muhlack S, Woitalla D, Glaubitz B, Schmidt-Wilcke T, Brüning T, (2018) Association of exposure to manganese and iron with relaxation rates R1 and R2*- magnetic resonance imaging results from the WELDOX II study. Neurotoxicology 64:68–77
358. Amankwaa EF, Tsikudo JB, Adovor K (2016) Recyclers at risk? Analysis of e-waste livelihoods and blood lead levels at Ghana's recycling hub, Agbogbloshie. https://www.theigc.org/wp-content/uploads/2016/07/Amankwaa-Tsikudo-2016-Policy-brief.pdf. Accessed: 30 March 2022
359. Ceballos DM, Dong Z (2016) The formal electronic recycling industry: challenges and opportunities in occupational and environmental health research. Environ Int 95:157–166
360. Terazono A, Oguchi M, Yoshida A, Medina RP, Ballesteros FC (2017) Material recovery and environmental impact by informal e-waste recycling site in the Philippines. In: Matsumoto M, Masui K, Fukushige S, Kondoh S (eds) Sustainability through innovation in product life cycle design. Springer Singapore, Singapore, pp 197–213
361. Duan H, Jiukun H, Tan Q, Liu L, Wang Y, Li J (2016) Systematic characterization of generation and management of e-waste in China. Environ Sci Pollut Res 23(2):1929–1943
362. D'Adamo I, Rosa P, Terzi S (2016) Challenges in waste electrical and electronic equipment management: a profitability assessment in three European countries. Sustainability 8(7):633
363. Tansel B (2017) From electronic consumer products to e-wastes: global outlook, waste quantities, recycling challenges. Environ Int 98:35–45
364. Zhang L, Xu Z (2016) A review of current progress of recycling technologies for metals from waste electrical and electronic equipment. J Clean Prod 127:19–36
365. Kumari A, Jha MK, Lee J-c, Singh RP (2016) Clean process for recovery of metals and recycling of acid from the leach liquor of PCBs. J Clean Prod 112:4826–4834
366. Roe FJC (1993) Book reviews. Review of: Ann Occup Hyg 37(3):341. Review of: Environ Health Criteria 134: Cadmium, World Health Organization (1992)
367. Pattee OH, Pain DJ (2003) Lead in the environment. Handbook of ecotoxicololgy, chap 15. CRC Press, Boca Raton, FL, pp 373–408
368. Marques AC, Cabrera J-M, de Fraga Malfatti C (2013) Printed circuit boards: a review on the perspective of sustainability. J Environ Manag 131:298–306
369. An T, Huang Y, Li G, He Z, Chen J, Zhang C (2014) Pollution profiles and health risk assessment of VOCs emitted during e-waste dismantling processes associated with different dismantling methods. Environ Int 73:186–194
370. Chen J, Huang Y, Li G, An T, Hu Y, Li Y (2016) VOCs elimination and health risk reduction in e-waste dismantling workshop using integrated techniques of electrostatic precipitation with advanced oxidation technologies. J Hazard Mater 302:395–403
371. Council of the European Union European Parliament (2017) Directive (EU) 2017/2102 of the European Parliament and of the Council of 15 November 2017 amending Directive 2011/65/EU on the restriction of the use of certain hazardous substances in electrical and electronic equipment. https://eur-lex.europa.eu/legal-content/EN/TXT/?qid=1512061986553&uri=CELEX:32017L2102. Accessed: 30 March 2022
372. Domínguez A, Geyer R (2019) Photovoltaic waste assessment of major photovoltaic installations in the United States of America. Renew Energy 133:1188–1200
373. Santos JD, Alonso-García MC (2018) Projection of the photovoltaic waste in Spain until 2050. J Clean Prod 196:1613–1628
374. Rocchetti L, Beolchini F (2015) Recovery of valuable materials from end-of-life thin-film photovoltaic panels: environmental impact assessment of different management options. J Clean Prod 89:59–64
375. Mahmoudi S, Huda N, Behnia M (2019) Photovoltaic waste assessment: forecasting and screening of emerging waste in Australia. Resour Conserv Recycl 146:192–205
376. Weckend S, Wade A, Heath G (2016) End-of-life management: solar photovoltaic panels. Technical Report T12-06:2016, International Renewable Energy Agency and International Energy Agency Photovoltaic Power Systems

377. Gautam A, Shankar R, Vrat P (2021) End-of-life solar photovoltaic e-waste assessment in India: a step towards a circular economy. Sustain Prod Consum 26:65–77
378. Zimmermann Y-S, Schäffer A, Corvini PFX, Lenz M (2013) Thin-film photovoltaic cells: long-term metal(loid) leaching at their end-of-life. Environ Sci Technol 47(22):13151–13159
379. Tanaka A, Hirata M, Shiratani M, Koga K, Kiyohara Y (2012) Subacute pulmonary toxicity of copper indium gallium diselenide following intratracheal instillations into the lungs of rats. J Occup Heal 54(3):187–195
380. Ramos-Ruiz A, Wilkening JV, Field JA, Sierra-Alvarez R (2017) Leaching of cadmium and tellurium from cadmium telluride (CdTe) thin-film solar panels under simulated landfill conditions. J Hazard Mater 336:57–64
381. Vellini M, Gambini M, Prattella V (2017) Environmental impacts of PV technology throughout the life cycle: importance of the end-of-life management for Si-panels and CdTe-panels. Energy 138:1099–1111
382. Belzile N, Chen Y-W (2015) Tellurium in the environment: a critical review focused on natural waters, soils, sediments and airborne particles. Appl Geochem 63:83–92
383. Merget R, Bauer T, Küpper H, Philippou S, Bauer H, Breitstadt R, Bruening T (2002) Health hazards due to the inhalation of amorphous silica. Arch Toxicol 75(11):625–634
384. Pickren G (2015) Making connections between global production networks for used goods and the realm of production: a case study on e-waste governance. Glob Networks 15(4):403–423
385. Echegaray F (2016) Consumers' reactions to product obsolescence in emerging markets: the case of Brazil. J Clean Prod 134:191–203
386. LeBel S (2016) Fast machines, slow violence: ICTs, planned obsolescence, and e-waste. Globalizations 13(3):300–309
387. Zeng X, Yang C, Chiang JF, Li J (2017) Innovating e-waste management. From macroscopic to microscopic scales. Sci Total Environ 575:1–5
388. Lundgren K (2012) The global impact of e-waste: addressing the challenge. International Labour Office, Programme on Safety and Health at Work and the Environment (SafeWork), Sectoral Activities Department (SECTOR), Geneva
389. Ryneal M (2016) Turning waste into gold: accumulation by disposal and the political economy of e-waste urban mining. Ph.D. thesis, Northern Arizona University, Flagstaff, AZ
390. Sabbaghi M, Cade W, Behdad S, Bisantz AM (2017) The current status of the consumer electronics repair industry in the U.S.: a survey-based study. Resour Conserv Recycl 116:137–151
391. Manzetti S, Ghisi R (2014) The environmental release and fate of antibiotics. Mar Pollut Bull 79(1–2):7–15
392. Fleming A (1944) The discovery of penicillin. Br Med Bull 2(1):4–5
393. Fleming A (1946) Penicillin. Its practical application, Blakiston, Philadelphia
394. Gaynes R (2017) The discovery of penicillin–new insights after more than 75 years of clinical use. Emerg Infect Dis 23(5):849–853
395. Renew JE, Huang CH (2004) Simultaneous determination of fluoroquinolone, sulfonamide, and trimethoprim antibiotics in wastewater using tandem solid phase extraction and liquid chromatography-electrospray mass spectrometry. J Chromatogr A 1042(1–2):113–121
396. Brown KD, Kulis J, Thomson B, Chapman TH, Mawhinney DB (2006) Occurrence of antibiotics in hospital, residential, and dairy effluent, municipal wastewater, and the Rio Grande in New Mexico. Sci Total Environ 366(2–3):772–783
397. Karthikeyan KG, Meyer MT (2006) Occurrence of antibiotics in wastewater treatment facilities in Wisconsin, USA. Sci Total Environ 361(1–3):196–207
398. Watkinson AJ, Murby EJ, Kolpin DW, Costanzo SD (2009) The occurrence of antibiotics in an urban watershed: from wastewater to drinking water. Sci Total Environ 407(8):2711–2723
399. Van Boeckel TP, Brower C, Gilbert M, Grenfell BT, Levin SA, Robinson TP, Teillant A, Laxminarayan R (2015) Global trends in antimicrobial use in food animals. Proc Natl Acad Sci U S A 112(18):5649–5654
400. Boy-Roura M, Mas-Pla J, Petrovic M, Gros M, Soler D, Brusi D, Menció A (2018) Towards the understanding of antibiotic occurrence and transport in groundwater: findings from the Baix Fluvià alluvial aquifer (NE Catalonia, Spain). Sci Total Environ 612:1387–1406

401. Rizzo L, Manaia C, Merlin C, Schwartz T, Dagot C, Ploy MC, Michael I, Fatta-Kassinos D (2013) Urban wastewater treatment plants as hotspots for antibiotic resistant bacteria and genes spread into the environment: a review. Sci Total Environ 447:345–360
402. Joakim Larsson DG, de Pedro C, Paxeus N (2007) Effluent from drug manufactures contains extremely high levels of pharmaceuticals. J Hazard Mater 148(3):751–755
403. Joakim Larsson DG (2014) Pollution from drug manufacturing: review and perspectives. Philos Trans R Soc B Biol Sci 369(1656)
404. Berkner S, Konradi S, Schönfeld J (2014) Antibiotic resistance and the environment–there and back again. EMBO Rep 15(7):740–744
405. Kraemer SA, Ramachandran A, Perron GG (2019) Antibiotic pollution in the environment: from microbial ecology to public policy. Microorganisms 7(6):1–24
406. Andersson DI, Hughes D (2012) Evolution of antibiotic resistance at non-lethal drug concentrations. Drug Resist Updat 15(3):162–172
407. Klaus Kümmerer (2009) Antibiotics in the aquatic environment–a review–Part II. Chemosphere 75(4):435–441
408. Bhullar K, Waglechner N, Pawlowski A, Koteva K, Banks ED, Johnston MD, Barton HA, Wright GD (2012) Antibiotic resistance is prevalent in an isolated cave microbiome. PLoS One 7(4):1–11
409. D'Costa VM, McGrann KM, Hughes DW, Wright GD (2006) Sampling the antibiotic resistome. Science 311(5759):374–377
410. D'Costa VM, King CE, Kalan L, Morar M, Sung WWL, Schwarz C, Froese D, Zazula G, CalmelsF Debruyne R, Golding GB, Poinar HN, Wright GD (2011) Antibiotic resistance is ancient. Nature 477(7365):457–461
411. Perron GG, Whyte L, Turnbaugh PJ, Goordial J, Hanage WP, Dantas G, Desai MM (2015) Functional characterization of bacteria isolated from ancient arctic soil exposes diverse resistance mechanisms to modern antibiotics. PLoS One 10(3):1–19
412. Lee SH, Lee JH (2011) Molecular characterization of TEM-type beta-lactamases identified in cold-seep sediments of Edison Seamount (South of Lihir Island, Papua New Guinea). In: Handbook of molecular microbial ecology II: metagenomics in different habitats, no May 2014, pp 545–552
413. Martínez JL (2008) Antibiotics and antibiotic resistance genes in natural environments. Science 321(5887):365–367
414. Barber M, Rozwadowska-Dowzenko M (1948) Infection by penicillin-resistant Staphylococci. Lancet 252(6530):641–644
415. Finland M (1979) Emergence of antibiotic resistance in hospitals, 1935–1975. Rev Infect Dis 1(1):4–21
416. Spellberg B (2008) Antibiotic resistance and antibiotic development. Lancet Infect Dis 8(4):211–212
417. Shoemaker NB, Vlamakis H, Hayes K, Salyers AA (2001) Evidence for extensive resistance gene transfer among *Bacteroides* spp. and among Bacteroides and other genera in the human colon. Appl Environ Microbiol 67(2):561–568
418. Knapp CW, Dolfing J, Ehlert PAI, Graham DW (2010) Evidence of increasing antibiotic resistance gene abundances in archived soils since 1940. Environ Sci Technol 44(2):580–587
419. Walsh C (2000) Molecular mechanisms that confer antibacterial drug resistance. Nature 406(775–781):775–781
420. Wright GD (2011) Molecular mechanisms of antibiotic resistance. Chem Commun 47(14):4055
421. Levy SB (1992) Active efflux mechanisms for antimicrobial resistance. Antimicrob Agents Chemother 36(4):695–703
422. Paulsen IT, Brown MH, Skurray RA (1996) Proton-dependent multidrug efflux systems. Microbiol Rev 60(4):575–608
423. Ross JI, Eady EA, Cove JH, Cunliffe WJ, Baumberg S, Wootton JC (1990) Inducible erythromycin resistance in Staphlyococci is encoded by a member of the ATP-binding transport super-gene family. Mol Microbiol 4(7):1207–1214

424. Philippon A, Labia R, Jacoby G (1989) Extended-spectrum beta-lactamases. Antimicrob Agents Chemother 33(8):1131–1136
425. Shen LL (1994) Molecular mechanisms of DNA gyrase inhibition by quinolone antibacterials. In: Andoh T, Ikeda H, Oguro M (eds) Advances in pharmacology, vol 29, chap 17. CRC Press, pp 285–304
426. Ferrero L, Cameron B, Crouzet J (1995) Analysis of gyrA and grlA mutations in stepwise-selected ciprofloxacin-resistant mutants of Staphylococcus aureus. Antimicrob Agents Chemother 39(7):1554–1558
427. Keeling PJ, Palmer JD (2008) Horizontal gene transfer in eukaryotic evolution. Nat Rev Genet 9(8):605–618
428. Norman A, Hansen LH, Sørensen SJ (2009) Conjugative plasmids: vessels of the communal gene pool. Philos Trans R Soc B Biol Sci 364(1527):2275–2289
429. Palmer KL, Kos VN, Gilmore MS (2010) Horizontal gene transfer and the genomics of enterococcal antibiotic resistance. Curr Opin Microbiol 13(5):632–639
430. Lederberg J (1956) Conjugal pairing in Escherichia coli. J Bacteriol 71(4):497–498
431. Bethke JH, Davidovich A, Cheng L, Lopatkin AJ, Song W, Thaden JT, Fowler VG, Xiao M, You L (2020) Environmental and genetic determinants of plasmid mobility in pathogenic *Escherichia coli*. Sci Adv 6(4):eaax3173
432. Lopatkin AJ, Meredith HR, Srimani JK, Pfeiffer C, Durrett R, You L (2017) Persistence and reversal of plasmid-mediated antibiotic resistance. Nat Commun 8(1):1689
433. Loftie-Eaton W, Bashford K, Quinn H, Dong K, Millstein J, Hunter S, Thomason MK, Merrikh H, Ponciano JM, Top EM (2017) Compensatory mutations improve general permissiveness to antibiotic resistance plasmids. Nat Ecol Evol 1(9):1354–1363
434. Vernikos G, Medini D (2014) Horizontal gene transfer and the role of restriction-modification systems in bacterial population dynamics. Evolutionary biology: genome evolution, speciation, coevolution and origin of life, no July 2014, chap 8. Springer International Publishing, Cham, pp 169–190
435. Springman AC, Lacher DW, Wu G, Milton N, Whittam TS, Davies H, Manning SD (2009) Selection, recombination, and virulence gene diversity among group B streptococcal genotypes. J Bacteriol 191(17):5419–5427
436. Fournier PE, Vallenet D, Barbe V, Audic S, Ogata H, Poirel L, Richet H, Robert C, Mangenot S, Abergel C, Nordmann P, Weissenbach J, Raoult D, Claverie JM (2006) Comparative genomics of multidrug resistance in Acinetobacter baumannii. PLoS Genet 2(1):62–72
437. Griffith F (1966) The significance of Penumococcal types. J Hyg (Lond) 64(2):129–175
438. Lorenz MG, Wackernagel W (1994) Bacterial gene transfer by natural genetic transformation in the environment. Microbiol Rev 58(3):563–602
439. Mazaheri Nezhad Fard R, Barton MD, Heuzenroeder MW (2011) Bacteriophage-mediated transduction of antibiotic resistance in enterococci. Lett Appl Microbiol 52(6):559–564
440. Colomer-Lluch M, Jofre J, Muniesa M (2011) Antibiotic resistance genes in the bacteriophage DNA fraction of environmental samples. PLoS One 6(3):1–11
441. Brüssow H, Hendrix RW (2002) Phage genomics: small is beautiful. Cell 108(1):13–16
442. Ochman H, Lawrence JG, Groisman EA (2000) Lateral gene transfer and the nature of bacterial innovation. Nature 405(6784):299–304
443. Baquero F, Martínez JL, Cantón R (2008) Antibiotics and antibiotic resistance in water environments. Curr Opin Biotechnol 19(3):260–265
444. Berendonk TU, Manaia CM, Merlin C, Fatta-Kassinos D, Cytryn E, Walsh F, Bürgmann H, Sørum H, Norström M, Pons MN, Kreuzinger N, Huovinen P, Stefani S, Schwartz T, Kisand V, Baquero F, Martinez JL (2015) Tackling antibiotic resistance: the environmental framework. Nat Rev Microbiol 13(5):310–317
445. Pitout JDD, Nordmann P, Laupland KB, Poirel L (2005) Emergence of Enterobacteriaceae producing extended-spectrum β-lactamases (ESBLs) in the community. J Antimicrob Chemother 56(1):52–59
446. Breathnach AS, Cubbon MD, Karunaharan RN, Pope CF, Planche TD (2012) Multidrug-resistant Pseudomonas aeruginosa outbreaks in two hospitals: association with contaminated hospital waste-water systems. J Hosp Infect 82(1):19–24

447. Wang SH, Sheng WH, Chang YY, Wang LH, Lin HC, Chen ML, Pan HJ, Ko HJ, Chang SC, Lin FY (2003) Healthcare-associated outbreak due to pan-drug resistant Acinetobacter baumannii in a surgical intensive care unit. J Hosp Infect 53(2):97–102
448. Valencia R, Arroyo LA, Conde M, Aldana JM, Torres M-J, Fernández-Cuenca F, Garnacho-Montero J, Cisneros JM, Ortiz C, Pachón J, Aznar J (2009) Nosocomial outbreak of infection with pan-drug-resistant Acinetobacter baumannii in a tertiary care university hospital. Infect Control Hosp Epidemiol 30(3):257–263
449. Kizny Gordon AE, Mathers AJ, Cheong EYL, Gottlieb T, Kotay S, Sarah Walker A, Peto TEA, Crook DW, Stoesser N (2017) The hospital water environment as a reservoir for carbapenem-resistant organisms causing hospital-acquired infections–a systematic review of the literature. Clin Infect Dis 64(10):1436–1444
450. Folkesson A, Jelsbak L, Yang L, Johansen HK, Ciofu O, Hoiby N, Molin S (2012) Adaptation of Pseudomonas aeruginosa to the cystic fibrosis airway: an evolutionary perspective. Nat Rev Microbiol 10(12):841–851
451. López-Causapé C, Rojo-Molinero E, Mulet X, Cabot G, Moyà B, Figuerola J, Togores B, Pérez JL, Oliver A (2013) Clonal dissemination, emergence of mutator lineages and antibiotic resistance evolution in Pseudomonas aeruginosa cystic fibrosis chronic lung infection. PLoS ONE 8(8):1–8
452. Leverstein-van Hall MA, Box ATA, Blok HEM, Paauw A, Fluit AC, Verhoef J (2002) Evidence of extensive interspecies transfer of integron-mediated antimicrobial resistance genes among multidrug-resistant Enterobacteriaceae in a clinical setting. J Infect Dis 186(1):49–56
453. Hocquet D, Muller A, Bertrand X (2016) What happens in hospitals does not stay in hospitals: antibiotic-resistant bacteria in hospital wastewater systems. J Hosp Infect 93(4):395–402
454. Reinthaler FF, Posch J, Feierl G, Wüst G, Haas D, Ruckenbauer G, Mascher F, Marth E (2003) Antibiotic resistance of *E. Coli* in sewage and sludge. Water Res 37(8):1685–1690
455. Ju F, Beck K, Yin Maccagnan A, McArdell C, Singer HP, Johnson DR, Zhang T, Bürgmann H (2019) Wastewater treatment plant resistomes are shaped by bacterial composition, genetic exchange, and upregulated expression in the effluent microbiomes. ISME J 13(2):346–360
456. Szczepanowski R, Linke B, Krahn I, Gartemann KH, Gützkow T, Eichler W, Pühler A, Schlüter A (2009) Detection of 140 clinically relevant antibiotic-resistance genes in the plasmid metagenome of wastewater treatment plant bacteria showing reduced susceptibility to selected antibiotics. Microbiology 155(7):2306–2319
457. Yang Y, Li B, Zou S, Fang HHP, Zhang T (2014) Fate of antibiotic resistance genes in sewage treatment plant revealed by metagenomic approach. Water Res 62:97–106
458. Chen H, Zhang M (2013) Occurrence and removal of antibiotic resistance genes in municipal wastewater and rural domestic sewage treatment systems in Eastern China. Environ Int 55:9–14
459. Munir M, Wong K, Xagoraraki I (2011) Release of antibiotic resistant bacteria and genes in the effluent and biosolids of five wastewater utilities in Michigan. Water Res 45(2):681–693
460. Joss A, Zabczynski S, Göbel A, Hoffmann B, Löffler D, McArdell CS, Ternes TA, Thomsen A, Siegrist H (2006) Biological degradation of pharmaceuticals in municipal wastewater treatment: proposing a classification scheme. Water Res 40(8):1686–1696
461. Bürgmann H, Frigon D, Gaze WH, Manaia CM, Pruden A, Singer AC, Smets BF, Zhang T (2018) Water and sanitation: an essential battlefront in the war on antimicrobial resistance. FEMS Microbiol Ecol 94(9):1–14
462. Pruden A, Pei R, Storteboom H, Carlson KH (2006) Antibiotic resistance genes as emerging contaminants: studies in Northern Colorado. Environ Sci Technol 40(23):7445–7450
463. Chee-Sanford JC, Mackie RI, Koike S, Krapac IG, Lin Y-F, Yannarell AC, Maxwell S, Aminov RI (2009) Fate and transport of antibiotic residues and antibiotic resistance genes following land application of manure waste. J Environ Qual 38(3):1086–1108
464. Woolhouse M, Ward M, Van Bunnik B, Farrar J (2015) Antimicrobial resistance in humans, livestock and the wider environment. Philos Trans R Soc B Biol Sci 370(1670)
465. Public Health Agency of Canada (2016) Canadian antimicrobial resistance surveillance system report 2016. https://www.canada.ca/en/public-health/services/publications/drugs-

health-products/canadian-antimicrobial-resistance-surveillance-system-report-2016.html. Accessed: 30 March 2022
466. Graham DW, Knapp CW, Christensen BT, McCluskey S, Dolfing J (2016) Appearance of β-lactam resistance genes in agricultural soils and clinical isolates over the 20th century. Sci Rep 6(February):1–8
467. Pruden A, Arabi M, Storteboom HN (2012) Correlation between upstream human activities and riverine antibiotic resistance genes. Environ Sci Technol 46(21):11541–11549
468. Peak N, Knapp CW, Yang RK, Hanfelt MM, Smith MS, Aga DS, Graham DW (2007) Abundance of six tetracycline resistance genes in wastewater lagoons at cattle feedlots with different antibiotic use strategies. Environ Microbiol 9(1):143–151
469. Storteboom H, Arabi M, Davis JG, Crimi B, Pruden A (2010) Identification of antibiotic-resistance-gene molecular signatures suitable as tracers of Pristine River, urban, and agricultural sources. Environ Sci Technol 44(6):1947–1953
470. Graham DW, Olivares-Rieumont S, Knapp CW, Lima L, Werner D, Bowen E (2011) Antibiotic resistance gene abundances associated with waste discharges to the Almendares river near Havana. Cuba. Environ Sci Technol 45(2):418–424
471. Pei R, Kim SC, Carlson KH, Pruden A (2006) Effect of river landscape on the sediment concentrations of antibiotics and corresponding antibiotic resistance genes (ARG). Water Res 40(12):2427–2435
472. Kristiansson E, Fick J, Janzon A, Grabic R, Rutgersson C, Weijdegård B, Söderström H, Joakim Larsson DG (2011) Pyrosequencing of antibiotic-contaminated river sediments reveals high levels of resistance and gene transfer elements. PLoS One 6(2):1–7
473. Li J, Cao J, Zhu YG, Chen QL, Shen F, Wu Y, Xu S, Fan H, Da G, Huang RJ, Wang J, De Jesus AL, Morawska L, Chan CK, Peccia J, Yao M (2018) Global survey of antibiotic resistance genes in air. Environ Sci Technol 52(19):10975–10984
474. Hu J, Zhao F, Zhang XX, Li K, Li C, Ye L, Li M (2018) Metagenomic profiling of ARGs in airborne particulate matters during a severe smog event. Sci Total Environ 615:1332–1340
475. Elias Dueker M, O'Mullan GD, Martínez JM, Juhl AR, Weathers KC (2017) Onshore wind speed modulates microbial aerosols along an urban waterfront. Atmosphere 8(11):215
476. Pirnay J-P, Bilocq F, Pot B, Cornelis P, Zizi M, Van Eldere J, Deschaght P, Mario Vaneechoutte, Jennes S, Pitt T, Daniel De Vos (2009) Pseudomonas aeruginosa population structure revisited. PLoS One 4(11):1–20
477. Hwang MSH, Morgan RL, Sarkar SF, Wang PW, Guttman DS (2005) Phylogenetic characterization of virulence and resistance phenotypes of Pseudomonas syringae. Appl Environ Microbiol 71(9):5182–5191
478. Christner BC, Morris CE, Foreman CM, Cai R, Sands DC (2008) Ubiquity of biological ice nucleators in snowfall. Science 319(5867):1214
479. Li J, Zhou L, Zhang X, Caijia X, Dong L, Yao M (2016) Bioaerosol emissions and detection of airborne antibiotic resistance genes from a wastewater treatment plant. Atmos Environ 124:404–412
480. Gao M, Jia R, Qiu T, Han M, Wang X (2017) Size-related bacterial diversity and tetracycline resistance gene abundance in the air of concentrated poultry feeding operations. Environ Pollut 220:1342–1348
481. McEachran AD, Blackwell BR, Delton Hanson J, Wooten KJ, Mayer GD, Cox SB, Smith PN (2015) Antibiotics, bacteria, and antibiotic resistance genes: aerial transport from cattle feed yards via particulate matter. Environ Health Perspect 123(4):337–343
482. Gilbert Y, Veillette M, Duchaine C (2010) Airborne bacteria and antibiotic resistance genes in hospital rooms. Aerobiologia (Bologna) 26(3):185–194
483. Schmitt H, Stoob K, Hamscher G, Smit E, Seinen W (2006) Tetracyclines and tetracycline resistance in agricultural soils: microcosm and field studies. Microb Ecol 51(3):267–276
484. Mariya Munir, Irene Xagoraraki (2011) Levels of antibiotic resistance genes in manure, biosolids, and fertilized soil. J Environ Qual 40(1):248–255
485. Avery LM, Williams AP, Killham K, Jones DL (2008) Survival of Escherichia coli O157:H7 in waters from lakes, rivers, puddles and animal-drinking troughs. Sci Total Environ 389(2–3):378–385

486. Jamieson RC, Gordon RJ, Sharples KE, Stratton GW, Madani A (2002) Movement and persistence of fecal bacteria in agricultural soils and subsurface drainage water: a review. Can Biosyst Eng/Le Genie des Biosyst au Canada 44:1–9
487. Andrews RE, Johnson WS, Guard AR, Marvin JD (2004) Survival of enterococci and Tn916-like conjugative transposons in soil. Can J Microbiol 50(11):957–966
488. Forsberg KJ, Reyes A, Wang B, Selleck EM, Sommer MOA, Dantas G (2012) The shared antibiotic resistome of soil bacteria and human pathogens. Science 337(6098):1107–1111
489. Götz A, Smalla K (1997) Manure enhances plasmid mobilization and survival of Pseudomonas putida introduced into field soil. Appl Environ Microbiol 63(5):1980–1986
490. Nielsen KM, Smalla K, Van Elsas JD (2000) Natural transformation of *Acinetobacter* sp. strain BD413 with cell lysates of *Acinetobacter* sp., *Pseudomonas fluorescens*, and *Burkholderia cepacia* in soil microcosms. Appl Environ Microbiol 66(1):206–212
491. Badger-Emeka LI, Emeka PM, Okosi M (2018) Evaluation of the extent and reasons for increased non-prescription antibiotics use in a University town. Nsukka Nigeria. Int J Health Sci (Qassim) 12(4):11–17
492. Fu J, Zhang Q, Huang B, Fan N, Jin R (2021) A review on anammox process for the treatment of antibiotic-containing wastewater: linking effects with corresponding mechanisms. Front Environ Sci Eng 15(1):17
493. Wang C, Zhao J, Xing B (2021) Environmental source, fate, and toxicity of microplastics. J Hazard Mater 407:124357
494. Guo X, Wang J (2019) The chemical behaviors of microplastics in marine environment: a review. Mar Pollut Bull 142:1–14
495. Andrady AL (2017) The plastic in microplastics: a review. Mar Pollut Bull 119(1):12–22
496. Plastics Europe and EPRO (2016) Plastics—the facts 2016. https://plasticseurope.org/wp-content/uploads/2021/10/2016-Plastic-the-facts.pdf. Accessed: 16 March 2022
497. Jambeck JR, Geyer R, Wilcox C, Siegler TR, Perryman M, Andrady A, Narayan R, Law KL (2015) Plastic waste inputs from land into the ocean. Science 347(6223):768–771
498. Dekiff JH, Remy D, Klasmeier J, Fries E (2014) Occurrence and spatial distribution of microplastics in sediments from Norderney. Environ Pollut 186:248–256
499. Woodall LC, Sanchez-Vidal A, Canals M, Paterson GLJ, Coppock R, Sleight V, Calafat A, Rogers AD, Narayanaswamy BE, Thompson RC (2021) The deep sea is a major sink for microplastic debris. R Soc Open Sci 1(4):140317
500. Xiong X, Zhang K, Chen X, Shi H, Luo Z, Chenxi Wu (2018) Sources and distribution of microplastics in China's largest inland lake–Qinghai Lake. Environ Pollut 235:899–906
501. Yu Q, Hu X, Yang B, Zhang G, Wang J, Ling W (2020) Distribution, abundance and risks of microplastics in the environment. Chemosphere 249:126059
502. Tang Y, Liu Y, Chen Y, Zhang W, Zhao J, He S, Yang C, Zhang T, Tang C, Zhang C, Yang Z (2021) A review: research progress on microplastic pollutants in aquatic environments. Sci Total Environ 766:142572
503. Browne MA, Galloway T, Thompson R (2007) Microplastic–an emerging contaminant of potential concern? Integr Environ Assess Manag 3(4):559–561
504. Gregory MR (1996) Plastic 'scrubbers' in hand cleansers: a further (and minor) source for marine pollution identified. Mar Pollut Bull 32(12):867–871
505. Lambert S, Wagner M (2016) Characterisation of nanoplastics during the degradation of polystyrene. Chemosphere 145:265–268
506. Kowalski N, Reichardt AM, Waniek JJ (2016) Sinking rates of microplastics and potential implications of their alteration by physical, biological, and chemical factors. Mar Pollut Bull 109(1):310–319
507. Liu L, Fokkink R, Koelmans AA (2016) Sorption of polycyclic aromatic hydrocarbons to polystyrene nanoplastic. Environ Toxicol Chem 35(7):1650–1655
508. Velzeboer I, Kwadijk CJAF, Koelmans AA (2014) Strong sorption of PCBs to nanoplastics, microplastics, carbon nanotubes, and fullerenes. Environ Sci Technol 48(9):4869–4876
509. Li J, Zhang K, Zhang H (2018) Adsorption of antibiotics on microplastics. Environ Pollut 237:460–467

510. Holmes LA, Turner A, Thompson RC (2012) Adsorption of trace metals to plastic resin pellets in the marine environment. Environ Pollut 160:42–48
511. Holmes LA, Turner A, Thompson RC (2014) Interactions between trace metals and plastic production pellets under estuarine conditions. Mar Chem 167:25–32
512. Hirai H, Takada H, Ogata Y, Yamashita R, Mizukawa K, Saha M, Kwan C, Moore C, Gray H, Laursen D, Zettler ER, Farrington JW, Reddy CM, Peacock EE, Ward MC (2011) Organic micropollutants in marine plastics debris from the open ocean and remote and urban beaches. Mar Pollut Bull 62(8):1683–1692
513. Jovanović B (2017) Ingestion of microplastics by fish and its potential consequences from a physical perspective. Integr Environ Assess Manag 13(3):510–515
514. Setälä O, Fleming-Lehtinen V, Lehtiniemi M (2014) Ingestion and transfer of microplastics in the planktonic food web. Environ Pollut 185:77–83
515. Neves D, Sobral P, Ferreira JL, Pereira T (2015) Ingestion of microplastics by commercial fish off the Portuguese coast. Mar Pollut Bull 101(1):119–126
516. D'Angelo S, Meccariello R (2021) Microplastics: a threat for male fertility. Int J Environ Res Public Health 18(5):1–11
517. Enyoh CE, Verla AW, Verla EN, Ibe FC, Amaobi CE (2019) Airborne microplastics: a review study on method for analysis, occurrence, movement and risks. Environ Monit Assess 191(11)
518. Banerjee A, Shelver WL (2021) Micro- and nanoplastic induced cellular toxicity in mammals: a review. Sci Total Environ 755:142518
519. Vianello A, Jensen L, Liu L, Vollertsen J (2019) Simulating human exposure to indoor airborne microplastics using a breathing thermal manikin. Sci Rep 9(1):8670
520. Etz BD, Mifkovic M, Vyas S, Shukla MK (2022) High-temperature decomposition chemistry of trimethylsiloxane surfactants, a potential fluorine free replacement for fire suppression. Chemosphere 308:136351
521. Jenness GR, Koval AM, Etz BD, Shukla MK (2022) Atomic insights into the hydrodefluorination of PFAS using silylium catalysts. Environ Sci Process Impacts 24:2085–2099
522. Luft CM, Schutt TC, Shukla MK (2022) Properties and mechanisms for PFAS adsorption to aqueous clay and humic soil components. Environ Sci Technol 56:10053–10061
523. Maza WA, Etz BD, Schutt TC, Chaloux BL, Breslin VM, Pate BB, Shukla MK, Owrutsky JC, Epshteyn A (2022) The impact of sub-micellar aggregation on the reduction kinetics of perfluorooctanoate (PFOA) by the hydrated electron. Environ Sci Technol Lett 9:226–232
524. Pisani WA, Wedgeworth DN, Roth MR, Newman JK, Shukla MK (2021) Multiscale modeling of polyamide 6 using molecular dynamics and micromechanics. Ind Eng Chem Res 60:13604–13613
525. Roth M, Pisani WA, Wedgeworth DN, Newman JK, Shukla MK (2022) Computational analysis on mechanical property reinforcement of nylon 6 polymer and nanofiller dispersion through addition of CNT/graphene/CNT-graphene nanofillers. J Polym Res 29:294
526. Pisani WA, Wedgeworth DN, Roth MR, Newman JK, Shukla MK (2021) Computational prediction of mechanical properties of PA6-graphene/carbon nanotube nanocomposites. J Phys Chem C 125:15569–15578
527. Buchanan JP, Reed-Gore ER, Jefcoat JA, Moser RD, Klaus KL, Peel HR, Buchanan RK, Barnes E, Alberts E, Shukla MK (2019) Increasing mechanical resilience and enhanced electrical conductivity through the incorporation of CNF reinforcing additives in PA6 nanocomposites. Struct Chem 30:341–349
528. Jha SK, Roth M, Todde G, Buchanan JP, Moser RD, Shukla MK, Subramanian G (2018) Role of Stone-Wales defects on the interfacial interactions among graphene, carbon nanotube, and nylon 6: A first-principles study. J Chem Phys 149:054703
529. Jha SK, Roth M, Todde G, Buchanan JP, Moser RD, Shukla MK, Subramanian G (2018) First-principles study of the interactions between graphene-oxide and amine-functionalized carbon nanotube. J Phys Chem C 122:1288–1298
530. Pramanik A, Jones S, Gao Y, Sweet C, Begum S, Shukla MK, Buchanan JP, Moser RD, Ray PC (2017) Development of chitosan wrapped CNT based 3D nanoporus membrane for separation and inactivation of rotavirus and shigella waterborne pathogens. J Mater Chem B 5:9522–9531

Chapter 2
A Generalized Force-Modified Potential Energy Surface (G-FMPES) for Mechanochemical Simulations

Sanjiv K. Jha and Gopinath Subramanian

Abstract We describe the modifications that a spatially varying external force produces on a Born-Oppenheimer potential energy surface (PES), and in this chapter, we present a formulation for describing a Generalized Force-Modified Potential Energy Surface (G-FMPES). Our formulation shows that the spatially varying force resembling hydrostatic pressure results in the G-FMPES having curvature different from that of the unmodified PES. Using electronic structure methods, the effect of pseudo-hydrostatic pressure on the PES is exemplified by calculating atomistic quantities (including transition states) for (i) conformational transitions in ethane (C_2H_6) and RDX (hexahydro-1,3,5-trinitro-*s*-triazine) molecules, (ii) the decomposition of RDX, and (iii) a Diels-Alder reaction between 1,3-butadiene and ethylene. The calculated transition states and Hessian matrices of stationary points of ethane and RDX molecules show that spatially varying external forces shift the stationary points and modify the curvature of the PES, thereby affecting the harmonic transition rates by altering both the energy barrier as well as the prefactor. The harmonic spectra of both molecules are blue-shifted with increasing compressive "pressure." Some stationary points on the RDX-PES disappear under the application of the external force, indicating the merging of an energy minimum with a saddle point. This change in the topology of the PES demonstrates that new reaction pathways may be introduced by the application of mechanical forces. Part of this chapter is reproduced with permission from Refs. (J Chem Phys 143(13):134109 [1]) Copyright 2015 AIP Publishing, (J Chem Phys 145(7):074307 [2]) Copyright 2016 AIP Publishing, and (Int J Quantum Chem 117(20):e25426 [3]) Copyright 2017 John Wiley & Sons.

S. K. Jha
Mathematics and Science Department, GateWay Community College, Maricopa Community Colleges, 108 N 40th Street, Phoenix, AZ 85034, USA
e-mail: sanjiv.jha@gatewaycc.edu

G. Subramanian (✉)
X-Computational Physics Division, Los Alamos National Laboratory, Los Alamos, NM 87545, USA
e-mail: gss@lanl.gov

M. Shukla et al. (eds.), *Emerging Materials and Environment*,
Challenges and Advances in Computational Chemistry and Physics 37,
https://doi.org/10.1007/978-3-031-39470-6_2

2.1 Introduction

Since the early days of humanity, mechanochemistry has been used to accelerate chemical reactions. The earliest example is very likely the striking of two pieces of stone in order to produce fire. In more recent times, mechanochemistry has been recognized as the use of mechanical forces to modify the potential energy surface of a system, and consequently, modify reaction pathways and rates. Recently, there has been a resurgence of interest in examining systems that have a subset of their atoms subjected to an external force. Examples include computing activation energies for defect diffusion in the vicinity of dislocations [4, 5], ring opening of cyclobutene [6], the design of mechanophores [7], inducing forbidden reactions by violating the Woodward-Hoffman rules [8], exploring the energy landscapes of mechanosensitive molecules [9], and mechanophore activation [10–12].

Compressive hydrostatic pressures are another practically useful control variable, and it has long been known that reactions conducted at pressures different from atmospheric can have mechanisms and rates different from their counterparts at atmospheric pressure [13–16]. Examples include developing superconducting behavior in condensed rare gases, and ionic compounds [17], polymorphic crystal-crystal phase transition in aluminum phosphate (berlinite) [18], bacterial spore inactivation [19], and the oxidation of hydrogen [20]. On the atomic level, compressive hydrostatic pressure manifests itself as an external force on the atoms that are participating in a reaction and can therefore be considered a subset of mechanochemistry [1, 6, 8, 12, 21–28].

There are numerous ways to compress a molecule in a simulation. The majority of studies to date assume, for simplicity, that the external force applied to the atoms is constant [6, 21, 23, 25], but, this need not always be the case. One example of a spatially varying force is seen in problems relevant to irradiation damage where the transition rates of defects (such as single interstitials) in metals and oxides have been demonstrated to be a function of the macroscopic stress [29–32]. In these cases, defects are embedded in a uniform stress field, but the transmission of the macroscopically uniform stress to the atomic level almost inevitably results in the individual atoms experiencing a force that is a complex function of the current atomic configuration. Consequently, atoms involved in a transition will experience an external force that is a complex function of the reaction progress. Another example of a spatially varying force is seen in energetic materials such as RDX in either the presence of a macroscopic stress field [33–35], or in the vicinity of defects [36–40]. Here, as with the case of irradiation damage, the force on the atoms involved in a transition is a function of their position within the crystal. A third example is seen in molecules embedded in a non-uniform external field (e.g., as seen in ion traps [41] where the confining external potential is quadratic) which results in a force that is non-uniform throughout the molecule, and therefore, the assumption of a constant force can be a poor approximation. A fourth example is seen in single-molecule force spectroscopy [42, 43] where the constant force assumption is valid only in the limit of "soft" handles through which the force is transmitted to the molecule. Therefore, a more realistic description of single-molecule pulling would also involve forces that are not constant.

The effects of external forces were recently investigated for several different systems [6–8, 23, 25, 27, 28]. Moreover, recent reviews were also published on the topic of mechanochemistry [22, 44, 45]. The effect of any external force (either constant or spatially varying) on a subset of atoms serves to modify the potential energy surface (PES) of the system being studied, and in this chapter, we present a formulation for describing a Generalized Force-Modified Potential Energy Surface (G-FMPES) [1]. This formulation can be used for both constant as well as spatially varying forces, and can be viewed as a generalization of earlier work by Ong et al. [6]. This formulation also shows that spatially varying forces result in the G-FMPES having a curvature different from that of the unmodified PES: this curvature change can be particularly important for studies that explore the G-FMPES by numerical integration [9], where the tacit assumption of unchanging curvatures is made. With the formulation of the G-FMPES in place, we show that standard atomistic techniques can be used to calculate quantities of practical interest, such as transition states and vibration spectra, under the influence of a spatially varying external force. We also discuss some results obtained for one particular mode of external load, which we refer to as pseudo-hydrostatic pressure, for conformation transitions in ethane (C_2H_6) and hexahydro-1,3,5-trinitro-*s*-triazine (RDX) [1], the decomposition of RDX [3], and a Diels-Alder reaction [2].

2.2 Methods

2.2.1 The Generalized Force-Modified Potential Energy Surface (G-FMPES)

Consider a 3-dimensional system comprised of N atoms with positions of all the atoms given by the $3N$-dimensional vector $\mathbf{R}$. Let the functional $V(\mathbf{R})$ denote the unmodified PES. We assume that the PES is smooth, and that third derivatives exist at all points. As is standard, the force on all the atoms is given by the negative gradient of the potential energy functional as

$$\mathbf{F}_{\text{grad}}(\mathbf{R}) = -\nabla V(\mathbf{R}) \tag{2.1}$$

In mechanochemical simulations, the external force, denoted by the $3N$-dimensional vector $\mathbf{F}_{\text{ext}}(\mathbf{R})$, is the result of an external potential $V_{\text{ext}}(\mathbf{R})$. These last two quantities are related by

$$\mathbf{F}_{\text{ext}}(\mathbf{R}) = -\nabla V_{\text{ext}}(\mathbf{R}) \tag{2.2}$$

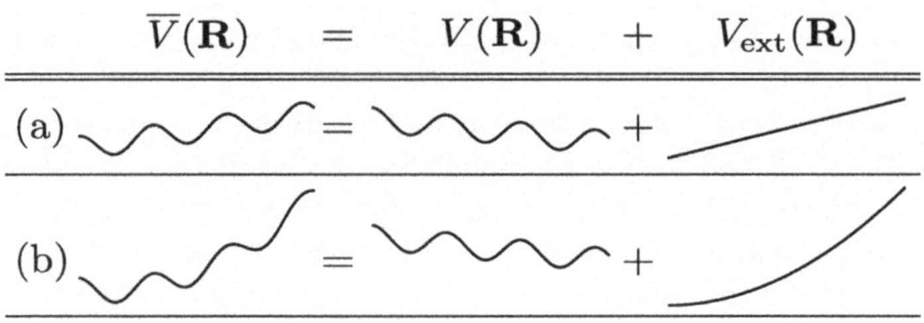

Fig. 2.1 A 1-dimensional generalized force-modified potential energy surface (G-FMPES), $\overline{V}(\mathbf{R})$, as the sum of an unmodified PES and an external potential. The schematic shows the results of adding two different kinds of external potentials: **a** linear, corresponding to a constant external force that preserves the curvature of the unmodified PES, and **b** quadratic, corresponding to a spatially varying force that changes the curvature of the unmodified PES. Reprinted (adapted) from [1], with the permission of AIP publishing

This external force, (form unspecified), modifies the PES and results in a G-FMPES:

$$\overline{V}(\mathbf{R}) = V(\mathbf{R}) + V_{ext}(\mathbf{R}) \tag{2.3}$$

This addition of functionals for a 1-dimensional PES is shown schematically in Fig. 2.1 for (a) linear, and (b) quadratic external potentials. We adopt the convention that a variable with an overline refers to a quantity computed on a G-FMPES. In mechanochemical simulations, the unmodified PES and external forces are prescribed, making it necessary to calculate the numerical value of $V_{ext}(\mathbf{R})$. As long as $\mathbf{F}_{ext}(\mathbf{R})$ is a conservative field, this is achieved by performing a path integral from a suitable reference configuration, say $\mathbf{R}_{ref}$, to the desired configuration $\mathbf{R}$ as

$$V_{ext}(\mathbf{R}) - V_{ext}(\mathbf{R}_{ref}) = \int_{\mathbf{s}=\mathbf{R}_{ref}}^{\mathbf{s}=\mathbf{R}} -\mathbf{F}_{ext}(\mathbf{s}) \cdot d\mathbf{s} \tag{2.4}$$

$$= \int_{\mathbf{s}=\mathbf{R}}^{\mathbf{s}=\mathbf{R}_{ref}} \mathbf{F}_{ext}(\mathbf{s}) \cdot d\mathbf{s} \tag{2.5}$$

Note that in the second integral, the integration path has been reversed by removing the negative sign. The actual path is unimportant (because of the conservative force field), and a straight line is chosen as it is possibly the simplest path. In most simulations, we are interested in the *relative* heights of stationary points, and therefore set $V_{ext}(\mathbf{R}_{ref}) = 0$. Equation (2.3) becomes

$$\overline{V}(\mathbf{R}) = V(\mathbf{R}) + \int_{\mathbf{s}=\mathbf{R}}^{\mathbf{s}=\mathbf{R}_{ref}} \mathbf{F}_{ext}(\mathbf{s}) \cdot d\mathbf{s} \tag{2.6}$$

and shows that external forces can, in general, produce stationary points that are different from those on the unmodified PES.

As discussed in the introduction, the external force vector is not necessarily a constant. However, if we set $\mathbf{F}_{\text{ext}}$ to be a constant, as has been the case with many mechanochemical studies [6–12], Eq. (2.6) for the G-FMPES reduces to

$$\overline{V}(\mathbf{R}) = V(\mathbf{R}) + \mathbf{F}_{\text{ext}} \cdot (\mathbf{R}_{\text{ref}} - \mathbf{R}) , \tag{2.7}$$

equivalent to the expression employed by previous studies. Thus, if we make the simplifying assumptions that previous researchers have made, our more general expression for the G-FMPES (Equation (2.6)) reduces to the simpler form employed in the past (Equation (2.7)).

2.2.2 *Curvature of the G-FMPES*

The local curvature of the G-FMPES is given by the Hessian matrix, which is in turn defined as the Jacobian of the gradient of the potential energy functional. Denoting the Hessian for the G-FMPES by $\overline{\mathbf{H}}(\mathbf{R})$, and applying the definition of the Hessian to the definition of a G-FMPES (Equation (2.3)), we obtain

$$\begin{aligned} \overline{\mathbf{H}}(\mathbf{R}) &= \mathbf{J}\left[\nabla \overline{V}(\mathbf{R})\right] \\ &= \mathbf{J}\left[\nabla \left\{V(\mathbf{R}) + V_{\text{ext}}(\mathbf{R})\right\}\right] \end{aligned} \tag{2.8}$$

Since the Jacobian and gradient operators are distributive, we get an expression for the Hessian matrix on the G-FMPES as

$$\begin{aligned} \overline{\mathbf{H}}(\mathbf{R}) &= \mathbf{H}(\mathbf{R}) - \mathbf{J}\left[\mathbf{F}_{\text{ext}}(\mathbf{R})\right] \\ &= \mathbf{H}(\mathbf{R}) + \mathbf{H}_{\text{ext}}(\mathbf{R}) \end{aligned} \tag{2.9}$$

where $\mathbf{H}(\mathbf{R})$ is the Hessian matrix of the unmodified PES, and we have designated

$$\mathbf{H}_{\text{ext}}(\mathbf{R}) = -\mathbf{J}\left[\mathbf{F}_{\text{ext}}(\mathbf{R})\right] \tag{2.10}$$

Equations (2.9) and (2.10) show that if the external force $\mathbf{F}_{\text{ext}}$ is a constant, then $\mathbf{H}_{\text{ext}}(\mathbf{R}) = 0$, the Hessian is unchanged, and the two surfaces have the same curvature (as shown in Fig. 2.1a). However, as external forces are not necessarily a constant, the curvatures of a G-FMPES are generally expected to be different from those of the unmodified PES. The effect of varying curvatures is discussed in Sect. 2.2.4.

2.2.3 *Pseudo-Hydrostatic Pressure*

As an example of a non-constant external force, we use a force vector referred to as "pseudo-hydrostatic pressure," where, the external force applied *on each atom*, $\mathbf{f}_{\text{ext}}^{(j)}$, is calculated as

$$\mathbf{f}_{\text{ext}}^{(j)} = P_{HP}\left[\mathbf{r}^{(j)} - \mathbf{c}\right]; \quad \forall \quad j = 1, 2, 3 \ldots N \tag{2.11}$$

where P_{HP} is a user-defined constant with units of kcal mol^{-1}Å^{-2}, $\mathbf{r}^{(j)}$ is the position vector of the jth atom, and $\mathbf{c}$ is the geometric centroid. Here, lowercase boldface symbols refer to 3-dimensional vectors. A negative value of P_{HP} implies compression, while a positive value implies expansion. Also, both the magnitude and the direction of the external force on each atom are a function of the configuration. Consequently, the external force vector on all atoms, $\mathbf{F}_{\text{ext}}$ is a function of the current configuration and is different, and not constant. This scheme for the application of external forces is equivalent to placing each atom of the system in the same parabolic potential centered at the geometric centroid $\mathbf{c}$. It results in a conservative force field, thereby satisfying the assumptions made in deriving Eq. (2.6). For the interested reader, a proof that the force field is conservative is available in Appendix 1.

2.2.4 Transition Rates on the G-FMPES

There are two key differences in calculating transition rates on a G-FMPES when compared with on an unmodified PES, and can be understood within the context of harmonic Transition State Theory (h-TST), where transition rates are computed as

$$\overline{k} = \overline{k}_0 \exp\left[-\frac{\overline{E}_a}{k_B T}\right] \tag{2.12}$$

where k_B is the Boltzmann constant, T is the temperature, $\overline{E}_a$ is the activation energy for a transition, and $\overline{k}_0$ is the prefactor (related to the curvature).

The activation energy, $\overline{E}_a$, is simply the difference in energies of the saddle point and the energy minimum on the G-FMPES. From Eq. (2.6), it is easy to see that stationary points on a G-FMPES (i.e., points with $\nabla\overline{V}(\mathbf{R}) = 0$) are in general different from those on an unmodified PES, and consequently, the activation energy for a transition can be expected to be different on a G-FMPES. Calculating the activation energy requires finding the saddle point configuration, which can be achieved using any of the various saddle point finding methods [46], and indeed, many researchers have accounted for this change [4, 29, 30] in the activation energy of a transition when computing transition rates.

However, these very studies have neglected the effects of curvature changes, which affect transition rates via the prefactor $\overline{k}_0$ with the justification that most significant changes in transition rates are due to changes in the activation energy, and therefore overwhelms any effects from curvature changes that manifest themselves in the prefactor. While this assumption may be acceptable in many situations, transitions with unusual prefactors such as the transformation of voids to stacking fault tetrahedra [47] with a prefactor tens of orders of magnitude higher than typical exist, and can be expected to show counterintuitive changes in transition rates when the external force is not a constant.

The effects of curvature change can be understood in the context of Vineyard's expression [48]:

$$\overline{k}_0 = \frac{\prod_{j=1}^{3N-m} \overline{\nu}_j^{(\text{min})}}{\prod_{j=1}^{3N-(m+1)} \overline{\nu}_j^{(\text{sad})}} \tag{2.13}$$

where the $3N - m$ different $\overline{\nu}_j^{(\text{min})}$ are the non-zero normal mode frequencies of the energy minimum, and the $3N - (m + 1)$ different $\overline{\nu}_j^{(\text{sad})}$ are the positive, non-zero normal mode frequencies of the saddle point. Subtraction of m in the numerator accounts for the m rigid body motion degrees of freedom, and subtraction of $(m + 1)$ in the denominator accounts for the rigid body motion degrees of freedom plus 1 imaginary frequency for a first-order saddle point. For systems with translational and rotational symmetry, $m = 6$. Application of external forces can break the symmetry, resulting in $m \leq 6$. These normal mode frequencies are in turn calculated by diagonalizing the Hessian matrix $\overline{\mathbf{H}}(\mathbf{R}) = \mathbf{H}(\mathbf{R}) + \mathbf{H}_{\text{ext}}(\mathbf{R})$.

For a constant external force resulting in $\mathbf{H}_{\text{ext}}(\mathbf{R}) = 0$, the prefactor $\overline{k}_0$ is calculated using the curvature of the unmodified PES at the locations of the new stationary points.

2.3 Applications

With the above understanding of how to compute energies, forces, and curvatures on a G-FMPES, standard atomistic techniques were used to compute quantities of practical interest. In this chapter, we will outline the effects of pseudo-hydrostatic pressure on reaction rates for (i) conformation transitions in ethane and RDX [1], (ii) the decomposition of RDX [3], and (iii) a Diels-Alder reaction [2].

Energies and their gradients (forces) on the unmodified PES (namely $V(\mathbf{R})$ and $\nabla V(\mathbf{R})$) were evaluated using an *Ab initio* electronic structure method with suitable basis sets, as implemented in the AIMS-MOLPRO code [49]. This combination of method and basis is a computationally inexpensive means of illustrating the main point of this chapter: the calculation of static quantities on a G-FMPES and their comparison with corresponding quantities on an unmodified PES. Transition states were calculated using the Nudged Elastic Band (NEB) method with the improved tangent estimate [50–52]. A more detailed description of our implementation is available in Appendix 2. Structure optimizations were performed using the Fast Inertial Relaxation Engine (FIRE) [53]. The integral in Eq. (2.6) was computed using the trapezoidal rule over a grid with resolution 5×10^{-3} Å along a straight line between $\mathbf{R}$ and $\mathbf{R}_{\text{ref}}$. Hessians of all stationary points on the G-FMPES had 6 numerically zero eigenvalues corresponding to translation and rigid body rotation. Hessians of all saddle points had only one negative eigenvalue, thereby ensuring that they were true first-order saddle points.

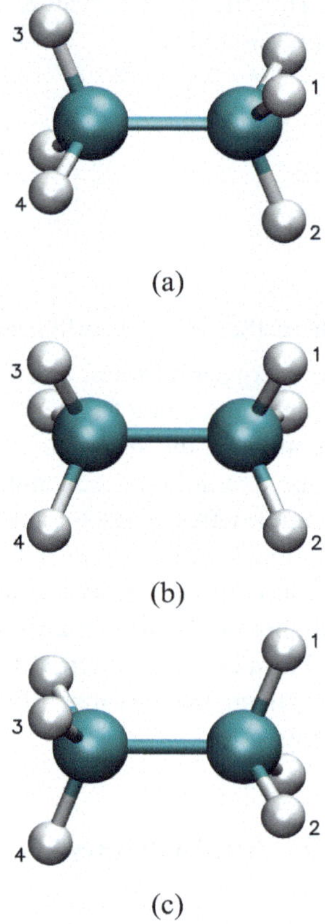

Fig. 2.2 Rotation of a CH_3 group by 120° about the C–C bond in ethane. **a** Start state—often referred to as the staggered conformation **b** Saddle point or eclipsed conformation (c) End state—also a staggered conformation. Reprinted (adapted) from [1], with the permission of AIP publishing

2.3.1 Rotational Barrier in Ethane

Using the Hartree-Fock (HF) method [54–56] with the 6-31G** basis set [57], we first examined the rotational energy barrier in ethane (C_2H_6): a 120° rotation of a CH_3 group about the C–C bond of an ethane molecule, as shown in Fig. 2.2. This transition is a prototypical example for understanding steric hindrance, first discovered by Kemp and Pitzer [58] who showed that a rotational barrier of 2.8 kcal mol^{-1} had to be present in order to obtain thermodynamic quantities in agreement with the experiment.

For this transition, the reference configuration was chosen to be the staggered conformation at $P_{HP} = 0$. Figure 2.3a shows the energy along the reaction coordinate for this transition at various pressures, and all energies are relative to the reference configuration. The nominal energy (no external pressure) was 3.02 kcal mol^{-1}, within 8% of the experimental value of 2.8 kcal mol^{-1} [58, 59]. With increasing compression (or decreasing values of P_{HP}), the saddle point energy of the saddle point progressively increases, while the energy of the staggered conformation decreases, thereby

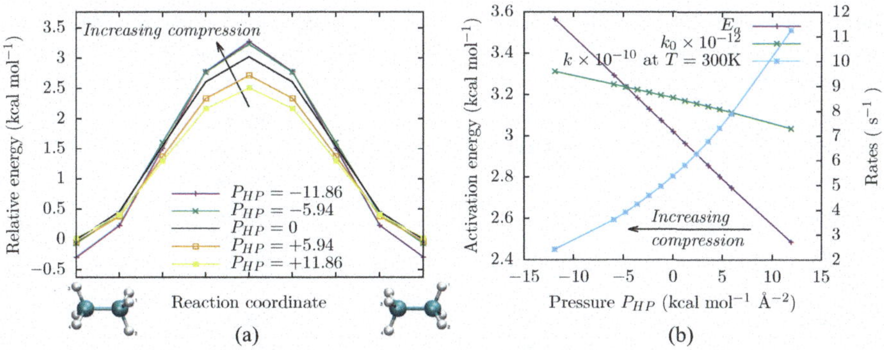

Fig. 2.3 For the rotational transition of ethane, **a** energy along the reaction coordinate as a function of pressure for the values of P_{HP} (in kcal mol^{-1} Å^{-2}) indicated in the legend, and **b** activation energies, prefactors, and transition rates as a function of pressure. All energies are relative to that of the staggered configuration at zero pressure. In both figures, an arrow indicates the direction of increasing compression. Reprinted (adapted) from [1], with the permission of AIP publishing

leading to a net increase in activation energy as a function of increasing compression. There is a corresponding increase in the prefactor with increasing compression (see Fig. 2.3), but this increase is overshadowed by the increase in activation energy, and consequently, there is a net decrease in the transition rate as a function of increasing compression.

Wavenumbers of the staggered conformation of ethane computed by diagonalizing the Hessian are given in Table 2.1 in units of inverse centimeters. The wavenumbers at zero pressure are in good agreement with the values available in the literature for calculations at the HF/6-31G** level of theory [60]. Compression shifts the spectrum to higher wavenumbers (blueshift) indicating that the modes become stiffer, while expansive pressures have the opposite effect, i.e., redshift the spectrum. This behavior is qualitatively consistent with the experimental behavior seen by Kurnosov et al. [61] who saw that the Raman spectra of both liquid and solid ethane shifted to higher wavenumbers with increasingly compressive pressure over a range of up to 6900 MPa.

2.3.2 Conformational Transformations in RDX

The second molecule examined was RDX, also known as hexahydro-1,3,5-trinitro-*s*-triazine, and is an energetic molecular crystal and is used in many explosive formulations. RDX has a chemical formula $C_3H_6N_6O_6$, and is shown in Fig. 2.4. Red atoms are oxygen and are part of the nitro (NO_2) groups, blue atoms are nitrogen, and cyan atoms are carbon. The ring formed by alternating C and N atoms is the triazine ring, and the nitro groups are on the side. The position of the nitro side groups with respect to the triazine ring as determined by the angle between the N−N bond and

Table 2.1 Calculated wavenumbers (in cm^{-1}) of the staggered conformation of ethane for the lowest, zero, and highest values of pseudo-hydrostatic pressure, P_{HP} (in kcal $mol^{-1}Å^{-2}$), studied

$P_{HP} = -11.86$	$P_{HP} = 0$	$P_{HP} = 11.86$
382.80	327.65	264.69
897.95	885.11	846.16
897.95	885.16	876.95
1208.96	1059.31	876.96
1366.90	1329.27	1281.82
1366.91	1329.31	1281.82
1510.56	1531.93	1538.96
1574.90	1564.79	1546.27
1648.04	1627.73	1601.37
1648.12	1627.81	1601.43
1651.56	1632.78	1612.73
1651.64	1632.86	1612.79
3344.70	3180.24	2992.31
3351.69	3185.88	2995.89
3423.52	3234.15	3020.84
3423.60	3234.22	3020.91
3445.54	3259.17	3047.84
3445.61	3259.23	3047.90

The wavenumbers of the rigid body modes of deformation are not reported
Adapted from [1]

the plane of the C–N–C triple (axial, intermediate, or equatorial, denoted by a, i, or e, respectively). Molecular dynamics simulations using a non-reactive and fully-flexible force fields have shown that conformational transitions play an important role in the stability of high-pressure phases [34], plastic deformation mechanisms of RDX [38, 62–64], and contribute to energy concentration in strained α-RDX crystals [33].

We studied the Caaa $\leftrightarrow$ Ceee transition (shown in Fig. 2.4) of an isolated RDX molecule because it has the highest activation energy of all the transitions between conformers for an RDX molecule with the Chair shaped triazine ring. The pseudo-hydrostatic pressures applied in this test case are particularly relevant because the deviatoric part of the stress tensor vanishes for the gas phase and the pressure ($-\mathrm{tr}(\boldsymbol{\sigma})/3$) becomes the critical term at high stresses and temperatures under which decomposition of shocked explosives takes place.

For this transition, the Caaa conformer at $P_{HP} = 0$ was chosen as the reference configuration $\mathbf{R}_{\text{ref}}$ with $V_{\text{ext}}(\mathbf{R}_{\text{ref}}) = 0$. Increasingly compressive pressures increased the energy barrier for the forward (Caaa $\rightarrow$ Ceee) transition while decreasing the barrier for the reverse (Caaa $\leftarrow$ Ceee) transition with a concomitant increase (or decrease) in the net rates (see Fig. 2.5). For high compressions, with pseudo-hydrostatic pressures with $P_{HP} \leq -2.37$ kcal mol^{-1} $Å^{-2}$, the Ceee conformer was

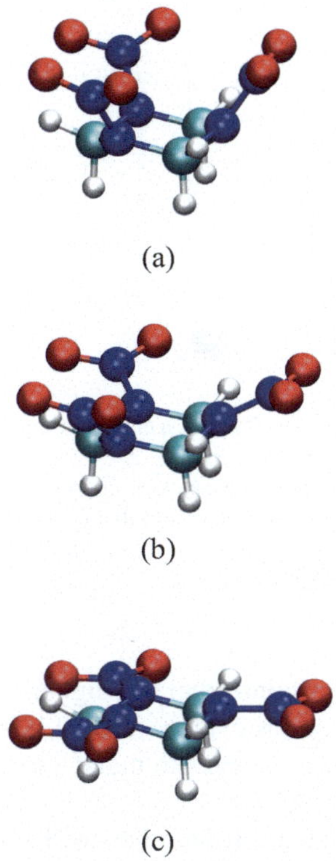

Fig. 2.4 Transition of an RDX molecule ($C_3H_6N_6O_6$) from the Caaa conformer to Ceee. **a** Caaa conformer, **b** saddle point, and **c** Ceee conformer. The red atoms are oxygen, and are part of the nitro (NO_2) groups, blue atoms are nitrogen, and cyan atoms are carbon. The ring formed by alternating C and N atoms is the triazine ring. Reprinted (adapted) from [1], with the permission of AIP publishing

no longer stable, and spontaneously transitioned into the Caaa conformer. Conversely, for low compressions with pseudo-hydrostatic pressures with $P_{HP} \geq 1.78$ kcal mol^{-1} Å^{-2}, the Caaa conformer was no longer stable and spontaneously transformed into the Ceee conformer.

This mechanism where external stresses lead to the disappearance of stationary points on the energy landscape has been observed before in the case of zirconium interstitials [29], and indicates that an energy minimum merges with a saddle point. It is entirely possible that new stationary points may be introduced by external stresses.

As with the case of ethane, we see that higher compressions cause the RDX molecule to adopt more compact conformations leading to enhanced steric effects. This observation agrees well with the fact that Caaa is present in the high-pressure polymorphs (β and ε) of RDX [67, 68]. For the Caaa $\rightarrow$ Ceee transition, we see that the transition rate decreases with increasing compression. However, the backward transition (Caaa $\leftarrow$ Ceee) rate increases with increasing compression, and this is

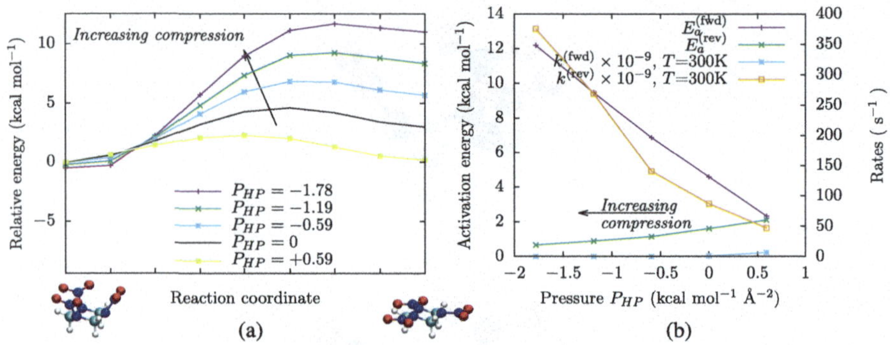

Fig. 2.5 For the Caaa to Ceee transition of RDX, **a** energy along the reaction coordinate as a function of pressure for the values of P_{HP} (in kcal mol^{-1} Å^{-2}) indicated in the legend, and **b** activation energies, prefactors, and transition rates as a function of pressure. Here, the forward transition is the Caaa → Ceee transition. All energies are relative to that of the Caaa conformation at zero pressure. In both figures, an arrow indicates the direction of increasing compression. Reprinted (adapted) from [1], with the permission of AIP publishing

thought to be because the Ceee conformer is a higher-energy conformer where steric effects are already pronounced (as evidenced by it being unstable at high compressions) and the extra compression serves to make transitions out of this high-energy conformer more accessible.

Table 2.2 shows the calculated wavenumbers with units of cm^{-1} of the two conformations for the lowest, zero, and highest values of pseudo-hydrostatic pressure used in this study. Again, as with the case of ethane, the harmonic vibrational spectrum of RDX shifts to higher wavenumbers with increasing compression. This is consistent with independent experimental studies [69–71], as well as molecular dynamics studies [72]. The softer modes of Ceee, from 54 to 122 cm^{-1}, however, behave in the opposite manner. Also shown in Table 2.2 are the wavenumbers obtained from DFT studies by Rice and Chabalowski [65] (at the B3LYP/6-31G** level), as well as experimental data from the vapor phase [66]. While there is general agreement between our harmonic frequencies and those of the experiment, there are significant differences between our frequencies and those of Rice and Chabalowski [65], but we suspect this is due to the differences in electronic structure theory techniques employed.

2.3.3 Decomposition Mechanism of RDX

We studied the three decomposition mechanisms of RDX shown in Fig. 2.6. In our study, all calculations were carried out with RDX in a chair conformation with all NO_2 groups in the axial position (Caaa), claimed to be the global minimum [65, 73–75] studies. All stationary points were characterized by performing frequency

Table 2.2 Calculated wavenumbers (in cm^{-1}) of the two conformers of RDX for the lowest, zero, and highest values of pseudo-hydrostatic pressure, P_{HP} (in kcal mol^{-1} $Å^{-2}$), studied

DFT, Ref. [65]	Experimental Ref. [66]	This work, Caaa			This work, Ceee		
		$P_{HP} = -1.78$	$P_{HP} = 0$	$P_{HP} = 0.59$	$P_{HP} = -1.78$	$P_{HP} = 0$	$P_{HP} = 0.59$
31		43.16	35.64	17.97	54.07	54.67	57.61
37		46.87	35.85	18.98	54.24	54.91	57.72
63		46.92	66.61	52.48	58.41	69.43	74.24
67		124.51	77.24	80.61	83.42	89.03	90.61
100		147.74	109.04	100.29	117.10	122.78	125.56
102		148.16	109.20	100.41	117.17	122.88	125.60
220		252.57	242.29	226.39	190.72	187.66	188.00
221		252.89	243.26	227.50	190.85	187.96	188.13
301		328.25	342.44	353.29	258.64	242.31	237.75
363		433.26	408.85	392.75	258.82	242.62	237.89
365		433.84	409.58	393.09	354.65	323.37	312.62
409		490.40	465.49	442.88	392.41	378.23	373.04
413		490.50	465.54	442.96	413.76	402.22	398.37
442		542.98	508.24	480.28	413.81	402.58	398.50
458		581.23	534.35	500.34	413.87	402.70	398.55
590		665.53	665.83	666.88	662.87	653.23	649.50
590		678.61	669.92	667.07	662.91	653.35	649.54
593		678.62	670.09	670.71	763.07	748.61	744.02
660		743.83	736.41	736.48	804.80	791.93	787.26
661	782	744.39	737.15	736.95	804.85	791.98	787.28
749	845	838.31	847.56	854.82	888.82	888.90	888.52
753	880	893.85	889.89	887.55	905.48	900.70	899.43
754		894.12	890.18	887.70	905.54	900.76	899.45
782	910	927.25	911.05	900.68	926.74	919.71	917.42
863		977.35	965.23	961.49	1025.01	1000.65	993.22
864		977.39	965.30	961.58	1025.04	1000.66	993.23
887		994.59	972.64	964.37	1039.98	1007.40	997.06
906	1014	1054.98	1048.14	1049.12	1132.72	1121.19	1118.04
907		1055.04	1048.31	1049.38	1132.73	1121.31	1118.09
935		1103.13	1096.06	1096.06	1158.32	1137.97	1131.37
1005		1183.59	1177.26	1181.35	1245.16	1222.61	1214.21
1005	1218	1183.95	1177.81	1181.77	1245.22	1222.61	1214.22
1141	1268	1299.40	1289.72	1291.61	1335.81	1319.68	1313.45
1242	1319	1414.78	1400.56	1399.42	1425.35	1403.41	1393.44
1250	1374	1415.18	1401.04	1399.72	1438.70	1407.28	1397.35
1252		1421.61	1414.65	1403.04	1464.11	1439.33	1430.03
1275	1420	1436.76	1418.06	1421.94	1464.18	1439.46	1430.07
1292	1444	1494.98	1494.98	1497.76	1476.04	1458.18	1451.63
1294		1520.75	1511.62	1505.00	1476.08	1458.33	1451.70
1345		1520.77	1511.83	1505.27	1525.41	1518.57	1516.90
1363		1541.13	1531.23	1528.75	1558.34	1537.31	1530.36

(continued)

Table 2.2 (continued)

DFT, Ref. [65]	Experimental Ref. [66]	This work, Caaa			This work, Ceee		
		$P_{HP} = -1.78$	$P_{HP} = 0$	$P_{HP} = 0.59$	$P_{HP} = -1.78$	$P_{HP} = 0$	$P_{HP} = 0.59$
1381		1541.77	1531.81	1528.99	1558.37	1537.38	1530.43
1384	1584	1585.01	1578.37	1577.65	1636.15	1612.26	1603.91
1402		1585.06	1578.44	1577.72	1636.37	1623.96	1619.76
1403		1598.31	1589.11	1585.93	1636.42	1624.01	1619.80
1464		1619.50	1616.34	1618.19	1657.76	1654.11	1653.79
1466		1619.53	1616.43	1618.31	1657.80	1654.11	1653.79
1482		1665.26	1655.60	1653.42	1670.68	1668.21	1667.74
1627		1854.72	1845.01	1839.91	1864.45	1848.85	1843.95
1658		1914.64	1890.13	1877.62	1889.74	1873.60	1868.46
1658		1914.99	1890.69	1878.07	1889.96	1873.61	1868.48
3064	3065	3311.08	3266.08	3245.22	3188.21	3172.70	3167.49
3064		3311.42	3266.52	3245.39	3188.29	3172.81	3167.52
3070		3323.14	3276.76	3254.68	3196.56	3179.52	3173.78
3192		3476.09	3434.69	3419.39	3484.77	3431.28	3412.98
3192		3476.20	3434.73	3419.44	3484.82	3431.33	3413.01
3194		3477.51	3436.30	3421.24	3487.38	3433.89	3415.59

The wavenumbers of the rigid body modes of deformation are not reported. For comparison, wavenumbers obtained by Rice and Chabalowski [65] for the Caaa conformer using a B3LYP functional, as well as experimental data from the vapor phase obtained by Karpowicz and Brill [66] are shown

Adapted from [3]

calculations. The zero-point energy (ZPE) correction to the electronic energy was included for all stationary points, and the ZPE is included in all energy values.

2.3.3.1 Concerted Mechanism

In the concerted mechanism, all three C−N bonds of RDX are broken simultaneously, yielding three molecules of CH_2−N−NO_2 (methylenenitramine, MN). The transition state (TS) is shown in Fig. 2.7a). The breaking C−N distances are between 2.00 Åand 2.14 Å. external forces modify the PES, altering both E_a and E_{rxn}. Furthermore, the sign of the applied external forces modifies the PES in different ways. A positive value of P_{HP} decreases the height of the reaction barrier and makes the reaction more exothermic. For the two highest values of P_{HP} we considered, the concerted reaction becomes exothermic, while for all other values of P_{HP} the reaction is endothermic. At all P_{HP} values, the mechanism remains concerted except for the two highest compressive values. In these last two cases, one of the C−N bonds breaks first followed by the simultaneous fission of the other two C−N bonds—a fundamental change in the reaction pathway (Table 2.3).

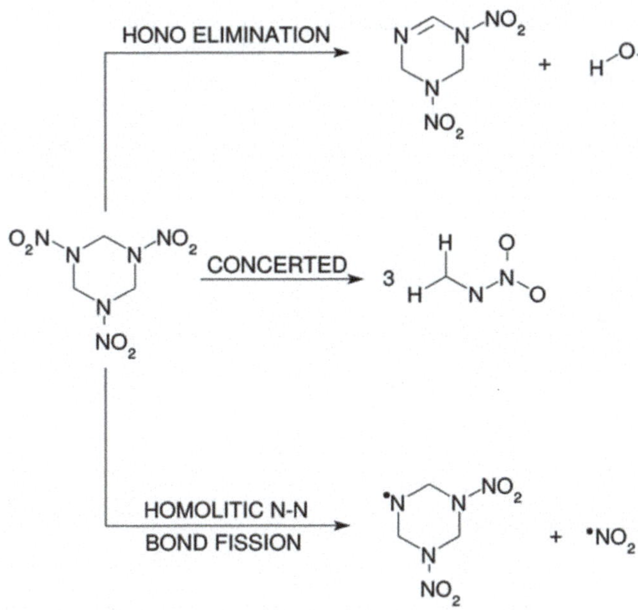

Fig. 2.6 Schematic of the initial decomposition reactions of RDX. Reprinted (adapted) from [3] with the permission of John Wiley & Sons

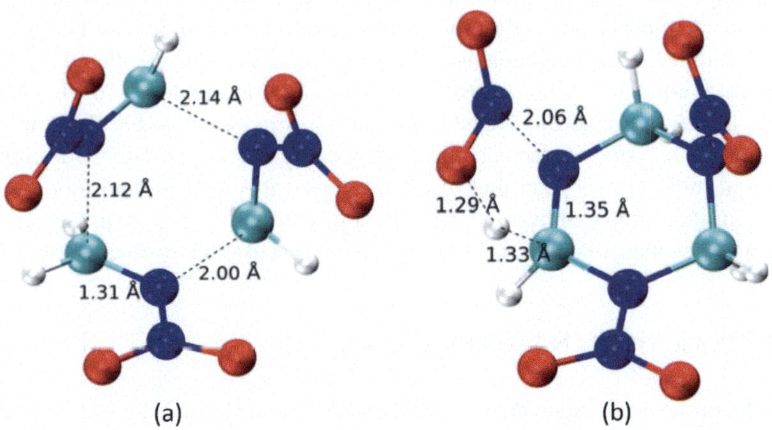

Fig. 2.7 Transition states of **a** concerted mechanism and **b** HONO elimination. Atom colors: C in cyan, N in blue, O in red, and H in white. Reprinted (adapted) from [3] with the permission of John Wiley & Sons

Table 2.3 Calculated activation (E_a) and reaction (E_{rxn}) energies at different values of pseudo-hydrostatic pressure, P_{HP} (in kcal mol^{-1} Å^{-2}), for concerted mechanism

P_{HP}	This work		Reference [73]	
	E_a	E_{rxn}	E_a	E_{rxn}
0	59.9	35.1	59.4	45.9
Expansion				
1.255	54.1	−16.9	……	……
0.941	57.2	−3.9	……	……
0.188	59.6	27.2	……	……
0.063	59.8	32.6	……	……
Compression				
−0.063	60.1	37.7	……	……
−0.188	60.6	42.8	……	……
−0.941	70.4	68.5	……	……
−1.255	77.1	76.0	……	……

All energies are expressed in kcal mol^{-1}
Adapted from [3]

2.3.3.2 HONO Elimination

In another possible decomposition mechanism of RDX, the close contact between one NO_2 group and one H leads to the elimination of one HONO molecule. The geometry of the TS found for this mechanism is shown in Fig. 2.7b. The calculated E_a for this mechanism is equal to 37.9 kcal mol^{-1}, while the E_{rxn} is equal to −10.1 kcal mol^{-1}. Both E_a and E_{rxn} make the HONO elimination a more favorable decomposition pathway than the concerted mechanism: the energy barrier for HONO elimination is ~20 kcal mol^{-1} lower than that of the concerted mechanism, and moreover, the HONO elimination is exothermic. The effect of external forces on this mechanism was minor (less than 1 kcal mol^{-1}), so we do not discuss this further (Table 2.4).

2.3.3.3 Homolytic N–N Bond Fission

A third decomposition mechanism of RDX starts via the homolytic fission of one of the N–N bonds. The N–N fission proceeds through a loose TS, associated with a large pre-exponential factor in the correlated rate constant [73, 76, 77]. We found a dissociation energy of 36.1 kcal mol^{-1} for this reaction. This value is in good agreement with previously calculated dissociation energies of RDX [73, 74, 78] ranging between 34 kcal mol^{-1} and 42 kcal mol^{-1}. The dissociation energy barrier for the N–N fission (36.1 kcal mol^{-1}) is very close to the energy barrier of the HONO elimination (37.9 kcal mol^{-1}). HONO elimination and N–N fission are very likely the favored starting routes for the decomposition of RDX. In a recent study using coupled-cluster theory, Molt et al. found a "true" TS for the N–N fission [79]. Molt et

Table 2.4 Calculated activation (E_a) and reaction (E_{rxn}) energies at different values of pseudo-hydrostatic pressure, P_{HP} (in kcal mol^{-1} Å^{-2}), for HONO elimination mechanism

P_{HP}	This work		Reference [73]	
	E_a	E_{rxn}	E_a	E_{rxn}
0	37.9	−10.1	39.2	−8.5
Expansion				
0.314	37.3	−28.7	......	......
0.188	37.5	−21.3	......	......
0.126	37.7	−17.8	......	......
0.063	37.7	−13.9	......	......
Compression				
−0.063	38.0	−6.4	......	......
−0.126	38.0	−3.2	......	......
−0.188	38.1	0.8	......	......
−0.314	38.4	7.9	......	......

All energies are expressed in kcal mol^{-1}
Adapted from [3]

al. calculated an energy barrier more than 10 kcal mol^{-1} higher than our and previous estimates [73, 74, 78], concluding that the N−N fission perhaps does not compete with the HONO elimination. The effect of external forces on the N−N fission (see Table 2.5) shows the same trend as for the two other mechanisms. The application of a compressive force works in opposition to the decomposition reaction, increasing the dissociation barrier, while the application of an expansive force lowers the barrier. Moreover, at the two highest positive values of P_{HP} that we considered, the N−N fission becomes exothermic.

2.3.4 A Diels-Alder Reaction

The Diels Alder (DA) reaction, where a diene reacts with a dienophile to form a cyclohexene derivative is an important route for synthesizing numerous cyclic compounds, and has been widely studied [80–95]. The simplest DA reaction, often called the parent DA reaction, is a [4 + 2] cycloaddition between 1,3-butadiene and ethylene which results in the formation of cyclohexene. DA reactions are used to synthesize a wide variety of compounds, such as pharmaceuticals [96–99], agrochemicals, as well as flavors, and fragrances [100].

In this section, we examine the effects of compressive external forces designed to mimic hydrostatic pressures on the mechanism of parent DA reaction between butadiene and ethylene using. As with other reactions outlined in this chapter, MEPs and transition states on the G-FMPES were calculated using a climbing-image nudged elastic band (CI-NEB) method [50, 51]. The levels of theory employed, how-

Table 2.5 Calculated reaction energies (E_{rxn}) at different values of pseudo-hydrostatic pressure, P_{HP} (in kcal mol^{-1} Å^{-2}), for homolytic N–N bond fission

P_{HP}	This work E_{rxn}	Reference [73] E_{rxn}
0	36.1	39.0
Expansion		
0.628	−58.6	……
0.314	−6.3	……
0.188	4.7	……
0.063	25.6	……
Compression		
−0.063	46.5	……
−0.188	67.5	……
−0.314	88.5	……
−0.628	135.6	……

All energies are expressed in kcal mol^{-1}
Adapted from [3]

ever, were CASSCF(6,6)/6-31G** [57, 101, 102], and B3LYP-DFT/6-311++G**, because while CASSCF is appropriate for studying organic chemical reactions [6, 103, 104], it often underestimates reaction exothermicity [105, 106], and energies calculated using the DFT method are known to provide a better match with experimental data. In this chapter, some experimentally testable predictions about the exothermicity, reaction rates, and the ratio of yield of the main reaction to the yield of the undesirable side reaction are made.

Experimentally, the activation energy of the parent DA reaction has been measured as 27.5 ± 2 kcal mol^{-1} [105, 107], and amongst much discussion, [103, 105, 108–121], is thought to have two mechanisms: a single-step concerted mechanism, and a two-step diradical mechanism. An undesirable side reaction is the formation of vinylcyclobutane [122]. These different pathways are depicted in Fig. 2.8. The concerted mechanism (Fig. 2.8, a $\rightarrow$ c) is a single-step process and results in the simultaneous formation of two new σ-bonds between the carbon atoms of butadiene and ethylene. The two-step mechanism (Fig. 2.8, a $\rightarrow$ b $\rightarrow$ c) involves a first transition to a stable diradical intermediate, followed by a second transition into cyclohexene. The third pathway (Fig. 2.8, a $\rightarrow$ d) is the undesirable side reaction that forms vinylcyclobutane.

2.3.4.1 Concerted Mechanism

In Fig. 2.9, we see the optimized geometries and bond lengths (in Å) of the reactants, transition state, and product on the unmodified PES. As the reaction progresses, the terminal methylene groups of the reactants rotate such that the protons (H) are

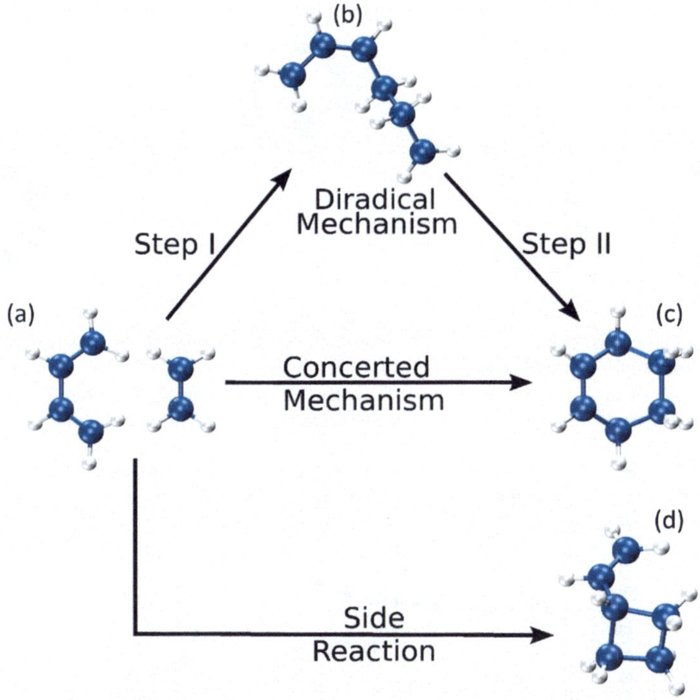

Fig. 2.8 Schematic of the parent DA reaction and the undesirable side reaction with structures of **a** butadiene and ethylene, **b** diradical, **c** cyclohexene, and **d** vinylcyclobutane. Atom colors: C in cyan, and H in white. Reprinted (adapted) from [2], with the permission of AIP publishing

repelled outwards. Geometries of the reactants, transition state, and product suggest that bond formation between C1–C5 and C4–C6 occurs to the same extent. The length of the C2–C3 bond shrinks from 1.48 Å to 1.36 Å, indicating a single → double bond transition, while that of the C5–C6 bond increases from 1.34 Å to 1.58 Å, indicating a double → single bond transition.

In Fig. 2.10, we see energies along the reaction pathway obtained using CASSCF (Fig. 2.10a) and DFT (Fig. 2.10b) methods. The energies of the reactants (butadiene + ethylene) have been shifted to zero. In the absence of any external pressure, the pathway obtained is shown by the black curve (with a filled circle) for both methods. Table 2.6 summarizes activation energies and reaction energies obtained in our calculations, and a comparison with values obtained by other simulations [105, 119] and experiments [107, 123]. We find that this reaction is exothermic with a reaction energy of -21.2 kcal mol^{-1}, and an energy barrier of 45.5 kcal mol^{-1}, and these values are comparable with other simulation results [119, 121]. Experimentally determined values [107, 123] are, however, significantly different and we believe that this discrepancy is because the CASSCF method is known to underestimate reaction exothermicity [105, 119, 121]. The B3LYP-DFT method, on the other hand, with a

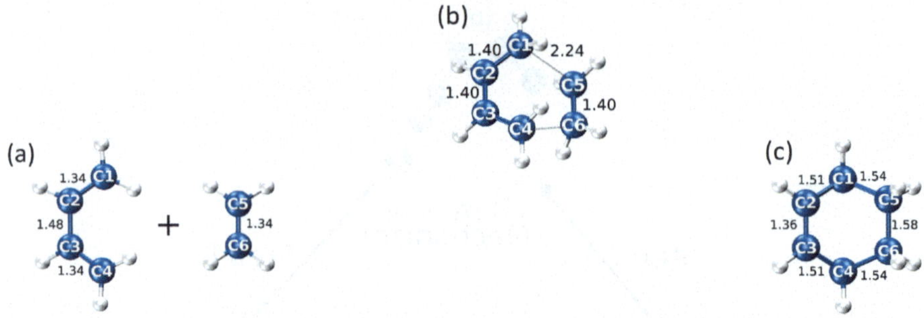

Fig. 2.9 Optimized structures of **a** reactants (butadiene and ethylene), **b** transition state, and **c** product (cyclohexene) of the parent DA reaction in the absence of external pressure, i.e., $P_{HP} = 0$. The shrinkage of the C2–C3 bond from 1.48 Å to 1.36 Å is consistent with the formation of a double bond. Reprinted (adapted) from [2], with the permission of AIP publishing

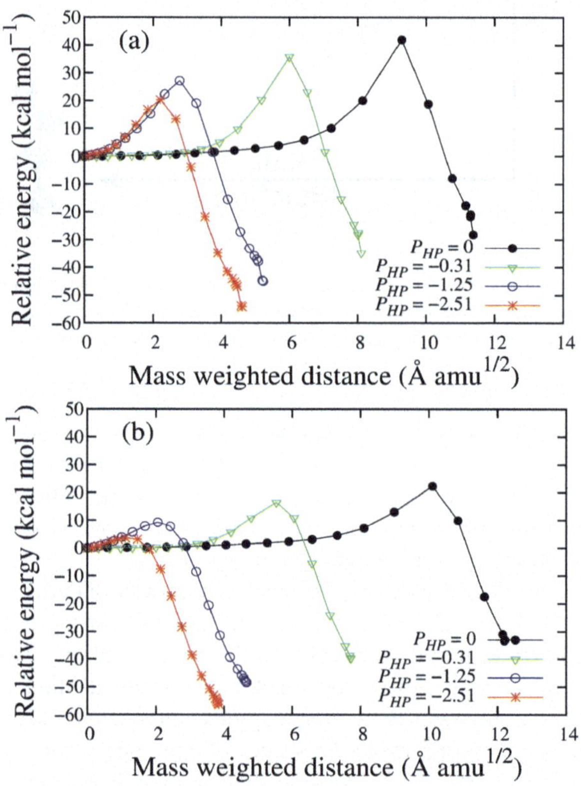

Fig. 2.10 Energy along the reaction coordinate as a function of pseudo-hydrostatic pressure, P_{HP}, for the concerted mechanism calculated at **a** CASSCF(6,6) and **b** B3LYP-density functional theory levels. The values of P_{HP} (in kcal mol^{-1} Å^{-2}) used in our simulation are indicated in the legend. Lower values of P_{HP} correspond to increasing compression due to the definition in Eq. (2.11). The energies of the reactants (butadiene + ethylene) are shifted to zero in all cases. The increasing compression lowers the energy barrier, and makes the reaction more exothermic. The vibrational ZPE correction has not been included in these plots. Reprinted (adapted) from [2], with the permission of AIP publishing

Table 2.6 Comparison of activation (E_a) and reaction (E_{rxn}) energies obtained in our calculations, calculations by other groups, as well as experimental values for the concerted mechanism

Energies	Previous studies			This work	
	Experimental	CASSCF(6,6)/ 6-31G** [119]	B3LYP/ 6-31G* [105]	CASSCF(6,6)/ 6-31G**	B3LYP/ 6-311++G**
E_a	27.5 ± 2 [107]	44.5	24.8	45.5	24.3
E_{rxn}	−40.0 [123]	−14.9	−36.6	−21.2	−27.6

All energies are expressed in kcal mol^{-1}
Adapted from [2]

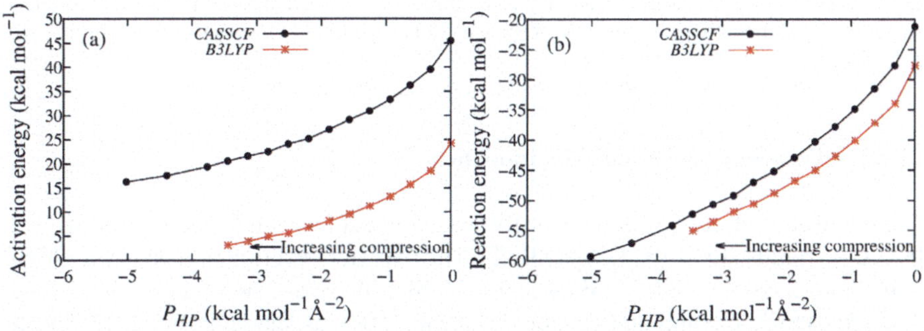

Fig. 2.11 **a** Activation energy, and **b** reaction energy as a function of compressive pseudo-hydrostatic pressure, P_{HP}, for the concerted mechanism between butadiene and ethylene. The dependence of both these quantities is a non-linear function of P_{HP}, implying that Bell's theory might not be valid for this system. Reprinted (adapted) from [2], with the permission of AIP publishing

higher basis set and diffuse functions (6-311++G**) results in computed activation energy of 24.3 kcal mol^{-1}, comparable with the experimental value of 27.5 ± 2 kcal mol^{-1} [107].

Effects of compressive pseudo-hydrostatic pressure on the systems are also shown in Fig. 2.10, with energies of the reactants (butadiene + ethylene) shifted to zero. As the magnitude of compression increases, the transition state shifts toward the reactants. Compression also lowered the activation energy for the concerted pathway, with higher compression resulting in lower activation energies. At the B3LYP level, reactants transitioned spontaneously to cyclohexene for $P_{HP} < -3.45$ kcal mol^{-1} Å^{-2}.

In Fig. 2.11a, we show the dependence of activation energy on compressive pseudo-hydrostatic pressure, P_{HP}. The relationship appears to be non-linear, and in violation of Bell's theory [124]. In Fig. 2.11b, we show shows the reaction energy as a function of P_{HP}, and it appears that reaction exothermicity increases with compression, and this dependence is also non-linear. These data indicate that the parent DA reaction becomes more exothermic with increasing compressive pseudo-hydrostatic pressure—an experimentally testable prediction.

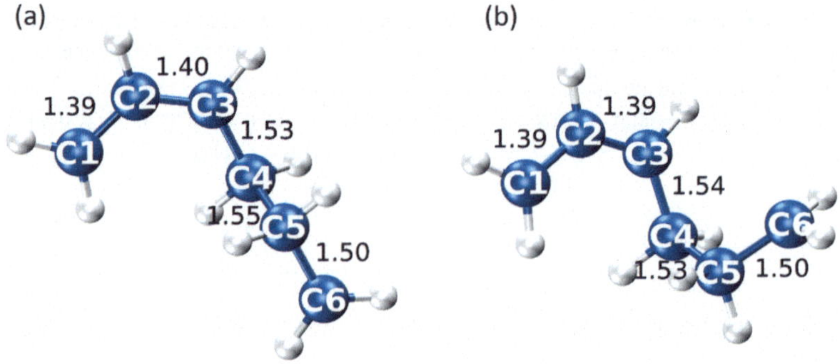

Fig. 2.12 Geometries of the **a** anti, and **b** gauche-diradical intermediates for the two-step diradical mechanism. C–C bond lengths are in Å. Reprinted (adapted) from [2], with the permission of AIP publishing

2.3.4.2 Two-Step Diradical Mechanism

This possible route for cyclohexene formation is depicted in Fig. 2.8, a → b → c. Here, C–C bonds are formed sequentially. The two-step reaction proceeds via a high-energy intermediate, called a diradical, of which there are two types. These types are referred to as the anti and gauche diradicals [105], as shown in Fig. 2.12. Unfortunately, the combination of DFT and NEB methods did not allow us to correctly describe the pathway on the PES for a system that switches spin state (singlet ↔ triplet) during the transition, viz., butadiene + ethylene → diradical → cyclohexene. Therefore, the diradical pathway was investigated only at the CASSCF level of theory.

We first discuss the anti-diradical pathway (Fig. 2.12a). Here, the reactants first form the anti-diradical, which then transitions into cyclohexene via torsional rotation about the C3−C4 and C4−C5 bonds. The energies along the reaction coordinate (expressed in the mass-weighted distance) as a function of P_{HP} are shown in Fig. 2.13. At $P_{HP} = 0$, the barrier for the formation of the anti-diradical is 44.0 kcal mol^{-1}, while the transition of the anti-diradical into cyclohexene was over a barrier of 3.9 kcal mol^{-1}. With increasing compression, the energy barrier for the formation of the anti-diradical decreased – this is likely because this process involves the translation of the molecules to form one bond, and the increasing compression facilitates this mechanism. The anti-diradical → cyclohexene transition involves rotation around C3−C4 and C4−C5 bonds, and we see that the energy barrier for this step increases with increasing compression. Here, increased compression results in the diradical becoming more compact, resulting in pronounced steric effects that increase the energy barrier. Beyond a critical value of compression ($P_{HP} \leq -2.51$ kcal mol^{-1}Å^{-2}), the anti-diradical becomes unstable, and is spontaneously converted into cyclohexene. The activation energies for the diradical mechanism as a function of pseudo-hydrostatic pressures, P_{HP}, are summarized in Table 2.7.

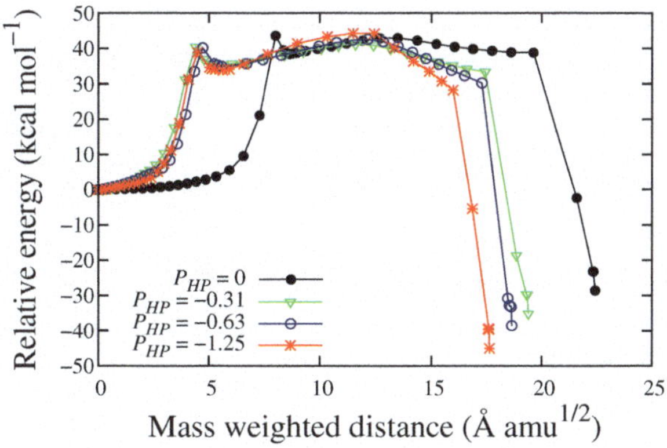

Fig. 2.13 Energy along the reaction coordinate as a function of pseudo-hydrostatic pressure, P_{HP}, for the two-step mechanism of DA reaction via the anti-diradical. The values of P_{HP} (in kcal mol^{-1} Å^{-2}) used in our simulation are indicated in the legend. The energies of the reactants (butadiene + ethylene) are shifted to zero in all cases. Increasing compression lowers the energy barrier for the first step, but raises the barrier for the second step. The vibrational ZPE correction has not been included in this plot. Reprinted (adapted) from [2], with the permission of AIP publishing

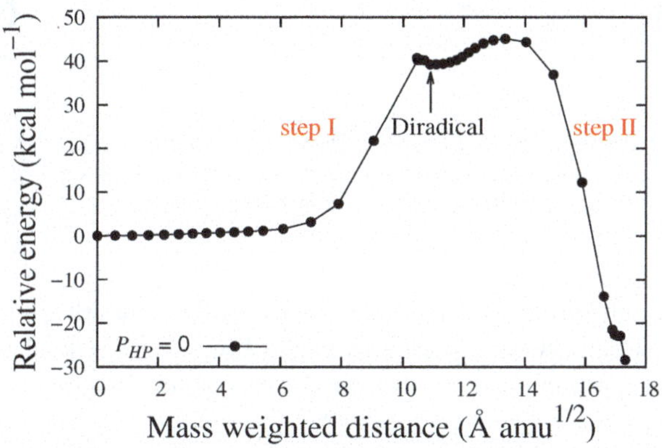

Fig. 2.14 Energy along the reaction coordinate for the two-step mechanism of DA reaction via the gauche-diradical. This reaction pathway was destabilized by the smallest value of compression that we attempted, viz., $P_{HP} = -0.31$ kcal mol^{-1} Å^{-2}, and the diradical spontaneously transitioned into vinylcyclobutane. The vibrational ZPE correction has not been included in this plot. Reprinted (adapted) from [2], with the permission of AIP publishing

Next, we discuss the gauche-diradical (Fig. 2.12b) pathway, whose MEP is shown in Fig. 2.14. Without compression, the barrier height for diradical formation (step I) is 41.4 kcal mol^{-1}, while the barrier for the diradical to transition into cyclohexene (step II) is 5.8 kcal mol^{-1}. The gauche-diradical is 0.9 kcal mol^{-1} below the transition state of step I. The activation energy of the overall reaction, defined as the highest peak along the MEP, is 46.3 kcal mol^{-1}. This activation energy is ~1 kcal mol^{-1} more than that of the concerted pathway, and ~2 kcal mol^{-1} more than that of the reaction

Table 2.7 Activation energies for diradical mechanism of DA reaction as a function of pseudo-hydrostatic pressure, P_{HP} (in kcal mol^{-1} Å^{-2}), calculated using CASSCF(6,6) level of theory

P_{HP}	Activation energy, E_a	
	Anti-diradical	Gauche-diradical
0	44.0	46.3
−0.31	43.6	.....
−0.63	43.0	.....
−1.25	44.3	.....

All energies are expressed in kcal mol^{-1}. For a non-zero value of P_{HP}, gauche-diradical is no longer a minimum in the PES
Adapted from [2]

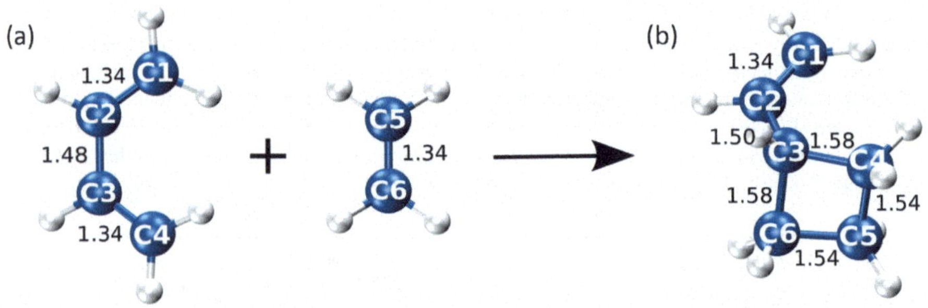

Fig. 2.15 Optimized structures of **a** reactants (butadiene and ethylene) and **b** product (vinylcyclobutane) for the side reaction between butadiene and ethylene in the absence of external pressure. Reprinted (adapted) from [2], with the permission of AIP publishing

pathway via the anti-diradical. Effects of compression were not studied, because, at even the smallest compressive pressures used (P_{HP} = −0.31 kcal mol^{-1}Å^{-2}), the gauche-diradical spontaneously transitioned into vinylcyclobutane.

2.3.4.3 Side Reaction: Formation of Vinylcyclobutane

Experimentally, vinylcyclobutane is produced in a side reaction when the parent DA reaction is being carried out [122], as we have observed in our calculations under compression, and therefore, we examined the side reaction mechanisms and the modifications produced by compressive pressure.

Figure 2.15 shows the optimized geometries and bond lengths (in Å) of the reactants (butadiene and ethylene) and the product (vinylcyclobutane) on the unmodified PES. The energies along the reaction coordinate as a function of compression, P_{HP}, for the reaction of butadiene + ethylene → vinylcyclobutane are shown in Fig. 2.16. For some values of $P_{HP} \neq 0$, a kink in the reaction pathway was seen. These kinks were present at a range of spring constants that spanned an order of magnitude. The activation energies, summarized in Table 2.8, are non-monotonic behavior with com-

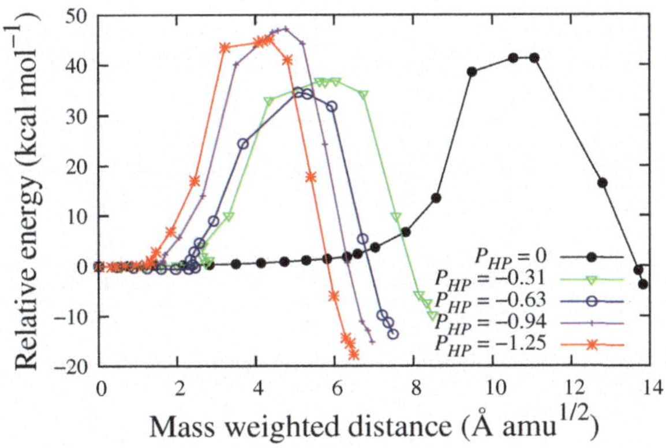

Fig. 2.16 Energy along the reaction coordinate for the side reaction that results in the formation of vinylcyclobutane, as a function of pseudo-hydrostatic pressures, P_{HP} (in kcal mol^{-1} Å^{-2}), indicated in the legend. The energies of the reactants (butadiene + ethylene) are shifted to zero in all cases. The vibrational ZPE correction has not been included in this plot. Reprinted (adapted) from [2], with the permission of AIP publishing

pression. At $P_{HP} = 0$, the energy barrier for this transition is 42.4 kcal mol^{-1}, and the reaction is slightly endothermic (by 2 kcal mol^{-1}). As the reaction proceeds, butadiene and ethylene approach each other, and yield vinylcyclobutane. The order in which these bonds form was observed to depend on the magnitude of the compressive pressure, and gives rise to different pathways. For lower compression ($-0.63 \leq P_{HP} \leq 0$ kcal mol^{-1} Å^{-2}), the optimized structures along the reaction path indicated the formation of C4−C5 bonds (see Fig. 2.15b) form first, followed by the C3−C6, and therefore the barrier decreases, suggesting that the increasing compression facilitates this process. On the other hand, higher compression ($P_{HP} \leq -0.94$ kcal mol^{-1} Å^{-2}) results in the formation of C3−C6 first, followed by C4−C5, and this pathway appears to present a higher barrier, therefore the barrier increases at $P_{HP} = -0.94$ kcal mol^{-1} Å^{-2}, however further increasing compression resulted in the decrease in the barrier by facilitating the side reaction. This switching of the order of bond formation seems to lead to the non-monotonic behavior of the energy barrier with compression.

2.4 Conclusions

In this chapter, we described a method for studying the influence of external forces on the Born-Oppenheimer potential energy surface (PES) of atomic systems, and demonstrated its applications by calculating reaction pathways for various reactions under the influence of external forces. This was done using a formulation for computing energies on a generalized force-modified potential energy surface (G-FMPES), and demonstrated that calculation of atomistic quantities (including transition states)

Table 2.8 Activation (E_a) and reaction (E_{rxn}) energies for butadiene + ethylene → vinylcyclobutane reaction as a function of pseudo-hydrostatic pressure, P_{HP} (in kcal mol^{-1} Å^{-2}), calculated using CASSCF(6,6) level of theory

P_{HP}	E_a	E_{rxn}
0	42.4	+2.0
−0.31	38.1	−4.0
−0.63	35.8	−7.8
−0.94	48.4	−9.4
−1.25	46.3	−11.9

The positive value of E_{rxn} implies the reaction is endothermic. All energies are expressed in kcal mol^{-1} Adapted from [2]

in the presence of a spatially varying force is easily accomplished using standard atomistic techniques. We have exemplified the formulation by calculating a variety of atomistic properties for (i) conformational transitions in ethane and RDX molecules, (ii) the decomposition of RDX, and (iii) the reaction between 1,3-butadiene and ethylene. For various transitions, we showed that the curvature of the G-FMPES is indeed different from that of the unmodified PES, and that it has an effect on the harmonic transition rates. Also seen were examples where the addition of an external pressure causes a state to become unstable, indicating the merging of a saddle point with the energy minimum, as a result of the increasing compressive pressures enhancing steric effects by forcing molecules into more compact conformations. For both the ethane and RDX molecules, increasing compressive pressures shifted the vibrational spectra toward stiffer mode: a behavior that is qualitatively consistent with independent experiments, as well as molecular dynamics studies. Our study of the parent Diels-Alder reaction between butadiene and ethylene indicated that the activation energy barrier of this reaction decreases with increasing compressive pressure. We also observed that one reaction pathway, the formation of cyclohexene by the gauche-diradical pathway, become inaccessible due to the compressive pressures. At sufficiently high compression, the barrier of the concerted pathway becomes zero and the reaction becomes spontaneous. Our estimates indicated that the compression used in our Diels-Alder reaction corresponds to a range of 68 MPa to 1410 MPa. This range can be accessed in laboratories by using shock-wave technologies or diamond-anvil cells, in order to test the predictions made in this work [61, 125].

Appendix 1: Proof that the external force field is conservative

For a 3-dimensional, N-atom system, the position vector of the jth atom, and the external force vector on it are given in component form as

$$\mathbf{r}^{(j)} = \left(x^{(j)}, y^{(j)}, z^{(j)}\right) \tag{2.14}$$

$$\mathbf{f}^{(j)}_{\text{ext}} = \left(f^{(j)}_{\text{ext},x}, f^{(j)}_{\text{ext},y}, f^{(j)}_{\text{ext},z}\right) \tag{2.15}$$

where various $\left\{\mathbf{f}_{\text{ext}}^{(j)}\ \forall\ j = 1, 2, 3 \ldots N\right\}$ make up the external force field $\mathbf{F}_{\text{ext}}(\mathbf{R})$.

Geometric centroid of the configuration is given by the mean position of all the atoms in the configuration as

$$\mathbf{c} = \left(c_x, c_y, c_z\right) = \left(\left\langle x^{(j)}\right\rangle, \left\langle y^{(j)}\right\rangle, \left\langle z^{(j)}\right\rangle\right) \tag{2.16}$$

where $\langle\cdot\rangle$ denotes an average taken over all N atoms.

According to our prescription of pseudo-hydrostatic pressure, external force on the jth atom is given as

$$\mathbf{f}_{\text{ext}}^{(j)} = P_{HP}\left[\mathbf{r}^{(j)} - \mathbf{c}\right]; \quad \forall \quad j = 1, 2, 3 \ldots N \tag{2.17}$$

where P_{HP} is a user-defined "pressure."

First derivatives of the external force vector are given by

$$\left\{\frac{\partial \mathbf{f}_{\text{ext}}^{(j)}}{\partial x^{(k)}}, \frac{\partial \mathbf{f}_{\text{ext}}^{(j)}}{\partial y^{(k)}}, \frac{\partial \mathbf{f}_{\text{ext}}^{(j)}}{\partial z^{(k)}}; \quad \forall \quad j, k = 1, 2, 3 \ldots N\right\} \tag{2.18}$$

The first of this set is

$$\frac{\partial \mathbf{f}_{\text{ext}}^{(j)}}{\partial x^{(k)}} = P_{HP}\left(\frac{\partial \mathbf{r}^{(j)}}{\partial x^{(k)}} - \frac{\partial \mathbf{c}}{\partial x^{(k)}}\right) \tag{2.19}$$

with

$$\frac{\partial \mathbf{r}^{(j)}}{\partial x^{(k)}} = \left(\frac{\partial x^{(j)}}{\partial x^{(k)}}, \frac{\partial y^{(j)}}{\partial x^{(k)}}, \frac{\partial z^{(j)}}{\partial x^{(k)}}\right) = \left(\delta_{jk}, 0, 0\right) \tag{2.20}$$

$$\frac{\partial \mathbf{c}}{\partial x^{(k)}} = \left(\frac{\partial \langle x\rangle}{\partial x^{(k)}}, \frac{\partial \langle y\rangle}{\partial x^{(k)}}, \frac{\partial \langle z\rangle}{\partial x^{(k)}}\right) = \left(\frac{1}{N}, 0, 0\right) \tag{2.21}$$

where δ_{jk} is the Kronecker delta. Using this, and similar results for the other derivatives, first derivatives of the external force vector are

$$\frac{\partial \mathbf{f}_{\text{ext}}^{(j)}}{\partial x^{(k)}} = P_{HP}\left(\delta_{jk} - \frac{1}{N}, 0, 0\right) \tag{2.22}$$

$$\frac{\partial \mathbf{f}_{\text{ext}}^{(j)}}{\partial y^{(k)}} = P_{HP}\left(0, \delta_{jk} - \frac{1}{N}, 0\right) \tag{2.23}$$

$$\frac{\partial \mathbf{f}_{\text{ext}}^{(j)}}{\partial z^{(k)}} = P_{HP}\left(0, 0, \delta_{jk} - \frac{1}{N}\right) \tag{2.24}$$

which are all constant for all values of j and k and therefore exist and are continuous everywhere, proving that the external force field is conservative.

Appendix 2: The NEB method on a G-FMPES

With the understanding of how to compute energies, forces, and curvatures on a G-FMPES, the NEB implementation for finding MEPs on a G-FMPES closely follows the original implementation, but with a few crucial modifications. For clarity, our implementation of a G-FMPES is outlined below.

The start and end point structures are re-optimized with the external force and a set of images is initialized between them. Consecutive images are connected by harmonic springs with an equilibrium length of zero and a user-specified spring stiffness k_{spring} (the actual value of which is not particularly important as long as it is greater than zero, but needs to be on the order of the system forces for efficient convergence). The band is iteratively optimized until the net force on each image is minimized to within a user-specified tolerance. At each iteration, the net force on the ith image located at $\mathbf{R}^{(i)}$ on the G-FMPES is given by the projected (nudged) forces as:

$$\overline{\mathbf{F}}^{(i)}_{\text{net,nudged}} = \left[\mathbf{F}^{(i)}_{\text{grad}} + \mathbf{F}^{(i)}_{\text{ext}}\right]_{\perp} + \left[\mathbf{F}^{(i)}_{\text{spring}}\right]_{\parallel} \tag{2.25}$$

where the subscript $\perp$ (or $\parallel$) on a vector indicates its component perpendicular (or parallel) to the unit tangent $\hat{\boldsymbol{\tau}}^{(i)}$. The un-normalized tangent $\boldsymbol{\tau}^{(i)}$ is computed using Henkelman and Jónsson's improved tangent estimate [50]. Defining

$$\boldsymbol{\tau}^{(i)+} = \mathbf{R}^{(i+1)} - \mathbf{R}^{(i)} \tag{2.26}$$

$$\boldsymbol{\tau}^{(i)-} = \mathbf{R}^{(i)} - \mathbf{R}^{(i-1)} \tag{2.27}$$

the tangent is estimated as

$$\boldsymbol{\tau}^{(i)} = \begin{cases} \boldsymbol{\tau}^{(i)+} & \text{if} \quad \overline{V}^{(i+1)} > \overline{V}^{(i)} > \overline{V}^{(i-1)} \\ \boldsymbol{\tau}^{(i)-} & \text{if} \quad \overline{V}^{(i+1)} < \overline{V}^{(i)} < \overline{V}^{(i-1)} \end{cases} \tag{2.28}$$

where the different $\overline{V}$ are computed using the line integral, as defined in the main article. In the event that the three consecutive values of $\overline{V}$ are neither strictly increasing nor strictly decreasing, i.e., if $\overline{V}^{(i+1)} \leq \overline{V}^{(i)} \geq \overline{V}^{(i-1)}$ or $\overline{V}^{(i+1)} \geq \overline{V}^{(i)} \leq \overline{V}^{(i-1)}$, in order to prevent abrupt switching between two possible tangents, the tangent is taken to be a weighted average as

$$\boldsymbol{\tau}^{(i)} = \begin{cases} \boldsymbol{\tau}^{(i)+} \Delta\overline{V}^{(i)+} + \boldsymbol{\tau}^{(i)-} \Delta\overline{V}^{(i)-} & \text{if} \quad \overline{V}^{(i+1)} > \overline{V}^{(i-1)} \\ \boldsymbol{\tau}^{(i)+} \Delta\overline{V}^{(i)-} + \boldsymbol{\tau}^{(i)-} \Delta\overline{V}^{(i)+} & \text{if} \quad \overline{V}^{(i+1)} < \overline{V}^{(i-1)} \end{cases} \tag{2.29}$$

with

$$\Delta\overline{V}^{(i)+} = \max\left(\left|\overline{V}^{(i+1)} - \overline{V}^{(i)}\right|, \left|\overline{V}^{(i-1)} - \overline{V}^{(i)}\right| \right) \tag{2.30}$$

$$\Delta\overline{V}^{(i)-} = \min\left(\left|\overline{V}^{(i+1)} - \overline{V}^{(i)}\right|, \left|\overline{V}^{(i-1)} - \overline{V}^{(i)}\right| \right) \tag{2.31}$$

The component of the spring force parallel to the tangent is computed as

$$\left[\mathbf{F}^{(i)}_{\text{spring}}\right]_{\parallel} = k_{\text{spring}} \left(\|\boldsymbol{\tau}^{(i)+}\| - \|\boldsymbol{\tau}^{(i)-}\| \right) \hat{\boldsymbol{\tau}}^{(i)} \tag{2.32}$$

At each iteration, the climbing-image method [51] is applied to the highest energy image along the band. This image, identified by $i = h$, is free from all spring forces and is assigned a climbing force computed by inverting the component of the gradient plus external forces along the tangent to this image which is expressed as

$$\overline{\mathbf{F}}^{(i)}_{\text{net,climb}} = \left(\mathbf{F}^{(i)}_{\text{grad}} + \mathbf{F}^{(i)}_{\text{ext}}\right) - 2\left[\left(\mathbf{F}^{(i)}_{\text{grad}} + \mathbf{F}^{(i)}_{\text{ext}}\right) \cdot \hat{\boldsymbol{\tau}}^{(i)}\right] \hat{\boldsymbol{\tau}}^{(i)} \tag{2.33}$$

and makes the highest image move uphill along the direction of the first eigenvector, and downhill along all other directions.

In the more recent variation with two climbing images [52], the highest image (with $i = h$) is not allowed to climb but is nudged in the usual manner by assigning its net force to be equal to $\overline{\mathbf{F}}_{\text{net,nudged}}$ as in Eq. (2.25). However, its two nearest neighbors (one from each side of the band, i.e., images with $i = h \pm 1$) are assigned climbing forces as in Eq. (2.33). This prescription results in a higher density of images near the saddle point and is particularly useful for MEPs with unusually high curvatures near the saddle point which would cause the NEB tangent direction to be different from the MEP tangent direction.

Having computed the net force on each image, the band is iteratively optimized until it is well-converged by moving each image toward the point of minimum net force using a suitable optimizer.

References

1. Subramanian G, Mathew N, Leiding J (2015) A generalized force-modified potential energy surface for mechanochemical simulations. J Chem Phys 143(13):134109
2. Jha SK, Brown K, Todde G, Subramanian G (2016) A mechanochemical study of the effects of compression on a diels-alder reaction. J Chem Phys 145(7):074307
3. Todde G, Jha SK, Subramanian G (2017) The effect of external forces on the initial dissociation of rdx (1,3,5-trinitro-1,3,5-triazine): a mechanochemical study. Int J Quantum Chem 117(20):e25426
4. Varvenne C, Bruneval F, Marinica M-C, Clouet E (2013) Point defect modeling in materials: coupling ab initio and elasticity approaches. Phys Rev B 88(13):134102
5. Clouet E, Garruchet S, Nguyen H, Perez M, Becquart CS (2008) Dislocation interaction with c in α-fe: a comparison between atomic simulations and elasticity theory. Acta Mater 56(14):3450–3460

6. Ong MT, Leiding J, Tao H, Virshup AM, Martínez TJ (2009) First principles dynamics and minimum energy pathways for mechanochemical ring opening of cyclobutene. J Am Chem Soc 131(18):6377–6379
7. Konda SSM, Brantley JN, Varghese BT, Wiggins KM, Bielawski CW, Makarov DE (2013) Molecular catch bonds and the anti-Hammond effect in polymer mechanochemistry. J Am Chem Soc 135(34):12722–12729
8. Wang J, Kouznetsova TB, Niu Z, Ong MT, Klukovich HM, Rheingold AL, Martinez TJ, Craig SL (2015) Inducing and quantifying forbidden reactivity with single-molecule polymer mechanochemistry. Nat Chem 7(4):323–327
9. Konda SSM, Avdoshenko SM, Makarov DE (2014) Exploring the topography of the stress-modified energy landscapes of mechanosensitive molecules. J Chem Phys 140(10):104114
10. Silberstein MN, Min K, Cremar LD, Degen CM, Martinez TJ, Aluru NR, White SR, Sottos NR (2013) Modeling mechanophore activation within a crosslinked glassy matrix. J Appl Phys 114(2):023504
11. Silberstein MN, Cremar LD, Beiermann BA, Kramer SB, Martinez TJ, White SR, Sottos NR (2014) Modeling mechanophore activation within a viscous rubbery network. J Mech Phys Solids 63:141–153
12. Bailey A, Mosey NJ (2012) Prediction of reaction barriers and force-induced instabilities under mechanochemical conditions with an approximate model: a case study of the ring opening of 1, 3-cyclohexadiene. J Chem Phys 136(4):01B613
13. Kumamoto K, Fukada I, Kotsuki H (2004) Diels–alder reaction of thiophene: dramatic effects of high-pressure/solvent-free conditions. Angew Chem Int Ed 43(15):2015–2017
14. Mianowski A, Robak Z, Tomaszewicz M, Stelmach S (2012) The boudouard-bell reaction analysis under high pressure conditions. J Therm Anal Calorim 110(1):93–102
15. Kiselev VD (2013) High-pressure influence on the rate of diels–alder cycloaddition reactions of maleic anhydride with some dienes. Int J Chem Kinet 45(9):613–622
16. Kiselev VD, Kornilov DA, Kashaeva EA, Potapova LN, Konovalov AI (2014) Effect of pressure on the rate of the diels-alder reaction of diethyl azodicarboxylate with 9, 10-dimethylanthracene. Russ J Org Chem 50(4):489–493
17. McMillan PF (2002) New materials from high-pressure experiments. Nat Mater 1(1):19–25
18. Gillet P, Badro J, Varrel B, McMillan PF (1995) High-pressure behavior in α-alpo 4: amorphization and the memory-glass effect. Phys Rev B 51(17):11262
19. J Torres A, Serment-Moreno V, Escobedo-Avellaneda ZJ, Velazquez G, Welti-Chanes J (2016) Reaction chemistry at high pressure and high temperature. In: High pressure processing of food. Springer, pp 461–478
20. Hashemi H, Christensen JM, Gersen S, Glarborg P (2015) Hydrogen oxidation at high pressure and intermediate temperatures: experiments and kinetic modeling. Proc Combust Inst 35(1):553–560
21. Ribas-Arino J, Shiga M, Marx D (2009) Understanding covalent mechanochemistry. Angew Chem Int Ed 48(23):4190–4193
22. Ribas-Arino J, Marx D (2012) Covalent mechanochemistry: theoretical concepts and computational tools with applications to molecular nanomechanics. Chem Rev 112(10):5412–5487
23. Lenhardt JM, Ong MT, Choe R, Evenhuis CR, Martinez TJ, Craig SL (2010) Trapping a diradical transition state by mechanochemical polymer extension. Science 329(5995):1057–1060
24. Kochhar GS, Bailey A, Mosey NJ (2010) Competition between orbitals and stress in mechanochemistry. Angew Chem Int Ed 49(41):7452–7455
25. Smalø HS, Rybkin VV, Klopper W, Helgaker T, Uggerud E (2014) Mechanochemistry: the effect of dynamics. J Phys Chem A 118(36):7683–7694
26. Avdoshenko SM, Makarov DE (2015) Finding mechanochemical pathways and barriers without transition state search. J Chem Phys 142(17):174106
27. Wang J, Kouznetsova TB, Craig SL (2015) Reactivity and mechanism of a mechanically activated anti-woodward–hoffmann–depuy reaction. J Am Chem Soc 137(36):11554–11557

28. Makarov DE (2016) Perspective: mechanochemistry of biological and synthetic molecules. J Chem Phys 144(3):030901
29. Subramanian G, Perez D, Uberuaga BP, Tomé CN, Voter AF (2013) Method to account for arbitrary strains in kinetic Monte Carlo simulations. Phys Rev B 87:144107
30. Goyal A, Phillpot SR, Subramanian G, Andersson DA, Stanek CR, Uberuaga Blas P (2015) Impact of homogeneous strain on uranium vacancy diffusion in uranium dioxide. Phys Rev B 91:094103
31. Vérité G, Domain C, Fu C-C, Gasca P, Legris A, Willaime F (2013) Self-interstitial defects in hexagonal close packed metals revisited: evidence for low-symmetry configurations in Ti, Zr, and Hf. Phys Rev B 87(13):134108
32. Garnier T, Manga VR, Bellon P, Trinkle DR (2014) Diffusion of Si impurities in Ni under stress: a first-principles study. Phys Rev B 90:024306
33. Mathew N, Picu RC (2011) Molecular conformational stability in cyclotrimethylene trinitramine crystals. J Chem Phys 135(2):024510
34. Munday LB, Chung PW, Rice BM, Solares SD (2011) Simulations of high-pressure phases in RDX. J Phys Chem B 115(15):4378–4386
35. Cawkwell MJ, Sewell TD, Zheng L, Thompson DL (2008) Shock-induced shear bands in an energetic molecular crystal: application of shock-front absorbing boundary conditions to molecular dynamics simulations. Phys Rev B 78(1):014107
36. Boyd S, Murray JS, Politzer P (2009) Molecular dynamics characterization of void defects in crystalline (1, 3, 5-trinitro-1, 3, 5-triazacyclohexane). J Chem Phys 131(20):204903
37. Ramos KJ, Hooks DE, Sewell TD, Cawkwell MJ (2010) Anomalous hardening under shock compression in (021)-oriented cyclotrimethylene trinitramine single crystals. J Appl Phys 108(6):066105
38. Munday LB, Solares SD, Chung PW (2012) Generalized stacking fault energy surfaces in the molecular crystal αRDX. Philos Mag 92(24):3036–3050
39. Mathew N, Picu CR, Chung PW (2013) Peierls stress of dislocations in molecular crystal cyclotrimethylene trinitramine. J Phys Chem A 117(25):5326–5334
40. Pal A, Picu RC (2014) Rotational defects in cyclotrimethylene trinitramine (rdx) crystals. J Chem Phys 140(4):044512
41. Austin DE, Peng Y, Hansen BJ, Miller IW, Rockwood AL, Hawkins AR, Tolley SE (2008) Novel ion traps using planar resistive electrodes: implications for miniaturized mass analyzers. J Am Soc Mass Spectrom 19(10):1435–1441
42. Dudko OK, Hummer G, Szabo A (2008) Theory, analysis, and interpretation of single-molecule force spectroscopy experiments. Proc Natl Acad Sci USA 105(41):15755–15760
43. Neuman KC, Nagy A (2008) Single-molecule force spectroscopy: optical tweezers, magnetic tweezers and atomic force microscopy. Nat Methods 5(6):491–505
44. Kochhar GS, Heverly-Coulson GS, Mosey NJ (2015) Theoretical approaches for understanding the interplay between stress and chemical reactivity. In: Polymer mechanochemistry. Springer, pp 37–96
45. Stauch T, Dreuw A (2016) Advances in quantum mechanochemistry: electronic structure methods and force analysis. Chem Rev 116(22):14137–14180
46. Henkelman G, Jóhannesson G, Jónsson H (2002) Methods for finding saddle points and minimum energy paths. In: Theoretical methods in condensed phase chemistry, vol 5. Springer, pp 269–302
47. Uberuaga BP, Hoagland RG, Voter AF, Valone SM (2007) Direct transformation of vacancy voids to stacking fault tetrahedra. Phys Rev Lett 99(13):135501
48. Vineyard GH (1957) Frequency factors and isotope effects in solid state rate processes. J Phys Chem Solids 3(1-2):121–127
49. Werner H-JKPJ, Knowles PJ, Knizia G, Manby FR, Schütz M, Celani P, Korona T, Lindh R, Mitrushenkov A, Rauhut G et al (2012) Molpro, version 2012.1, a package of ab initio programs. see http://www.molpro.net
50. Henkelman G, Jónsson H (2000) Improved tangent estimate in the nudged elastic band method for finding minimum energy paths and saddle points. J Chem Phys 113(22):9978–9985

51. Henkelman G, Uberuaga BP, Jónsson H (2000) A climbing image nudged elastic band method for finding saddle points and minimum energy paths. J Chem Phys 113(22):9901–9904
52. Zarkevich NA, Johnson DD (2015) Nudged-elastic band method with two climbing images: finding transition states in complex energy landscapes. J Chem Phys 142(2):024106
53. Bitzek E, Koskinen P, Gähler F, Moseler M, Gumbsch P (2006) Structural relaxation made simple. Phys Rev Lett 97:170201
54. Hartree DR (1928) The wave mechanics of an atom with a non-coulomb central field. Part I. theory and methods. In: Mathematics proceedings Cambridge, vol 24. Cambridge University Press, pp 89–110
55. Fock V (1930) Näherungsmethode zur lösung des quantenmechanischen mehrkörperproblems. Zeitschrift für Physik 61(1–2):126–148
56. Slater JC (1930) Note on Hartree's method. Phys Rev 35:210–211
57. Krishnan R, Binkley JS, Seeger R, Pople JA (1980) Self-consistent molecular orbital methods. xx. A basis set for correlated wave functions. J Chem Phys 72(1):650–654
58. Kemp JD, Pitzer KS (1936) Hindered rotation of the methyl groups in ethane. J Chem Phys 4(11):749–749
59. Eyring H, Grant DM, Hecht H (1962) The rotational barrier in ethane. J Chem Educ 39(9):466
60. Johnson RD III (ed) (2013) NIST computational chemistry comparison and benchmark database, NIST standard reference database number 101, release 16a, august 2013. see http://cccbdb.nist.gov/
61. Kurnosov AV, Ogienko AG, Goryainov SV, Larionov EG, Manakov AY, Lihacheva AY, Aladko EY, Zhurko FV, Voronin VI, Berger IF et al (2006) Phase diagram and high-pressure boundary of hydrate formation in the ethane- water system. J Phys Chem B 110(43):21788–21792
62. Cawkwell MJ, Ramos KJ, Hooks DE, Sewell TD (2010) Homogeneous dislocation nucleation in cyclotrimethylene trinitramine under shock loading. J Appl Phys 107(6):063512
63. Mathew N, Picu RC (2013) Slip asymmetry in the molecular crystal cyclotrimethylenetrinitramine. Chem Phys Lett 582:78–81
64. Munday LB, Mitchell RL, Knap J, Chung PW (2013) Role of molecule flexibility on the nucleation of dislocations in molecular crystals. Appl Phys Lett 103(15):151911
65. Rice BM, Chabalowski CF (1997) Ab initio and nonlocal density functional study of 1, 3, 5-trinitro-s-triazine (RDX) conformers. J Phys Chem A 101(46):8720–8726
66. Karpowicz RJ, Brill TB (1984) Comparison of the molecular structure of hexahydro-1, 3, 5-trinitro-s-triazine in the vapor, solution and solid phases. J Phys Chem 88(3):348–352
67. Davidson AJ, Oswald IDH, Francis DJ, Lennie AR, Marshall WG, Millar DIA, Pulham CR, Warren JE, Cumming AS (2008) Explosives under pressure-the crystal structure of γ-RDX as determined by high-pressure x-ray and neutron diffraction. Cryst Eng Comm 10(2):162–165
68. Millar DIA, Oswald IDH, Barry C, Francis DJ, Marshall WG, Pulham CR, Cumming AS (2010) Pressure-cooking of explosives-the crystal structure of ε-RDX as determined by x-ray and neutron diffraction. Chem Commun 46:5662–5664
69. Dreger ZA, Gupta YM (2010) Raman spectroscopy of high-pressure- high-temperature polymorph of hexahydro-1, 3, 5-trinitro-1, 3, 5-triazine (ε-RDX). J Phys Chem A 114(26):7038–7047
70. Zheng X, Zhao J, Tan D, Liu C, Song Y, Yang Yanqiang (2011) High-pressure vibrational spectroscopy of hexahydro-1, 3, 5-trinitro-1, 3, 5-triazine (RDX). Propellants Explos Pyrotech 36(1):22–27
71. Ciezak JA, Jenkins TA, Liu Z, Hemley RJ (2007) High-pressure vibrational spectroscopy of energetic materials: hexahydro-1, 3, 5-trinitro-1, 3, 5-triazine. J Phys Chem A 111(1):59–63
72. Pereverzev A, Sewell TD, Thompson DL (2013) Molecular dynamics study of the pressure-dependent terahertz infrared absorption spectrum of α-and γ-RDX. J Chem Phys 139(4):044108
73. Chakraborty D, Muller RP, Dasgupta S, Goddard WA (2000) The mechanism for unimolecular decomposition of rdx (1, 3, 5-trinitro-1, 3, 5-triazine), an ab initio study. J Phys Chem A 104(11):2261–2272

74. Harris NJ, Lammertsma K (1997) Ab initio density functional computations of conformations and bond dissociation energies for hexahydro-1, 3, 5-trinitro-1, 3, 5-triazine. J Am Chem Soc 119(28):6583–6589
75. Shishkov IF, Vilkov LV, Kolonits M, Rozsondai B (1991) The molecular geometries of some cyclic nitramines in the gas phase. Struct Chem 2(1):57–64
76. Chakraborty D, Muller RP, Dasgupta S, Goddard WA (2001) A detailed model for the decomposition of nitramines: RDX and HMX. J Comput Aided Mater 8(2-3):203–212
77. Schweigert IV (2015) Ab initio molecular dynamics of high-temperature unimolecular dissociation of gas-phase RDX and its dissociation products. J Phys Chem A 119(12):2747–2759
78. Wu CJ, Fried LE (1997) Ab initio study of RDX decomposition mechanisms. J Phys Chem A 101(46):8675–8679
79. Molt RW, Watson T, Bazanté AP, Bartlett RJ, Richards NGJ (2016) Gas phase RDX decomposition pathways using coupled cluster theory. Phys Chem Chem Phys 18(37):26069–26077
80. Diels O, Alder K (1928) Syntheses in the hydroaromatic series. I. addition of "diene" hydrocarbons. Justus Liebigs Ann Chem 460:98–122
81. Carruthers W (1978) Some Mod Methods Org Synth, 2nd edn. Cambridge University Press, Cambridge
82. Cycloaddition reactions in organic synthesis. Pergamon Press, Oxford (1990)
83. Boger DL, Weinberg SN (1987) Hetero-Diels methodology in organic synthesis. Academic Press Inc, San Diego
84. Houk KN, Gonzalez J, Li Y (1995) Pericyclic reaction transition states: passions and punctilios, 1935-1995. Acc Chem Res 28(2):81–90
85. Lee GY, Kim HY, Han IS (1999) DFT studies of the diels-alder reaction of 1, 4-diaza-1, 3-butadiene with acrolein. Korean Chem Soc 20(5):621–623
86. Domingo LR, Andrés J (2003) Enhancing reactivity of carbonyl compounds via hydrogen-bond formation. A DFT study of the hetero-diels- alder reaction between butadiene derivative and acetone in chloroform. J Org Chem 68(22):8662–8668
87. Zhang X, Haifeng D, Wang Z, Yun-Dong W, Ding K (2006) Experimental and theoretical studies on the hydrogen-bond-promoted enantioselective hetero-diels-alder reaction of danishefsky's diene with benzaldehyde. J Org Chem 71(7):2862–2869
88. Akiyama T, Morita H, Fuchibe K (2006) Chiral brønsted acid-catalyzed inverse electron-demand AZA diels-alder reaction. J Am Chem Soc 128(40):13070–13071
89. Esquivias J, Arrayás RG, Carretero JC (2007) Catalytic asymmetric inverse-electron-demand diels- alder reaction of n-sulfonyl-1-AZA-1, 3-dienes. J Am Chem Soc 129(6):1480–1481
90. Liu D, Canales E, Corey EJ (2007) Chiral oxazaborolidine- aluminum bromide complexes are unusually powerful and effective catalysts for enantioselective diels-alder reactions. J Am Chem Soc 129(6):1498–1499
91. Domingo LR, Sáez JA (2009) Understanding the mechanism of polar diels-alder reactions. Org Biomol Chem 7(17):3576–3583
92. Jiang H, Cruz DC, Li Y, Lauridsen VH, Jørgensen KA (2013) Asymmetric organocatalytic thio-diels–alder reactions via trienamine catalysis. J Am Chem Soc 135(13):5200–5207
93. Sato S, Maeda Y, Guo J-D, Yamada M, Mizorogi N, Nagase Shigeru, Akasaka Takeshi (2013) Mechanistic study of the diels-alder reaction of paramagnetic endohedral metallofullerene: reaction of la@ c82 with 1, 2, 3, 4, 5-pentamethylcyclopentadiene. J Am Chem Soc 135(15):5582–5587
94. Yuan C, Biao D, Yang L, Liu B (2013) Bioinspired total synthesis of bolivianine: a diels-alder/intramolecular hetero-diels-alder cascade approach. J Am Chem Soc 135(25):9291–9294
95. Dell'Amico L, Vega-Peñaloza A, Cuadros S, Melchiorre P (2016) Enantioselective organocatalytic diels–alder trapping of photochemically generated hydroxy-o-quinodimethanes. Angew Chem Int Ed 55(10):3313–3317
96. Andrews DR, Barton DHR, Hesse RH, Pechet MM (1986) Synthesis of 25-hydroxy-and 1. alpha., 25-dihydroxy vitamin d3 from vitamin d2 (calciferol). J Org Chem 51(25):4819–4828

97. Lygo B, Bhatia M, Cooke JWB, Hirst DJ (2003) Synthesis of (±)-solanapyrones a and b. Tetrahedron Lett 44(12):2529–2532
98. Burns AC, Forsyth CJ (2008) Intramolecular diels-alder/Tsuji allylation assembly of the functionalized trans-decalin of salvinorin A. Org Lett 10(1):97–100
99. Chackalamannil S, Davies RJ, Wang Y, Asberom T, Doller D, Wong J, Leone D, McPhail AT (1999) Total synthesis of (+)-himbacine and (+)-himbeline. J Org Chem 64(6):1932–1940
100. Funel JA, Abele S (2013) Industrial applications of the diels–alder reaction. Angew Chem Int Ed 52(14):3822–3863
101. Roos BO (1987) The complete active space self-consistent field method and its applications in electronic structure calculations. Adv Chem Phys 69:399–445
102. Roos BO (1980) The complete active space SCF method in a fock-matrix-based super-ci formulation. Int J Quantum Chem Symp 18(S14):175–189
103. Bernardi F, Bottoni A, Field MJ, Guest MF, Hillier IH, Robb MA, Venturini A (1988) MC-SCF study of the diels-alder reaction between ethylene and butadiene. J Am Chem Soc 110(10):3050–3055
104. Wiest O, Montiel DC, Houk KN (1997) Quantum mechanical methods and the interpretation and prediction of pericyclic reaction mechanisms. J Phys Chem A 101(45):8378–8388
105. Goldstein E, Beno B, Houk KN (1996) Density functional theory prediction of the relative energies and isotope effects for the concerted and stepwise mechanisms of the diels-alder reaction of butadiene and ethylene. J Am Chem Soc 118(25):6036–6043
106. Goodrow A, Bell AT, Head-Gordon M (2009) Transition state-finding strategies for use with the growing string method. J Chem Phys 130(24):244108
107. Rowley D, Steiner H (1951) Kinetics of diene reactions at high temperatures. Discuss Faraday Soc 10:198–213
108. Evans MG (1939) The activation energies of reactions involving conjugated systems. Trans Faraday Soc 35:824–834
109. Robert Burns Woodward and Roald Hoffmann (1969) The conservation of orbital symmetry. Angew Chem Int Ed 8(11):781–853
110. Dewar MJ, Griffin AC, Kirschner S (1974) Mindo/3 study of some diels-alder reactions. J Am Chem Soc 96(19):6225–6226
111. Dewar MJS, Olivella S, Rzepa HS (1978) Ground states of molecules. 49. mindo/3 study of the retro-diels-alder reaction of cyclohexene. J Am Chem Soc 100(18):5650–5659
112. Frey HM, Pottinger R (1978) Thermal unimolecular reactions of vinylcyclobutane and isopropenylcyclobutane. J Chem Soc Faraday Trans 1 74:1827–1833
113. Dewar MJS, Pierini AB (1984) Mechanism of the diels-alder reaction. studies of the addition of maleic anhydride to furan and methylfurans. J Am Chem Soc 106(1):203–208
114. Dewar MJS (1984) Multibond reactions cannot normally be synchronous. J Am Chem Soc 106(1):209–219
115. Dewar MJS, Olivella S, Stewart JJP (1986) Mechanism of the diels-alder reaction: reactions of butadiene with ethylene and cyanoethylenes. J Am Chem Soc 108(19):5771–5779
116. Houk KN, Lin YT, Brown FK (1986) Evidence for the concerted mechanism of the diels-alder reaction of butadiene with ethylene. J Am Chem Soc 108(3):554–556
117. Dewar MJS, Jie C (1992) Mechanisms of pericyclic reactions: the role of quantitative theory in the study of reaction mechanisms. Acc Chem Res 25(11):537–543
118. Townshend RE, Ramunni G, Segal G, Hehre WJ, Salem L (1976) Organic transition states. V. the diels-alder reaction. J Am Chem Soc 98(8):2190–2198
119. Sakai S (2000) Theoretical analysis of concerted and stepwise mechanisms of diels-alder reaction between butadiene and ethylene. J Phys Chem A 104(5):922–927
120. Cui C-X, Liu Y-J (2014) A thorough understanding of the diels-alder reaction of 1, 3-butadiene and ethylene. J Phys Org Chem 27(8):652–660
121. Li Y, Houk KN (1993) Diels-alder dimerization of 1, 3-butadiene: an ab initio CASSCF study of the concerted and stepwise mechanisms and butadiene-ethylene revisited. J Am Chem Soc 115(16):7478–7485

122. Bartlett PD, Schueller KE (1968) Cycloaddition. VIII. ethylene as a dienophile. a minute amount of 1, 2-cycloaddition of ethylene to butadiene. J Am Chem Soc 90(22):6071–6077
123. Lide DR (eds) (2009) CRC handbook of chemistry and physics, 89th edn. CRC/Taylor & Francis, Boca Raton, FL
124. Bell GI (1978) Models for the specific adhesion of cells to cells. Science 200(4342):618–627
125. Ivanov AN, Litvin DF, Savenko BN, Smirnov LS, Voronin VI, Teplykh AE (1995) High pressure cell for neutron diffraction investigations. High Pressure Res 14(1–3):209–214

Chapter 3
Chemometric Modeling of Emerging Materials for the Removal of Environmental Pollutants

Supratik Kar and Jerzy Leszczynski

Abstract Widespread usage of pharmaceuticals, personal care products (PPCPs), and agrochemicals followed by the release of household waste, industrial and hospital wastes has affected the environment and ecosystems immensely. These toxic chemicals are primarily classified under contaminants of emerging concerns (CEC) and/or environmental pollutants (EPs). Due to their harmful effects, timely removal of these EPs is an utmost requirement under risk management of the environment. A series of traditional techniques are accepted by the environmental organization to remove these products from the environment. Adsorption is one of the low-budget, easy to perform, and efficient approaches. With the advancement of nanotechnology, materials like carbon nanotubes (CNTs), magnetic nanoparticles, modified activated carbons/biochar, clay polycations, polyamide nanofilters (PNF), etc. have emerged as the materials of interest at present time. Along with the existing hazardous chemicals in the ecosystems, every day thousands of new chemicals are introduced to the environment. As a result, there is a continuous requirement for efficient materials which are capable of adsorbing these contaminants from the environment. In this perspective, chemometric-based modeling and machine learning (ML) models are shown to be capable of predicting important structural and physicochemical features that are responsible for the efficient adsorption property of these emerging materials. Once these features are identified, further modification in the structure of these materials can be performed to make them much more efficient adsorbers than the existing materials. The present chapter discusses the CECs and EPs, emerging materials in the present time, along with details about the chemometric and ML models which can be employed for modeling of the adsorption of EPs. Finally, successful case

---

S. Kar
Chemometrics & Molecular Modeling Laboratory, Department of Chemistry, Kean University, Union, New Jersey, USA
e-mail: skar@kean.edu

J. Leszczynski (✉)
Interdisciplinary Center for Nanotoxicity, Department of Chemistry, Physics and Atmospheric Sciences, Jackson State University, Jackson, MS 39217, USA
e-mail: jerzy@icnanotox.org

M. Shukla et al. (eds.), *Emerging Materials and Environment*,
Challenges and Advances in Computational Chemistry and Physics 37,
https://doi.org/10.1007/978-3-031-39470-6_3

studies for the prediction of adsorption of EPs onto different emerging materials are meticulously discussed with mechanistic interpretations.

3.1 Environmental Pollutants

Pharmaceuticals and personal care products (PPCPs), by-product waste, solvents, and contaminated gases from the chemical and pharmaceutical industry, oil refineries, and bulk industries are termed as EPs when they are released into the environment [1]. The hazardous effects of these pollutants are a serious matter of concern over the decades. In many cases, these organic chemicals are treated up to manifold to reduce their toxic effects before release, but still, even a minimal concentration can be dangerous and sometimes fatal to aquatic bodies, can cause soil erosion, and air pollution, which can subsequently cause health hazards to humans [2]. Pharmaceuticals are more dangerous due to their intended effects on non-target organs of human systems [3]. Among the EPs, persistent organic pollutants (POPs) are a group of hazardous chemicals which don't break down easily in the environment, and are deliberately prepared and utilized in the health system, agriculture and pest control, manufacturing, or industry. The POPs can bioaccumulate in the fat tissue of living systems, biomagnify in food chains, and are persistent in the ecosystem. The most frequently defined EPs are illustrated in Fig. 3.1.

3.1.1 Pharmaceuticals

Extensive and irrational use of pharmaceuticals resulted in significant negative outcomes for the environment and health systems over the last decades [4]. Over 600 active pharmaceutical ingredients (APIs) and their metabolites have been detected in surface water, groundwater, sewage effluent as well as in soil samples for more than 71 countries all over the world in a concentration ranging from few ng/L to 1000 μg/L [5]. The most frequently detected therapeutic classes in aquatic and terrestrial compartments are painkillers, cardiovascular drugs, antibiotics, and central nervous system-acting drugs.

3.1.2 Personal Care Products (PCPs)

Apart from pharmaceuticals, PCPs are, worldwide, the most commonly detected pollutants in the ecosystem at a substantial degree of concentration that majorly contributes to the toxicity to aquatic system [6]. The PCPs include a range of chemicals used in day-to-day life like soaps, cosmetics, toothpaste, preservatives (ethylparaben (EPB), methylparaben (MPB), propylparaben (PPB), celestolide (ADBI)),

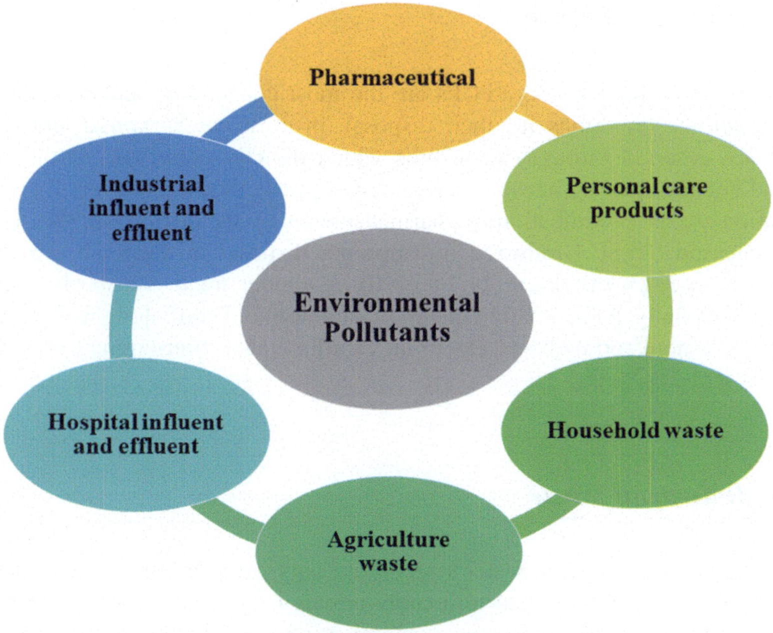

Fig. 3.1 Types of most common environmental pollutants (EPs)

fragrance/musk (tonalide (AHTN), galaxolide (HHCB), 4-methyl-benzylidene-camphor (4-MBC), sunscreens/UV filters (2-ethyl-hexyl-4-trimethoxycinnamate (EHMC), octyl-triazone (OC), octyl-methoxycinnamate (OMC), insect repellents (1,4-dichlorobenzene, N,N-diethyl-m-toluamide (DEET)), disinfectants (methyltriclosan, triclosan (TCS), and triclocarbon (TCC))) [7]. This is likely because the majority of the PCPs are meant for external use and thus do not undergo metabolic changes like pharmaceuticals would inside the body. Therefore, a vast quantity of unchanged PCPs are released into the surface water of water bodies and wastewater treatment plants (WWTPs) followed by oceans and rivers [8].

3.1.3 Agrochemicals

Antibiotics like oxytetracycline with streptomycin are very frequently used in plant agriculture in monitoring bacterial diseases of flowers and fruits. Studies have reported that antibiotics employed for agrochemicals account for $< 0.5\%$ of total antibiotic use in the USA. Again, agrochemicals like herbicides, insecticides, fungicides, and growth hormones for plants are very frequently detected in surface and groundwater over the years. It is a serious threat to the ecosystem as many of them are unaltered after usage [9].

3.1.4 Household Waste

Among the household waste, PPCPs are the most concerning contaminants. Due to inappropriate directions for their disposal, in most cases expired and unused PPCPs are deserted through waste bins which directly contribute to the terrestrial and aquatic toxicity hazards. Literature suggests that the reasons for unused medication are the modification of pharmaceuticals by the doctor (48.9%), or self-discontinuation (25.8%) followed by improper disposal in the trash (76.5%) or flushing them down the drain (11.2%) [10]. A similar trend is also observed for PCPs. Additionally, paint, coating materials of household stuff along with polymer-based toys, nanomaterials-based electronics stuff are also contributing to household waste [3].

3.1.5 Industrial Waste

Most industrial chemicals in water systems are only partially treated in wastewater treatment plants (WWTPs) and ultimately released into the aquatic environment, leading to pervasive and continuous contamination [11]. Although the majority of industries are required to follow the good manufacturing practices (GMP), still many reports suggest the release of harmful contaminants from industrial waste over time throughout the world. Concentrations up to several mg/L of pharmaceutical wastes have been identified in industrial effluents for many generic drugs, specifically in Asian countries [12].

3.1.6 Hospital Waste

Hospital wastes in the form of influents and effluents consisting of therapeutics, biologicals, and excreta are another important source of contamination. As per multiple studies, 16 pharmaceuticals were detected in hospital waste [13]. Moreover, iodinated X-ray contrast media like iohexol, iopromide, and iomeprol, which are frequently employed as diagnostic tools for capturing X-ray images of soft tissues, are frequently eliminated as waste without apposite treatment contributing to the persistence of these wastes in the ecosystem for a long period [14].

3.2 Emerging Material Concerning the Removal of EPs

To perform risk management of the environment, removal of the above-mentioned EPs and/or POPs is the first step. A good number of EPs have been detected in different water bodies at concentrations of ng l^{-1} to μg l^{-1} in surface water, rivers, lakes, groundwater, marine biota, sewage treatment plants (STPs), and wastewater treatment plants (WWTPs) over the years throughout the world [5]. Nowadays, considerable attention has been provided to the water quality of water bodies, as most of them are contaminated with an enormous number of EPs, especially PPCPs. Modified activated carbons/biochar, clay polycations, polyamide nanofilters (PNF), magnetic nanoparticles (MNPs), single-walled carbon nanotubes (SWCNTs), and multiwalled carbon nanotubes (MWCNTs) are emerging materials that could be applied to remove the organic contaminants from different areas of the environment [15]. Thus, the study of their removal capacity is very much important to a reduction in toxicity, risk assessment, as well as risk management of the environment. Additionally, an in-depth understanding of their structural and physiochemical features will be helpful to develop much more efficient materials. In this background, the development of in silico models for predicting the extent to which EPs can be removed from different waste sources through treatment unit operations would be beneficial. A series of emerging materials is portrayed in Fig. 3.2.

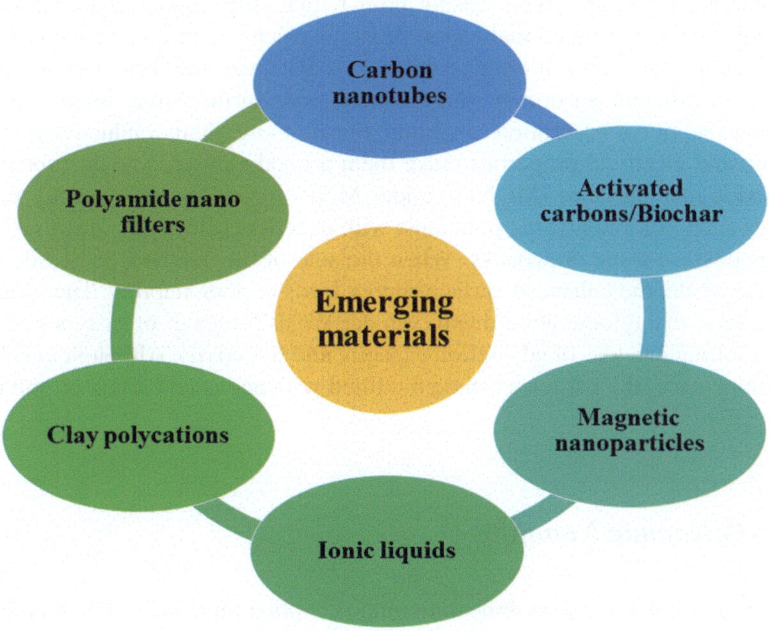

Fig. 3.2 Types of emerging materials for the removal of EPs

3.2.1 Carbon Nanotube (CNT)

One of the most talked-about innovations of nanotechnology is carbon nanotubes (CNTs) which can be used as a new generation adsorbent due to their high adsorption affinity toward organic contaminants and high surface area. The CNTs have acquired an extensive interest as alternative adsorbents for pollution management due to their ability to remove inorganic and organic pollutants from the environment. Another significant advantage of CNTs is that they can be functionalized easily to enhance their selectivity toward specific target pollutants. In rationalizing the targeted modification or functionalization of CNTs, computational studies along with chemometric tools showed significant promise over the last decade. Several studies were performed to predict the adsorption property of organic chemicals, radioactive elements, heavy metal ions, etc. for both single-walled carbon nanotubes (SWCNTs) and multiwalled carbon nanotubes (MWCNTs). Such studies are usually followed by introspection and identification of major structural features responsible for the interactions [15].

3.2.2 Magnetic Nanoparticles (MNPs)

Magnetic nanoparticles (MNPs) have proven their impact in the fields of nanomedicine, imaging, and biosensing along with multiple uses in material science. The modification and functionalization of nanoparticles with diverse biomolecules provide enormous application possibilities [16]. Over the last decade, MNPs appeared as efficient adsorbents due to their large surface area, tunable particle size, and exceptional adsorption capability. Furthermore, their conductivity, charge, magnetic and electrical properties make them a good choice as an adsorbent. The metal oxide nanoparticles (MONPs) under MNPs have a high propensity to bind toxicants such as ionic organic pollutants with excellent selectivity. However, these particles possess some drawbacks. When the size of the MONPs decreases to the nanometer scale, the enhanced surface energy leads to poor stability. Therefore, the particles tend to agglomeration due to van der Waals forces or other types of interaction resulting in a loss of adsorption capacity and selectivity. Although to enhance their competence, the particles can be modified with activated carbon or polymeric support [17].

3.2.3 Polyamide Nanofilters

Nanofiltration (NF) based on thin-film-composite polyamide (TFC-PA) membranes is an established and energy-efficient approach for extracting small solutes from liquid solvents [18]. The selective layer of polyamide-based NF membranes is prepared by interfacial polymerization (IP) on a porous support. These filters are extremely useful for waste management in recent times. The functionalization of the thin film may improve their adsorption and provide superior separation efficiency with low energy consumption. Here, in silico models become increasingly important for designing efficient coated filters for the management of organic and inorganic wastes.

3.2.4 Clay Polycations

The cation exchange capability, large surface area, flexibility, simple production, easy regeneration, and low cost of clay and/or clay minerals make them an imperative choice in the removal of organic compounds and wastewater treatment [19]. Among clay minerals, hybrid clay polyelectrolyte nanocomposites (CPN) received attention for the removal of organic pollutants from wastewater. The most frequently used clay polycations are nanocomposites of poly (4-vinylpyridine-costyrene) (PVPcoS), chitosan/montmorillonite (MMT) composite, composite of magnetite/bentonite clay composite, polydiallyl dimethylammonium chloride (PDADMAC) with MMT, and PVPcoS/MMT clay composite. Based on the efficiency and enhanced adsorption, the protonated PVPcoS-MMT (HPVPcoS-MMT) is better than the normal PVPcoS/MMT. The HPVPcoS-MMT system has proven its capability of delivering highly efficient removal of atrazine and pyrene due to its large, functionalized surface.

3.2.5 Activated Carbon and Biochar

Biochar is a stable carbon-rich material, derived from biomass that is heated in a limited oxygen environment. Biochar shows incredible potential to handle water/wastewater contaminants [20]. Similarly, activated carbon is a carbon-rich solid which is derived from biomass or carbonaceous substances such as coal or tar pitch and is also obtained through pyrolysis. The process helped to activate the carbon material that significantly increases the surface area of the material, helping it to adsorb a greater number of molecules. Higher adsorption capability helps these materials to be effective at eliminating pollutants and contaminants from water and air. Biochar has been observed as an efficient adsorbent for heavy metals, such as arsenic (As). Depending on the initial conditions, duration of the formation process, and temperature, as well as the starting material, the products contain dissimilar functional groups which can adsorb different contaminants. Therefore, structural modification of biochar can enhance the adsorption property to some extent.

3.2.6 *Ionic Liquids (ILs)*

Ionic liquids (ILs) have widespread usage in industrial processes. ILs typically consist of organic cations (e.g., imidazolium, pyrrolidinium, pyridinium, and ammonium) and inorganic anions (likely bromide, chloride, hexafluorophosphate, and tetrafluoroborate). Under 100 °C, ILs behave as non-volatile liquids. Due to this non-volatile nature, null flammability, and thermal stability, ILs behave as environmentally friendly solvents and are defined as "designer solvents". ILs have the potential for the removal of several organic pollutants [21].

3.3 Chemometric Modeling and Machine Learning Techniques

Chemometrics (also termed cheminformatics) is the scientific area that applies mathematical and statistical techniques to design and/or choose optimal procedures and experiments, and to deliver crucial chemical information by analyzing available chemical data. Chemometric methods employ both linear and nonlinear techniques for the development of the models. The linear models are built based on the simple correlation between the descriptors (X variables) and the response (Y variable). The nonlinear models are built based on spline functions, quadratic terms, and other higher-order polynomials. One of the most employed modeling approaches under chemometrics is quantitative structure–activity relationship (QSAR) [22, 23]. To develop a QSAR model, a series of classification, regression, and machine learning approaches can be utilized. In this present chapter, we focus our discussion mainly to QSAR models.

The ideology of QSAR analysis assumes that the molecules available in nature contain information related to their physical, chemical, or biological (including toxicological) activity which can be suitably described in terms of a mathematical correlation. Equation 3.1 describes the response elicited by chemicals to be a mathematical function of chemical features.

$$\text{Chemical Response} = f(\text{Chemical attributes}) = f(\text{Structure, Property}) \quad (3.1)$$

Here, response refers to any type of physicochemical property, activity, or toxicity shown by the chemicals, while the chemical attributes correspond to the quantitative information obtainable about the chemicals using suitable theoretical/ experimental techniques. The nomenclature of the technique is usually done depending upon the nature of the response. Accordingly, three broad types can be identified, namely QSAR, QSPR, and QSTR denoting the response to be a biological activity (e.g., anticancer, antibacterial), physicochemical property (molar refractivity, aqueous solubility, etc.), and toxicity (non-systemic such as ecotoxicity, systemic such as cardiotoxicity, nephrotoxicity), respectively. However, we shall use the term QSPR in this present study as modeling of property is the interest of the present chapter. The major chemometric approaches can be classified under two classes: regression- and classification-based methods [24–32]. The most commonly used approaches are following under regression-based techniques: multiple linear regression (MLR), stepwise regression, partial least squares (PLS), genetic function approximation (GFA), and genetic partial least square analysis (G/PLS). While, classification-based techniques consist of approaches like principal component analysis (PCA), factor analysis (FA), factor analysis followed by MLR (FA-MLR) and PLS (FA-PLS), and linear discriminant analysis (LDA).

Machine learning (ML) models are another popular in silico technique which is quite popular for the prediction of diverse characteristics starting from toxicity to physicochemical properties of emerging materials [33–35]. The most frequently used ML approaches are artificial neural network (ANN), support vector machine (SVM), decision tree (DT), and random forest (RF).

3.4 Why Are Chemometric Modeling and ML Models Required?

Chemometric modeling is an important tool for designing and enhancing the adsorption properties of emerging materials within a short period while minimizing reliance on experimental efforts. Final proposed materials need to be checked experimentally, but individually checking the adsorption properties of hundreds or thousands of compounds would involve millions of dollars, time, and labor and would ultimately be impractical. Thus, chemometric modeling can narrow the field of promising materials that need to be experimentally validated. Due to the advancement of material science, 100 new compounds are introduced daily and many of them possess multiple toxic effects on the animal as well as ecosystems. Therefore, removal of these materials after their intended use from ecosystem is of the utmost importance, and each kind of organic pollutants has specific materials which can remove them. Chemometric modeling can offer faster designing, modeling, and predictions of adsorption properties of thousands of emerging materials specific to organic pollutants.

3.5 Predictive in Silico Models on the Adsorption of Organic Pollutants by EMs

3.5.1 Modeling of EP's Adsorption by CNT

Multiple in silico modeling experiments have been performed over time employing SWCNT, MWCNT, and sometimes a combination of both. For a better understanding of these models, we have divided the section into three subsections.

3.5.1.1 Modeling of EP's Adsorption by SWCNT

Ghosh et al. [36] developed PLS-based QSPR models to identify imperative structural and physicochemical properties contributing to the adsorption coefficient of 40 synthetic organic environmental pollutants (majorly PPCPs, contrasting agents, dyes) onto SWCNTs. The developed QSPR models suggested that the hydrophobic surface of the molecules (MLOGP2 has the highest contribution to the adsorption coefficient), molecular shape and branching, number of H atoms attached with the α-C atom (tylosin and sulfamethoxazole), presence of two carbon atoms at the topological distance 4 (ethinyl estradiol and phenanthrene), and the presence of C and O atoms at the topological distance 6 (ethinyl estradiol and diuron) may increase the adsorption of hazardous chemicals to the SWCNTs. Conversely, the presence of C and S atoms at topological distance 7 (lincomycin and methyl orange) and the number of tertiary aliphatic amines (lincomycin) could be unfavorable for adsorption. If we summarize the critical features, hydrophobic interactions (ethinyl estradiol and 1,2,4-trichlorobenzene showed better adsorption due to higher MLOGP2 values), hydrogen bonding, and electrostatic interactions are responsible for better adsorption of the studied pollutants onto SWCNTs (Fig. 3.3). This model offers a better understanding of the necessary features of a chemical to have better adsorption onto SWCNTs which may be useful for ERM.

Lata and Vikas [37] reported QSPR models for the adsorption coefficient of 40 organic compounds by SWCNTs. These characteristics were modeled by QSPR

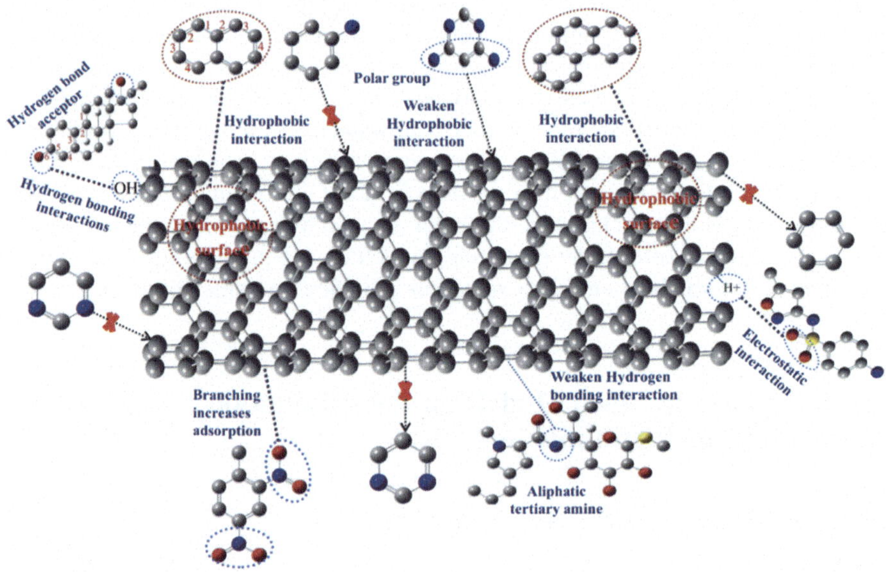

Fig. 3.3 Mechanistic interpretation of adsorption of organic environmental pollutants onto SWCNTs/functionalized SWCNTs. The figure is adapted with permission from Elsevier [36]

employing quantum–mechanical descriptors to analyze major structural information that could be imperative for functionalizing of SWCNTs to increase the adsorption of these organic chemicals. The authors also checked the predictivity of the linear solvation energy relationship (LSER) models for the adsorption prediction. The obtained models offered crucial insights and reported that the mean polarizability was the key feature which most affected the chemicals' adsorption to SWCNTs. The authors also reported that models prepared with specific descriptors are inferior to the models obtained from the mixture of solvatochromic and quantum mechanical descriptors. In conclusion, the authors indicated that the developed models could be used to predict the adsorption efficiency of steroid hormones, nucleobases, and agrochemicals like herbicides, insecticides, pesticides, and endocrine disrupting chemicals to SWCNTs. As per the authors' prediction, they claimed progesterone and guanine will be strongly adsorbed by the SWCNTs. The developed in silico models could be used to predict the nanotoxicity along with the adsorption of major environmental pollutants and biomolecules to SWCNTs.

The adsorption data (logK) of 61 organic pollutants onto SWCNTs were modeled employing linear and nonlinear modeling techniques named MLR and SVM by Wang et al. [38]. Both models were validated extensively utilizing classical regression and error-based validation metrics to check the robustness, goodness-of-fit, and predictive ability of the model. Both models were built using the same dataset and molecular structural descriptors for comparison. Based on the model's interpretations, the authors concluded that:

(a) Lipophilic or hydrophobic chemicals choose to interact with SWNTs than with hydrophilic ones,
(b) Cyclic compounds with substituents might be well adsorbed by SWNTs,
(c) Chemicals with higher α values tend to have stronger dispersion interactions and dipole-induced dipole forces with SWNTs, where $\alpha = \frac{(100 * \text{Molecular Polarizability})}{\text{Volume}}$
(d) F, O, Cl atoms were influential toward the adsorption of organic pollutants by SWCNTs,
(e) The electrostatic interactions between H atoms and π electrons in SWCNTs as well as the H bonding interactions between H atoms and O atoms of water also affected the adsorption.
(f) The molecular dipole moment (μ) encoding the dipole–dipole interactions had the slightest effect on the adsorption.
(g) Among the van der Waals, hydrophobic, hydrogen bonding, and electrostatic interactions, the hydrophobic and van der Waals interactions contributed most to the adsorption.

Liu et al. [39] demonstrated density-functional theory (DFT) study followed by development of a QSPR model of 25 benzene derivatives (BzD) to identify the major structural attributes of the adsorption characteristics and vital interactions pattern to SWCNTs. The authors designed 6–10 interaction modes to identify the probable interaction configuration for each BzD–SWCNT system in the gas phase where all geometries were optimized using DFT at the M062X/6-31G(d) level of theory. The performed DFT simulations helped to optimize the structural, dynamic, and energetic characteristics of the complexes for determining the adsorption mechanisms. Later a QSPR model was prepared containing 8 descriptors using stepwise regression approach utilizing a total of 79 three-dimensional surface area, volume, and shape (SurVolSha) descriptors. Based on the statistical evaluations, the model had promising predictive competence and could be employed to predict the studied systems. The introspection of the study illustrated that π-π stacking was governs the adsorption of BzD onto the SWCNTs, while the substituents played a secondary role on the adsorption. The final optimized BzD–SWCNT configurations demonstrated that the parallel modes are more stable than the perpendicular modes.

3.5.1.2 Modeling of EP's Adsorption by MWCNT

The surface area normalized adsorption coefficients of 69 aromatic organic compounds on MWCNTs were modeled using the Monte Carlo method by Ahmadi and Akbari [40]. Simplified molecular-input line-entry system (SMILES) and hydrogen-suppressed molecular graphs (HSGs) of each chemical were considered to generate the descriptors for the QSPR model employing the CORAL (CORrelation and Logic) software [41]. The authors reported that the results were highly acceptable and predictable, and the QSPR model could be applied for the prediction of the adsorption coefficient of aromatic compounds on MWCNTs with only the SMILES string for chemical query.

Wang et al. [42] prepared QSPR models of adsorption coefficients data of diverse organic compounds for two different datasets. The first dataset contained 59 organic compounds with adsorption coefficient data (logK1), and the second one contained 69 organic compounds with adsorption coefficient data related to specific surface area (logKSA) for MWCNTs. First, the authors optimized the geometry of all compounds through DFT method at STO-3G level. Thereafter, the norm index descriptors based on a series of distance matrices, including Euclidean distance matrix, adjacent matrix, and step matrix, were employed to develop the models to predict the adsorption affinity to MWNTs. The authors reported highly acceptable goodness-of-fit, robustness, and predictivity outcomes for both models (R^2 for the training set and the test set of 0.95 and 0.98 for logK1; and 0.93 and 0.98 for logKSA, respectively). Each model was also checked for applicability domain to find outliers in the developed model that would result in unreliable predictions in the test set. Additionally, the authors performed a Y-randomization analysis and found that the random models showed lower validation metrics values than the actual models. A thorough mechanistic interpretation of the models concluded that the chemicals get adsorbed to the MWCNTs due to physicochemical interactions such as π-π interaction, hydrophobic, electrostatic interactions, and hydrogen bonding.

Rahimi-Nasrabadi et al. [43] developed multiple linear and nonlinear QSPR models to predict the adsorption coefficient data (logK1) of 59 aromatic organic compounds by MWCNTs. The authors split the entire dataset into three different sets where the training, validation, and test set, and consist of 43, 11, and 5 chemicals, respectively. First, Austin Model1 (AM1) was used to optimize the molecular geometry of studied compounds. Then, the descriptors were computed by HyperChem program (ver. 7) and Dragon descriptors. Four Abraham descriptors were also included in the pool of descriptors to evaluate the probable nonlinear relationship between logK1 and the mentioned descriptors. Authors employed linear stepwise MLR and nonlinear multilayered perceptron neural network (MLP-NN) models and selected five descriptors by using self-organizing map (SOM) for model development. Both the linear and nonlinear models offered statistically acceptable results both internally and externally. The MLP-NN model offered better statistical quality compared to simple MLR model. Based on the outcomes, the authors proposed that there is a nonlinear relationship that exists between the structures of the compounds and their adsorption onto MWCNTs. The authors also interpreted that hydrogen bond accepting ability, molar volume, molecular mass and size, and the amount of branching were the crucial features to characterize and regulate the adsorption of the organic chemicals by MWCNTs.

Adsorption coefficients of 40 heterogeneous aromatic organic chemicals to MWCNTs were modeled through QSPR by Heidari and Fatemi using the CORAL software [44]. This is one of the first reported QSPR models of adsorption coefficients of chemicals on CNTs using the CORAL software tool. Two different models were developed employing hydrogen-filled graph-based descriptors (model 1) and hybrid descriptors (model 2) which were the amalgamation of SMILES and hydrogen filled graph-based (HFG) descriptors. The authors utilized 3D response surfaces of all subsets to optimize the Monte Carlo (MC) parameters of models. The obtained results depicted a good robustness, goodness-of-fit, and acceptability of both models. The statistical results supported that the HFG descriptor-based model was superior compared to the hybrid-based descriptor model.

Hassanzadeh et al. [45] modeled adsorption of two datasets consists of 40 and 69 organic chemicals by MWCNTs. The first dataset contained 40 diverse chemicals with defined adsorption affinity properties (k1) for MWCNTs and modeled with solvatochromic descriptors. Dataset 2 included adsorption affinity of 69 organic pollutants connected to specific surface area (KSA) for MWCNTs and modeled with 3D descriptors. The authors generated nonlinear models utilizing a blend of radial basis function neural network (RBFN) and GA called as whole space GA-RBFN (wsGA-RBFN). All models were validated both internally and externally using numerous strict statistical parameters. The statistical results showed that the wsGA-RBFN-based model was significant to predict adsorption coefficients (logK1) of dataset 1 and logKSA for dataset 2 as compared to simple MLR and GA-RBFN models. Based on these results, the authors claimed that models with solvatochromic descriptors were superior to the physicochemical descriptors in terms of the statistical quality of the model.

Wang et al. [46] reported 3D QSPR models employing three techniques, namely MLR, ANN, and SVM for modeling the adsorption coefficient of 39 aromatic organic chemicals onto MWCNTs. The statistical quality of the models suggested that the nonlinear ANN and SVM models performed better than the MLR model. Close examination of the models revealed that the number of N and O atoms, octanol/water, number of atoms in rings of a ring molecule, and partition coefficient

($logK_{ow}$) contributed positively toward the adsorption of chemicals to MWCNTs. On the contrary, dipole moment and the assessed number of H-bonds that would be accepted by the solute from water molecule in an aqueous solution (accptHB) had disadvantageous effects toward the adsorption to MWCNTs.

3.5.1.3 Modeling of EP's Adsorption by Both SWCNT and MWCNT

Roy et al. [47] developed statistically and mechanistically interpretable QSPR models for adsorption affinity (2 datasets consist of 59 and 69 organic contaminants including PPCPs as well as industrial chemicals) and dispersibility index (3rd dataset includes 29 organic solvents significant for PPCPs manufacture) of organic chemicals to MWCNTs and SWCNTs, respectively employing simple 2D features. The obtained consensus MLR models from datasets 1 and 2 suggested the following insights related to adsorption affinity of the studied compounds to MWCNT: (a) π–π interactions between organic pollutants and π-system of the MWCNTs influence the adsorption affinity positively and the interaction increases with an increasing number of aromatic rings; (b) electron-donating groups like $-NH_2$, –OH and electron-withdrawing groups like $-NO_2$ and –Cl support the π–π interactions; (c) electrostatic interactions (present features: nArOH, B03[O–O], F03[O–O], F04[N–O], nHBint, and O%) also favor increased adsorption affinity; d) hydrogen bonding interactions (compounds with features like nArOH, F03[O–O], B03[O–O], nHBint, F04[N–O] and O%) also contribute positively along with hydrophobic interactions (minsCH3, X0v, X2A, and X1A) for aliphatic chemicals. The PLS model for dispersibility (dataset 3) interpreted that the organic solvents having features like tertiary amine, polar solvents with a low number of electron donors or solvents with medium polarity, and lower ionization potential (chemicals with smaller ionization potential supports easy interaction between π-system of the SWCNT and π-cation of solvents due to easy conversion of the chemical into cationic form) might be good solvents to improve the dispersibility of SWCNTs (Fig. 3.4). As dispersibility is directly correlated with adsorption affinity, all features are directly related to adsorption capability of the studied chemicals to both forms of CNTs.

3.5.2 Modeling of EP's Adsorption by Clay Polyelectrolyte Nanocomposites (CPN)

Protonated montmorillonite–poly-4-vinylpyridine-co-styrene (Mt–HPVPcoS), a hybrid clay polyelectrolyte nanocomposite (CPN), has a proven efficiency as adsorption material for organic pollutants. To predict the adsorption coefficient ($logk_d$) of organic pollutants including largely PPCPs, Kar et al. [48] proposed a statistically acceptable and mechanistically interpretable QSPR model. The three-descriptor-based QSPR model showed a correlation coefficient of 0.75. The model suggested

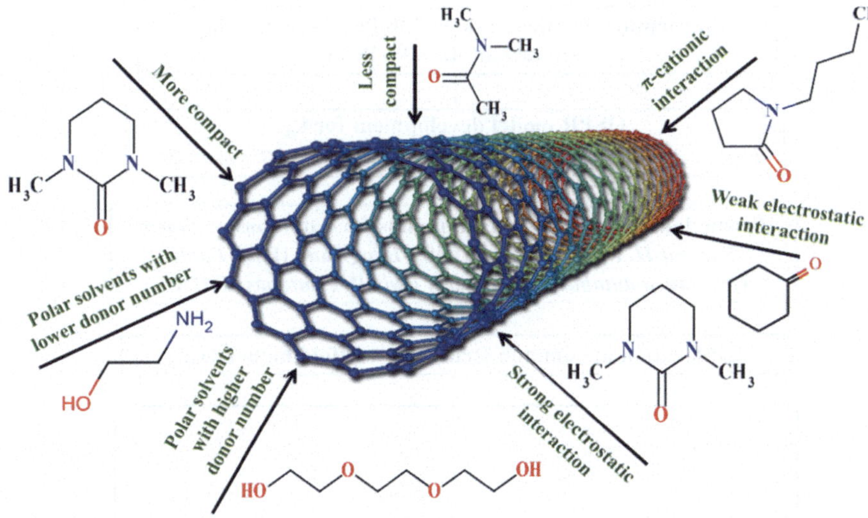

Fig. 3.4 Major features responsible for adsorption of organic environmental pollutants to SWCNTs. The figure is adapted with permission from The Royal Society of Chemistry [47]

that the conformational restriction/flexibility and measures of unsaturation positively contributed to the increase in the adsorption coefficient. Conversely, higher electrophilicity decreases the adsorption of specific compounds which means that the existence of nucleophiles increases the adsorption value to Mt–HPVPcoS. To apply the model in a broader perspective, the authors predicted the $\log k_d$ value of 900 thousand chemicals from five databases (Super Natural II, Interbioscreen (IBS), DrugBank (DB), Paola Gramatica database (PGD), and Hair Dye database (HDD)) covering a huge number of PPCPs. To check the prediction reliability of a huge number of data, the authors utilized two layers screening with the AD algorithm and PRI tool. From the complete prediction and maximum common substructure-based (MCS) analysis, the authors successfully extracted the important structural scaffold for higher $\log k_d$ to the mentioned CPN for an individual database. The authors identified that the most common scaffolds, considering both modeled and prediction compounds, for higher $\log k_d$ are derivatives of nitrobenzene, methoxy benzene, diphenylmethane, sulfonate/sulfonic acid, as well as primary, secondary, and tertiary amines (Fig. 3.5). The top 5 chemicals concerning $\log k_d$ to Mt–HPVPcoS for each database were also reported by the authors in Fig. 3.6.

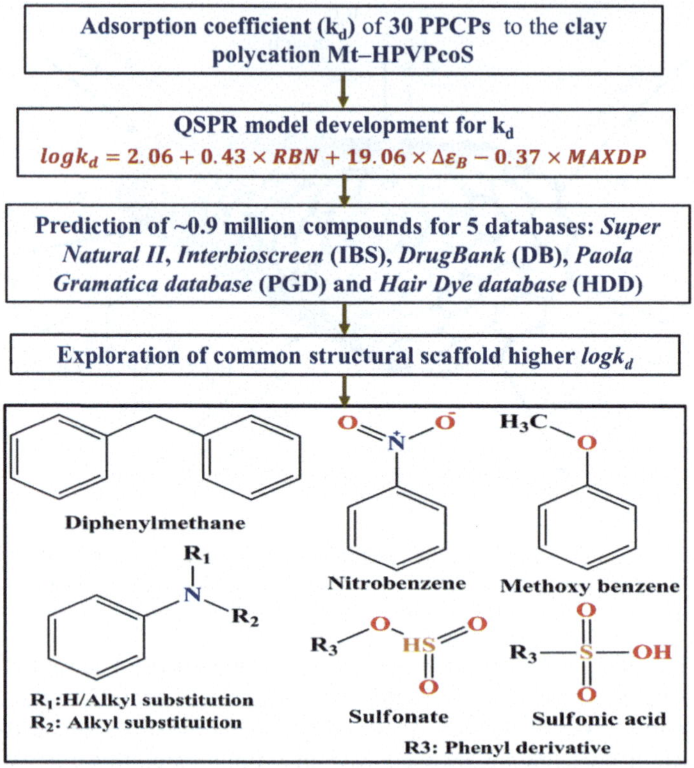

Fig. 3.5 Complete workflow along with the identified mechanistic interpretation from QSPR models

3.5.3 Modeling of EP's Adsorption by Polyamide Nanofiltration

Yangali-Quintanilla [49] developed a QSPR model for the prediction of the rejection of 106 emerging contaminants, including pharmaceuticals, agrochemicals, and other organic compounds, by polyamide nanofiltration (NF-90, NF-200) membranes. Multiple statistical tools including PLS, PCA, and MLR were employed to predict rejection by NF using combined interactions between membrane characteristics, compound properties, and filtration operating conditions as descriptors. The generated QSPR model identified a molecule's hydrophobicity (log D), equivalent width related to the size of the compound, salt rejection as membrane characteristics, and the depth and length of a compound as the most significant features that impact rejection by NF.

$$\begin{aligned}\text{Rejection} = {} & 265.15 \times \text{eqwidth} - 117.36 \times \text{depth} + 81.662 \times \text{length} \\ & - 5.229 \log D - 0.272 \times \text{MWCO} - 62.57 \end{aligned} \tag{3.2}$$

Super Natural ID	Pred. $logK_d$	IBS Natural ID	Pred. $logK_d$
SN00132349	7.37	STOCK1N-25773	6.22
SN00139488	7.27	STOCK1N-33117	6.18
SN00140779	7.22	STOCK1N-25583	5.94
SN00122334	7.16	STOCK1N-70175	5.86
SN00121058	7.14	STOCK1N-26068	5.84
IBS Synthetic ID	**Pred. $logK_d$**	**DB ID**	**Pred. $logK_d$**
STOCK1S-62119	7.36	DB00941	6.658
STOCK1S-72088	7.19	DB07251	6.564
STOCK1S-31833	7.10	DB11864	6.508
STOCK7S-65481	7.07	DB12147	6.442
STOCK2S-10058	7.05	DB01616	6.423
PGD ID	**Pred. $logK_d$**	**HDD ID**	**Pred. $logK_d$**
Alverine	6.42	C.I. Direct Red 81	6.02
Clomifene	6.08	C.I. Acid Orange 3	5.80
Prenylamine	6.04	C.I. Food Black 2	5.78
Quinacrine	5.99	C.I. Direct Yellow 12	5.52
Aprindine	5.93	C.I. Acid Violet 49	5.25

Fig. 3.6 Top 5 compounds from each database based on the predicted $logk_d$ value

Based on this model, the authors proposed that rejection increased by size and steric hindrance and decreased by solute hydrophobicity related to adsorption and partitioning mechanisms. The steric hindrance followed by electrostatic repulsion effects is the result of major membrane characteristics, i.e., salt rejection. The authors, however, note that for the NF membranes with a broad range of pore size and distribution, it may be difficult to estimate the rejection of contaminants. Further, the authors employed an external dataset consisting of 89 compounds (gathered from three different datasets corresponding to membrane Filmtec NF-90, Trisep TS-80, and Desal HL) and found an acceptable correlation coefficient of 0.75. Based on an external set analysis, the rejection cases by NF-90 for trichloroethene (TCE) and bromoform (BF) may be considered as an uncharacteristic outcome as their rejections did not match to their size and hydrophobicity when compared to other molecules. The proposed approach is related to indirect toxicity modeling of PPCPs for risk assessment and management basis.

3.5.4 Modeling of EP's Adsorption by Activated Carbons and Biochar

De Ridder et al. [50] developed a multivariate linear regression model to predict equilibrium carbon loading (CL) on a specific activated carbon (F400) for solutes with a wide range of properties. The authors developed separated models employing groups (bins) of solutes with similar properties under a specific dataset and solute removals were predicted for each bin separately. The models reported squared regression values ranging from 0.61 to 0.84. Based on the interpretations of each model, the authors concluded the following:

- The model(s) achieved 0.5 log unit deviation of the predicted carbon loading of the studied solutes compared to experimental values.
- Hydrophobic partitioning is applicable for all solutes, as linear relationships were observed between log*D* and carbon loading for each bin.
- The CL can only be predicted precisely once the set of solutes with a wide-ranging variant in solute properties is divided into four bins: aliphatic without H-bond d/a groups, aliphatic with H-bond d/a groups, aromatic with H-bond d/a groups, and aromatic without H-bond d/a groups.
- London dispersion forces only had a dominant influence for the bin containing aliphatic solutes without H-bond d/a groups.
- Carbon loading was higher at lower log D values for solutes which were able to form H-bonds, and aromatic solutes which could form pi–pi bonds.
- At a log*D* of 3.7, the solutes in all bins showed similar removal which indicates that hydrophobic partitioning is the leading removal mechanism at that log*D* value, superseding the effects of H-bond formation and pi–pi interaction.
- The effects of the presence or absence of H-bond d/a groups were independent of the amount of H-bond acceptor or H-bond donor groups. Due to the limited number of H-bond d/a groups on the carbon surface, a solute can only form a H-bond with a single functional group on the carbon.

Zhang et al. [51] employed a cosine similarity approach that focused on developing a model on the adsorption of organic pollutants by biochars, activated carbons, and carbon nanotubes. Previously developed models can only predict for particular adsorption isotherm models or under specific equilibrium concentrations with a simple MLR approach. In this study, the authors built a machine learning process based on neural networks (NNs). The authors considered grouped adsorption data for adsorbent–adsorbate pairs under different equilibrium concentrations, and polyparameter linear free energy relationships (pp-LFERs). The dataset consists of aqueous adsorption for 165 organic compounds onto 50 biochars, 35 granular activated carbons, 34 carbon nanotubes, and 30 polymeric resins. Later, the NN-LFER models were effectively applied to numerous equilibrium concentrations irrespective of the adsorption isotherm models and reported fewer prediction deviations. The predictions were improved by the inclusion of two key descriptors: pore volume and BET surface area for the adsorbents. Finally, interpreting the NN-LFER models based on the Shapley values, the authors concluded that consideration of equilibrium concentration and multiple critical properties of the adsorbents are the key reason for better prediction with fewer deviations from the experimental data.

3.6 Future Direction and Conclusion

The characterization and knowledge of environmental pollutants in the form of organic pollutants and wastewater contaminants are the foundation for selecting best the possible emerging material for use in the treatment and/or removal of the pollutants. In many cases, specific material is good for a particular pollutant irrespective of its efficiency. Thus, the choice of the right adsorbent is crucial. Once the properties of the adsorbent are known, the functionalization or modification of its chemical structure can be made easier based on computational and experimental exercises. Electronic and structural features of studied materials have already demonstrated their effectiveness for building predictive chemometric models from the enormous data available in the literature. A good number of chemometric models are available for the adsorption of compounds from organic pollutants to diverse forms of CNTs, but the number of models for MNPs, clay polycations, biochars, and ILs are sparse or non-existent. Moving forward, additional models to capture the adsorption properties to MNPs, clay polycations, biochars, ILs, etc. are required. These models may be beneficial not only in the design of efficient materials for the future but also work as a useful tool for prediction regarding any new compound before its synthesis. Another crucial point is the lack of a good modeling database for these emerging materials. The development of these databases will be enormously helpful to future researchers to both easy access available models and to avoid unintentionally re-developing an existing model. To that end, open access models in a single platform are also required for easy access for both experimental and computational researchers.

Acknowledgements SK wants to thank the administration of Dorothy and George Hennings College of Science, Mathematics and Technology (HCSMT) of Kean University for providing research opportunities through research release time and resources. JL thankful to the National Science Foundation (NSF/CREST HRD-1547754 and NSF/RISE HRD-1547836) for financial support.

Conflicts of Interest
The authors declare no conflict of interest.

References

1. Ferguson PJ, Bernot MJ, Doll JC, Lauer TE (2013) Detection of pharmaceuticals and personal care products (PPCPs) in near-shore habitats of souther Lake Michigan. Sci Total Environ 458–460:187–196
2. Faulwetter JL, Gagnon V, Sundberg C, Chazarenc F, Burr MD, Brisson J, Camper AK, Stein OR (2009) Microbial processes influencing performance of treatment wetlands: a review. Ecol Eng 35:987–1004
3. Kar S, Sanderson H, Roy K, Benfenati E, Leszczynski J (2020) Ecotoxicological assessment of pharmaceuticals and personal care products using predictive toxicology approaches. Green Chem 22:1458–1516
4. Busfield J (2015) Assessing the over use of medicines. Soc Sci Med 131:199–206

5. Kümmerer K (2013) Pharmaceuticals in the environment: sources, fate, effects and risks. Springer Science & Business Media, Berlin
6. Yang X, Flowers RC, Weinberg HS, Singer PC (2011) Occurrence and removal of pharmaceuticals and personal care products (PPCPs) in an advanced wastewater reclamation plant. Water Res 45:5218–5228
7. Montes-Grajales D, Fennix-Agudelo M, Miranda-Castro W (2017) Occurrence of personal care products as emerging chemicals of concern in water resources: a review. Sci Total Environ 595:601–614
8. Cortez FS, Seabra P, Santos AR, Cesar A, Choueri RB, Martini GDA, Bohrer-Morel MB (2012) Biological effects of environmentally relevant concentrations of the pharmaceutical Triclosan in the marine mussel Perna perna (Linnaeus, 1758). Environ Pollut 168:145–150
9. Kar S, Roy K, Leszczynski, J (2017) On application QSARs in food and agricultural sciences: history and recent developments. In: Roy K (ed) Advances in QSAR modeling with applications in pharmaceutical, chemical, food, agricultural and environmental sciences, under the book series. Challenges and Advances in Computational Chemistry and Physics (Series Ed: Leszczynski J), vol 24. Springer, pp 203–302
10. Persson M, Sabelström E, Gunnarsson B (2009) Handling of unused prescription drugs-knowledge, behaviour and attitude among Swedish people. Environ Int 35:771–774
11. Fernandez C, Gonzalez-Doncel M, Pro J et al (2010) Occurrence of pharmaceutically active compounds in surface waters of the henares-jarama-tajo river system (Madrid, Spain) and a potential risk characterization. Sci Total Environ 408:543–551
12. Li D, Yang M, Hu J et al (2008) Determination and fate of oxytetracycline and related compounds in oxytetracycline production wastewater and the receiving river. Environ Toxicol Chem 27:80–86
13. José Gómez M, Petrovic M, Fernández-Alba AR et al (2006) Determination of pharmaceuticals of various therapeutic classes by solid-phase extraction and liquid chromatographye-tandem mass spectrometry analysis in hospital effluent wastewaters. J Chromatogr 1114:224–233
14. Ternes TA, Hirsch R (2000) Occurrence and behavior of X-ray contrast media in sewage facilities and the aquatic environment. Environ Sci Technol 34:2741–2748
15. Jorio A, Dresselhaus G, Dresselhaus MS (2008) Carbon nanotubes: advanced topics in the synthesis, structure, properties and applications. Springer, Berlin
16. Kudr J, Haddad Y, Richtera L, Heger Z, Cernak M, Adam V, Zitka O (2017) Magnetic nanoparticles: from design and synthesis to real world applications. Nanomat (Basel, Switzerland) 7:243
17. Saxena R, Saxena M, Lochab A (2020) Recent progress in nanomaterials for adsorptive removal of organic contaminants from wastewater. Chem Select 5:335–353
18. Huang S, Wu M-B, Zhu C-Y, Ma M-Q, Yang J, Wu J, Xu Z-K (2019) Polyamide nanofiltration membranes incorporated with cellulose nanocrystals for enhanced water flux and chlorine resistance. ACS Sustainable Chem Eng 7:12315–12322
19. Shabtai IA, Mishael YG (2017) Catalytic polymer-clay composite for enhanced removal and degradation of diazinon. J Hazard Mater 335:135–142
20. Enaime G, Baçaoui A, Yaacoubi A, Lübken M (2020) Biochar for wastewater treatment—conversion technologies and applications. Appl Sci 10:3492
21. Hashemi B, Zohrabi P, Dehdashtian S (2018) Application of green solvents as sorbent modifiers in sorptive-based extraction techniques for extraction of environmental pollutants. TrAC Trends Anal Chem 109:50–61
22. Roy K, Kar S, Das RN (2015) Understanding the basics of QSAR for applications in pharmaceutical sciences and risk assessment. Academic Press, New York
23. Roy K, Kar S, Das RN (2015) A primer on (Q)SAR/QSPR modeling: fundamental concepts. Springer, UK
24. Draper NR, Smith H (1998) Applied regression analysis. Wiley-VCH, Weinheim
25. Darlington RB (1990) In Regression and linear models. McGraw- Hill, New York
26. Fan Y, Shi LM, Kohn KW, Pommier Y, Weinstein JN (2001) Quantitative structure-antitumor activity relationships of camptothecin analogues: cluster analysis and genetic algorithm-based studies. J Med Chem 44:3254–3263

27. Wold S, Sjostrom M, Eriksson L (2001) PLS-regression: a basic tool of chemometrics. Chemom Intell Lab Syst 58:109–130
28. Rogers D, Hopfinger AJ (1994) Application of genetic function approximation to quantitative structure-activity relationships and quantitative structure-property relationships. J Chem Inf Comput Sci 34:854–866
29. Rogers D (1999) Rational drug design, Truhlar DG (ed). Springer-Verlag, New York
30. Wold S, Eriksson L (1995) Validation tools. In: Chemometric methods in molecular design, van de Waterbeemd H (ed). VCH, Weinheim, pp 309–317
31. Franke R, Gruska A (1995) Principal component and factor analysis. In Van de Waterbeemd H (ed) Chemometric methods in molecular design. VCH, Weinheim, pp 113–158
32. Fisher RA (1936) The use of multiple measurements in taxonomic problems. Ann Eugen 7:179
33. Zupan J, Gasteiger J (1999) Neural networks in chemistry and drug design. Wiley-VCH, Weinheim
34. Cristianini N, Shawe-Taylor J (2000) An Introduction to Support Vector Machines and other kernel-based learning methods. Cambridge University Press, UK
35. Breiman L (2001) Random forests. Mach Learn 45:5–32
36. Ghosh S, Ojha PK, Roy K (2019) Exploring QSPR modeling for adsorption of hazardous synthetic organic chemicals (SOCs) by SWCNTs. Chemosphere 228:545–555
37. Lata S (2018) Concentration dependent adsorption of aromatic organic compounds by SWCNTs: Quantum-mechanical descriptors for nano-toxicological studies of biomolecules and agrochemicals. J Mol Graph Model 85:232–241
38. Wang Y, Chen J, TangW XD, Liang Y, Li X (2019) Modeling adsorption of organic pollutants onto single-walled carbon nanotubes with theoretical molecular descriptors using MLR and SVM algorithms. Chemosphere 214:79–84
39. Liu Y, Zhang J, Chen X, Zheng J, Wang G, Liang G (2014) Insights into the adsorption of simple benzene derivatives on carbon nanotubes. RSC Adv 4:58036–58046
40. Ahmadi S, Akbari A (2018) Prediction of the adsorption coefficients of some aromatic compounds on multi-wall carbon nanotubes by the Monte Carlo method. SAR QSAR Environ Res 29:895–909
41. http://www.insilico.eu/coral
42. Wang Y, Yan F, Jia Q, Wang Q (2017) Assessment for multi-endpoint values of carbon nanotubes: quantitative nanostructureproperty relationship modeling with norm indexes. J Mol Liq 248:399–405
43. Rahimi-Nasrabadi M, Akhoondi R, Pourmortazavi SM, Ahmadi F (2015) Predicting adsorption of aromatic compounds by carbon nanotubes based on quantitative structure property relationship principles. J Mol Struct 1099:510–515
44. Heidari A, Fatemi MH (2017) A theoretical approach to model and predict the adsorption coefficients of some small aromatic molecules on carbon nanotube. JCCS 64:289–295
45. Hassanzadeh Z, Kompany-Zareh M, Ghavami R, Gholami S, Malek-Khatabi A (2015) Combining radial basis function neural network with genetic algorithm to QSPR modeling of adsorption on multi-walled carbon nanotubes surface. J Mol Struct 1098:191–198
46. Wang QL, Apul OG, Xuan P, Luo F, Karanfil T (2013) Development of a 3D QSPR model for adsorption of aromatic compounds by carbon nanotubes: comparison of multiple linear regression, artificial neural network and support vector machine. RSC Adv 3:23924–23934
47. Roy J, Ghosh S, Ojha PK, Roy K (2019) Predictive quantitative structure–property relationship (QSPR) modeling for adsorption of organic pollutants by carbon nanotubes (CNTs). Environ Sci Nano 6:224–247
48. Kar S, Ghosh S, Leszczynski J (2019) Is clay-polycation adsorbent future of the greener society? in silico modeling approach with comprehensive virtual screening. Chemosphere 220:1108–1117
49. Yangali-Quintanilla V, Sadmani A, McConville M, Kennedy M, Amy G (2010) A QSAR model for predicting rejection of emerging contaminants (pharmaceuticals, endocrine disruptors) by nanofiltration membranes. Water Res 44:373–384

50. de Ridder DJ, Villacorte L, Verliefde ARD, Verberk JQJC, Heijman SGJ, Amy GL, van Dijk JC (2010) Modeling equilibrium adsorption of organic micropollutants onto activated carbon. Water Res 44:3077–3086
51. Zhang K, Zhong S, Zhang H (2020) Predicting aqueous adsorption of organic compounds onto biochars, carbon nanotubes, granular activated carbons, and resins with machine learning. Environ Sci Technol 54:7008–7018

Chapter 4
How Environmental Chemicals of Concern Emerge: ILs and HNTs

William M. Nelson

Abstract Chemicals of concern may emerge in our environment in many ways. Two distinct ways are through the design of the compound or by molecules/functionalities added to the original compound. In this chapter, we will use ionic liquids (ILs) and halloysite nanotubes (HNTs) to illustrate these two ways. ILs are salts that are typically composed of organic cations and inorganic anions. While they were initially promoted as "green" solvents, this characterization is not necessarily true. The potential number and uses for ILs seem almost limitless. HNTs are a form of naturally occurring clay, possessing mostly a nanotubular structure and a similar chemical composition to kaolin. The tubular morphology and the aluminum silicate structure of HNTs have allowed them extensive use in science, engineering, and medicine. The growth of the biorefinery and green chemistry, circular chemistry, and sustainability will provide increasing access of these compounds into the environment. This extended use of both ILs and HNTs increases their presence in the environment. While the ILs which may be toxic depending upon their composition, HNTs may become toxic due to what is loaded in/on them. While the utility of these chemicals is unquestionable, understanding their environmental behavior, fate, and effects is critical.

4.1 Introduction

Ionic liquids (ILs) and halloysite nanotubes (HNTs) can be used to illustrate two ways in which environmental chemicals of concern (ECC) emerge or become more prevalent in the environment. These classes of compounds are radically different: ILs are touted as designer compounds, while HNTs are basically nanoscale clay containers that can be modified. Both classes of chemicals are receiving increasing interest (and, therefore, usage.) Due to their chemical properties and wide range of

W. M. Nelson (✉)
Engineer Research and Development Center, Environmental Laboratory,
US Army Corps of Engineers, 3909 Halls Ferry Road, Vicksburg, MS 39180, USA
e-mail: William.M.Nelson@usace.army.mil

M. Shukla et al. (eds.), *Emerging Materials and Environment*,
Challenges and Advances in Computational Chemistry and Physics 37,
https://doi.org/10.1007/978-3-031-39470-6_4

possible modifications/derivatizations, both ILs and HNTs are finding application in an increasing variety of processes. This chapter will introduce these chemicals and review some of the current areas of research and potential applications, while showing how they become ECC.

ILs are salts which consist of anions and cations with low melting points, often producing liquids at room temperature with negligible vapor pressure. These component anions and cations can be designed to produce the desired characteristics of the resulting salt. The current literature on the compounds exceed 110,000 articles (2022, Scopus).

HNTs are naturally occurring aluminosilicate clay related to kaolinite. The individual clay compounds often possess a hollow tubular structure. The clay is found globally and is mined from natural deposits in countries such as the USA, China, New Zealand, France, and Belgium. The HNT structure allows physical entrapment of chemicals and/or chemical modification of the inside or outside of the nanotube's surfaces. Analogous to IL, the types of entrapment molecules and extent of chemical modification are increasing and current literature on the compounds exceed 6,000 articles (2022, Scopus).

4.2 Ionic Liquids (ILs)

Solvent use is ubiquitous and poses threats to both human and environmental health. A solvent is a substance that dissolves a solute, resulting in a solution. The solutions can be found in research (e.g., chemical synthesis), in industry (e.g., purification processes), or in commercial products (e.g., paints, paint removers, inks, and dry cleaning). A solvent is usually a liquid but can also be a solid, a gas, or a supercritical fluid. ILs, especially salts that are liquid at room temperature, can be useful solvents in all of these applications. ILs, through extensive research efforts, can now find uses beyond solvents (e.g., catalysts, medicine, and environmental applications.)

The reduction of solvent use, or the use of more environmentally benign solvents, has been one of the ways to reduce the amount and/or toxicity of chemicals entering the environment. The appropriate design and/or selection of a solvent for a process can greatly improve the efficiency of a chemical production process. ILs provide potential solutions in this area. As the commercial performance of ILs improve, and their usage increases, these chemicals will also have a greater likelihood of becoming an ECC [1].

4.2.1 IL Overview

ILs are liquid salts with a wide range of potential properties. The nature of the elements, functional groups composing the salt, and their chemical arrangements in the ions (in either the anion or cation) will strongly affect the resulting properties. Over the past four decades, ILs have been prepared and have had impacts in science and technologies. They have been used in academic and industrial applications, and their applications go far beyond use as solvents. Initially, their very low vapor pressure made them potential candidates for relieving some of the major challenging facing society as clean, efficient, and eco-friendly alternative replacements for volatile organic chemicals and solvents. However, their chemical nature and numerous possible designs have led to other applications. The yields from reactions conducted in ILs at large scales are promising, but questions arise concerning the recovery of these compounds and their regeneration. Furthermore, the steps involved in their synthesis and their toxicity raise questions about these compounds. While they are still referred as ILs, these ionic compounds may not be liquid at room temperature and may have value as ionic chemicals due to their properties. The potential applications of ILs in electro-chemistry, solvent, engineering, catalysis, biological aid, physical chemistry, analysis, and many more are indicative of how broadly their use could impact the environment [2]. This will be further elaborated in the ensuing sections.

The increased interest in the development and use of ILs is measured in several ways. Total publishing activity (mentioned earlier) and patent count serve to mark research activity, industrial usage, and toxicological knowledge of ILs. The volume of these measurements provides the clearest measure of the impact of ILs, and they also provide information on their ready availability. The chemical structure of ILs reveals that existing knowledge of the toxicity characteristics of the anions/cations will provide insight into the toxicity of the salt. This information should be used to guide the design and use of greener, less toxic, and more benign ILs [3].

Outside of negligible vapor pressure, the chemical characteristics of these salts will depend heavily upon the component ions. The term "ionic liquid" is very broad, and multiple structures can be converted into ILs, which introduces complexity and diversity to IL structure and properties. Novel ILs may contain more than one cation or anion. There is no single model that could be comprehensive enough to describe the entire family of ILs. For example, there are magnetic ionic liquids (MILs), polymeric ILs (PILs), and energetic ILs (EILs).

4.2.2 How Do ILs Become ECCs?

While the utility of these chemicals is unquestionable, understanding their environmental behavior, fate, and effects is critical. As solvents, ILs can reduce the effects of other coexisting contaminants by serving as important solvent substitutes and providing techniques that might be adopted for the removal of toxic compounds from the environment [4]; their own environmental impact must be studied. When ILs are being used for any chemical purpose, they must be regarded as environmental contaminants when released. As the use of ILs has expanded in recent years, more studies are appearing about their environmental impact. Since these compounds are salts of varying size and complexity, a way to begin the assessment of these chemicals is through environmental life cycle assessment (LCA) studies on their component ions [5]. However, there is a shortage of the characterization factors of ILs for human toxicity and ecotoxicity impact categories. Therefore, complete life cycle inventory data for ILs and precursor chemicals are essential for inventory analysis; and the LCIA needs to be clearly defined about the level of detail on the release of ILs into the environment.

Where in the ecosystem will ILs appear as ECC? While ILs can be designed as solvents or active chemicals to dissolve most solutes, they themselves can also be engineered to be soluble in other media (from water to organic media to other solvents). Given that, these compounds can potentially be found in any environmental sector. Toxicological research on microalgae and cyanobacteria can help to establish a solid foundation for aquatic ecotoxicological assessments [4]. New information on emerging pollutants such as ionic liquids is gained by studying the toxicological effects of pollutants on microalgae and cyanobacteria. A significant area where information is lacking is in environmental lifetimes of types of ILs.

A second way to approach the environmental effects of ILs is to integrate the most recent and pertinent data available on the (bio)degradability and toxicity of ionic liquids, and then couple this information with an analysis on the conscious use of these compounds on a large scale [4]. The integrated data should focus on the recognition of toxicophores in the individual ILs, and on the way, the ILs will be used. This information will shed light on probably areas where ILs will emerge as ECC.

4.2.3 Examples of Use of ILs

In addition to being used as solvents in chemical processes and biomass refining ILs are employed as auxiliaries and catalysts in chemical synthesis. They are used in analytical equipment. They make up electrolytes in lithium-ion batteries, supercapacitors, and metal plating baths. They can be found as lubricants and coolants. This range of applications of ILs shows both its value and the potential for its appearance in the environment. The examples that follow will highlight both sides of this issue.

4.2.3.1 Environmental Remediation via Extraction

Gas extraction: In the fuel industry, sulfur-containing compounds (SCC) are a growing issue with standards becoming more stringent. Sulfur leads to the release of sulfur dioxide, during the combustion process. The capture of SO_2 from flue gases, while the most common solution, is a remediation process that will allow businesses to keep their emissions within limits. ILs are a useful technology here. By acting as an acidic gas absorbent for removal of SO_2, IL-mediated SO_2 capture provides promising method for better SO_2 absorption capacity and a greener technology [6].

The green improvements result from the ease to regenerate the absorbing medium, which also proves to be more economical. Similarly, ILs are also a promising alternative for H_2S capture, a greenhouse gas. Through the proper design of ionic liquid anions, cations, and functional groups on the H_2S absorption medium, separation and oxidation are both improved and made more economical [7]. New strategies to maximize the extractive desulfurization rely on the catalyzed breakdown of the captured sulfur-containing species [5]. Lastly, ILs can be incorporated into supported ionic liquid membranes, polymer/ionic liquid composite membranes, gelled ionic liquid membranes, and poly(ionic liquid)-based membranes. These membranes and lattices can be used in environmental cleanups and separations [8]. An important question in this technology is the lifetime of the catalysts and the final disposal.

Poly(ionic liquid)s (PILs): PILs are a subclass of ionic polyelectrolytes with broad range of structural and functional properties [9]. These extended electrolytes have unique energy transfer characteristics and electron transfer properties. Stimuli-responsive PILs have applications in systems, due to their smart responsive behavior toward different chemical, physical, and biochemical stimuli such as pH, redox, CO_2, temperature, light, and enzymes. They can be used as switches, batteries, and signaling. The pH-responsive PILs have potential applications in the biomedical area including therapy, drug delivery, and diagnostics. CO_2-responsive PILs are useful in environmental monitoring of this gas.

Metal recovery: The importance of precious metals due to batteries, communication, and computers underscores the need for reliable isolation, recovery, and purification of metals. Hydrometallurgy is an efficient way of recovering metal ions from mineral ores and waste where the most important steps are the leaching of the minerals and the recovery and purification of metal ions by solvent extraction (SX). ILs are being extensively applied in this process technology with the aim to make it environmentally friendly by replacing organic volatile solvents. The ILs can be reutilized due to their negligible vapor pressure. While promising, this technology still requires further improvements before it can be employed on an industrial scale [10]. Additionally, consideration for the end-of-lifetime disposal of these ILs is critical.

Contaminant extraction: Related to metal extraction is the environmental need for extraction of contaminants. The contaminants differ, so the ability to design specific extraction templates involving ILs is possible. ILs are useful in the microextraction of pesticide residues in fruits and vegetables. The advantages and limitations of current applications of these ILs in microextraction techniques must be carefully weighed [7]. The unique physicochemical properties of ionic liquids make them ideal sorbents in separation techniques, but present challenges for IL recoveries. Further, because they can trap chemicals due to their chemical functional groups, these same groups might indiscriminately interact with environmental chemicals. The number of applications in analytical separation has increased considerably over the past 5 years, which also increases the likelihood of their appearance in the environment.

The need for water to ensure sustainability in the twenty-first century is critical. Efficient technologies to address water scarcity through the treatment of wastewater to facilitate its reuse are needed. ILs have been applied to many industrial and analytical separation processes, but their applications in the wastewater treatment, especially in the removal of organic pollutants, can be invaluable here [2]. The potential applications of ILs include solvent extraction, solvent membrane technologies, and ionic liquid-modified materials that are mainly used as adsorbents. Implementation of IL-based technologies for wastewater treatment has many challenges, including the selection of non-hazardous ionic liquids, technological applications, high testing requirements for individual uses and scaling-up of the entire pollutant removal, disposal, and IL regeneration process. The studies necessary to provide this information are currently ongoing, but there remains a lot of work. Incorporating recovery as one of the design criteria for the ILs brings this into the new emphases on circular chemistry [11].

4.2.3.2 ILs in Syntheses

Synthesis and synthetic chemistry provide paths to both use and design ILs. Initially, the use of ILs in synthesis seemed unlimited, as they were called "designer" solvents because of unlimited potential for design of their physical, chemical, and even biological properties by selection of cation, anion, and their associated functional groups during the synthetic reactions [12]. It was found that ILs may affect the kinetics of chemical reactions proceeding in these solvents. As a direct result of environmental concerns and sustainability, designing ILs for the environment must also be a critical criterion.

Chemical reactions: One of the most exciting discoveries early in the work with ILs was that the chemical properties of ILs could greatly influence the chemical reactions conducted in them. Specific interactions between components of ILs resulted in relatively high viscosity of this media and caused microviscosity changes in the close vicinity of reacting species. This resulted in greater catalytic conversions and heightened specificity. Specific non-directional interactions such as electrostatic or dispersion forces and directional interactions, such as hydrogen bonding between components of ILs and reactants, affected both translational and rotational mobility of reaction species as well as activation parameters [13]. The challenge continues to be how to direct these effects toward better and more efficient reactions.

As an example of how ILs effect chemical reactions can lead to remediation technologies for the environment, it is shown that alkylation of isoalkane and alkene are important techniques for octane chemical conversions. Widely adopted catalysts (H_2SO_4 and HF) for alkylation at industrial level possess environmental concerns, are difficult in recycling, and are corrosive. If ILs can serve as green catalysts for obtaining alkylates, and various ILs are effective for alkylation reactions, then this could provide a path to remove some of the more harsh chemical procedures [14]. More is needed to be learned about the chemistry involved in this process.

In traditional organic/synthetic chemistry procedures, the use of simple, efficient, and inexpensive catalysts is highly desirable. The development of efficient and environmentally benign synthetic protocols involving the catalytic activity of ILs has attracted increasing interest in modern organic syntheses. This might provide a solution to the growing concern over the environment. ILs have shown their usefulness as environmentally benign sustainable alternative to the conventional organic solvents in synthetic chemistry, as they increase efficiency of organic transformations [15]. It will be important to continue to evaluate ILs both as solvents and as catalysts for future reactions.

4.2.3.3 ILs in Industry

Textiles: The wastewater found in the textile industry contains dyes that result in downstream pollution and socioeconomic issues. Photocatalysis has been an efficient remediation technique used to address this issue. Newer developments in photocatalyst immobilization and photocatalytic reactor design, coupled with methods such as microwave irradiation, sonication incorporate ILs as the reaction medium and catalyst. These novel applications are a valuable asset for the textile industry [6]. Ecotoxicological studies are required in real environments to validate the potential applications of the designed nanostructured doped photocatalysts.

As the previous paragraph illustrated, the use of ILs can assist the textile industry (and any major industry facing similar waste-stream issues) as it attempts to reduce its environmental impact by creating, designing, and developing novel and sustainable industrial processes, more aligned with ecology. A valuable way this is accomplished is by the substitution of various textile wet processes by new cleaner processes using ILs [16].

Metalworking: A cutting fluid is a type of coolant or lubricant designed specifically for metalworking processes, such as machining and stamping. These fluids typically include oils, oil–water emulsions, pastes, gels, aerosols (mists), and air or other gases. Cutting fluids are made from petroleum distillates, animal fats, plant oils, water and air, or other raw ingredients. These fluids can be used as cutting fluids, cutting oils, cutting compounds, coolants, or lubricants. When these are spent and discharged, contaminated effluents can pollute the water, air and soil. ILs are being developed for use as replacement fluids in these cooling technologies [17].

The metal working industry is large, and achieving sustainability in machining processes is critical. ILs can be synthesized for use as lubricants that have the potential to replace the mineral oils in machining. ILs can also be a potential additive aiding better performance in terms of cutting force, tool wear, and surface roughness [8]. As the use of ILs increase, however, they will appear in greater concentration in the environment [18].

Agriculture: Loss and waste of food in the agri-food industry is a matter of great concern, causing negative economic, social, and environmental impacts. The wine, fruit juice, and vegetable oil-processing industries generate significant amounts of wastes and side streams containing potentially valuable bioactive compounds. ILs, through extraction processes, can lead to the recovery of these valuable wastes. This is accomplished by the design of ILs to recover desirable compounds from residual matrices. The recovered materials have further applications in food, pharmaceutical, and cosmetic industries. Mostly, this is accomplished with liquid–liquid extraction of phenolic compounds from liquid agri-food matrices [19].

Beyond extractions, ILs can be used as reaction solvents in agriculture. The manufacture of green materials using natural raw materials has gained extensive interest, due to its renewability, biodegradability, abundance, and low cost. ILs can be used to process starch, which can replace synthetic polymers. Replacing volatile organic solvents in polysaccharide processing shows that there are more possibilities for ILs in starch dissolution, gelatinization, modification and plasticization, and their industrial applications [13].

A final example involves the design of ILs bearing bioactive components. These can be used as chemical tools in many agricultural applications, as well as in the preservation of agricultural products [20]. Appropriately designed ILs can be used as bactericides, fungicides, herbicides, wood preservation agents, plants stimulants, and growth regulators. Their possible advantages in comparison with conventional, presently used formulations will have to be weighed based upon their environmental and human toxicity characteristics. While these compounds have unique properties and serve many functions, more needs to be known about their biodegradability, toxicity, and the methods for their determination and prediction.

4.2.4 Biorefinery and Green Chemistry

Using the tools emerging from green chemistry, the biorefinery converts biomass to energy and other beneficial byproducts. Biorefining uses the sustainable processing of biomass to provide a spectrum of biobased products. One of the principles of sustainable development is to replace polluting petroleum polymers with alternatives derived from renewable materials; these efforts are important and are expanding within the biorefinery. ILs have emerged as an ideal solvent, catalyst, or additive to process many biopolymers, as their chemical structure presents challenges to routine industrial chemical methods [21].

Protein-rich biomass provides a valuable feedstock for the emerging biomass-chemical industry [12]. In order to fully utilize these new streams of raw materials, new methods are needed for the extraction of proteins from biomass, which will become raw materials for future reactions. Moreover, the development of physical and (bio)chemical technologies for synthesis of polymers, processing of commodity chemicals, development of pharmaceuticals, and production of other fine chemicals will be necessary for the biorefinery to replace the current petroleum-based products. One of the targets using these biomass products will be the synthesis of fine chemicals and ionic liquids. Once again, increased usage will potentially generate ECC.

4.2.4.1 Biorefinery (BF)

Biorefineries are refineries that can provide multiple chemicals by fractioning biomass into multiple intermediates (carbohydrates, proteins, triglycerides) that can be further converted into desired products. The use of biomass as feedstock can provide a benefit by being a renewable resource, reducing the impacts on the environment, and lowering pollutants emissions and reducing the emissions of hazard products [22]. While the intent of biorefineries includes supplying the current fuels and chemical building blocks, the actual processes it uses must be evaluated for the effects of their processes/products on the environment.

Because of the efficacy of ILs in pretreatment and production of renewable or platform chemicals from biomass, they will be used in many BF processes [14]. At this juncture, it will be important to design and synthesize more efficient, low cost and less toxic ionic liquids that will accomplish the goals of the biorefinery. This will also include assessment of ILs in how they are used in the biorefinery, their stability, recycling, and environmental hazards.

4.2.4.2 Green Chemistry (GC)

Sustainable and green chemistry (GC) are a different way of practicing chemistry and chemical engineering. This approach involves the design, development, and implementation of chemical products and processes that have minimal effects on the environment. The goal of this effort is to protect and benefit the economy, humans, and the environment by designing and implementing ways to preserve resources, conserve energy, and discover replacements for hazardous substances. ILs emerged from efforts in GC, and there are still many areas of active research involving ILs. Current work ranges from novel ILs to their potential use in research and industry. What makes an IL "green" is its use in context [23]. None of the mentioned properties of ILs necessarily result in clean, or green processes. Green comes from increasing the efficiency, reducing waste and losses of products/processes.

Within GC, there are increasing numbers of studies on the design and evaluation of functional ILs. They have potential application as electrolyte solution substituents not only for batteries but also other electrochemical devices. Also, the use of ILs as solvents for biomolecules, especially for dissolution of proteins, is increasing. Zwitterionic ILs and polymerized ILs have also seen greater use [24]. Recent research on hydrogen synthesis by dark fermentation of galactose from macroalgal biomass involving IL-assisted catalysis extends the utility of ILs to help achieve a maximum recovery of sugars while detoxifying the byproducts [20]. This is critical work in the biorefinery

The combination of ionic liquids (ILs) and nanotechnologies gives more attractive and efficient possibilities for further industrially applicable nanomaterials. Nanotechnology manipulates matter on an atomic, molecular, and supramolecular scale and is a technology that will enable enormous changes in today's industries and markets while potentially creating whole new industries ranging from sensors to catalysis. Using ILs in nanocomposites would facilitate an interaction of molecules and could also help in electron transfer reactions. New materials can be used as electrosensors, catalysts, electronics, and photonics. The advantages of using ILs in nanomaterials include improved thermal and electrochemical stability, controlled particle size, increased conductivity, reduced toxicity, and increased reactivity in an eco-friendly way. In essence, ILs can enhance the properties of existing materials, thereby enabling a wider application area for nanotechnology [14]. This increased use also opens up more ways for ILs to become ECC [25].

A final example involves biocompatible ionic liquids (Bio-ILs). These ILs are potentially useful in applications ranging from the electrochemical to the biomedical fields. The most promising strategies for their synthesis involve using molecules from biorenewable sources as a basis for both the anionic and cationic components of the Bio-ILs structure. If successful, the resultant ILs would be "accepted" in biological systems. This strategy bodes well for reducing the environmental toxicity of these ILs. Several studies have shown reduced impact on the environment and improved health safety [26]. There appears to be the potential for great impact on ensuring the safety of these materials.

4.2.5 Medicine

The application of ILs in the fields of medicine clearly provides instances of designing molecules for the environment. ILs with their tunable and remarkable properties can be candidates in multiple domains such as biotechnology, therapeutics as well as pharmaceutical sciences. As more studies emerge for uses in this area, the classification, properties, and toxicity of ILs will also improve. Their use in medical fields will illustrate the types of biological activity shown by ILs, which includes antimicrobial and anticancer properties. Key to this is the identification of the Active Pharmaceutical Ingredient-Ionic Liquids (API-ILs) for explaining the medical effects of these designer ILs. This new area will greatly assist the development of ILs that possess useful potential in the field of medicine, biology, and chemistry [27]. For example, microbiological activities of ILs have applications in medicine and pharmacology [28]. The biological activity of ILs including their antimicrobial properties (depending on the type of a cation or an anion) has a vast array of medical applications. Some ILs with antibacterial efficacy can be considered for use as topical and general medications and may even find applications in the treatment of wounds and infections [29].

4.3 Halloysite Nanotubes (HNT)

Halloysite nanotubes (HNTs) are types of naturally occurring clays with similar chemical composition to kaolin. Its nanoscale interior cavity from its tubular morphology and has hydroxyl groups on its surfaces, which allows derivatives to be prepared. While not as numerous as ILs, many applications have been discovered for this unique, cheap, and abundant clay [30]. It is suitable for a number of interesting applications [31]. The functionalized halloysite nanotubes can be employed as reinforcing fillers for polymers as well as carriers for the sustained release of active molecules, such as antioxidants, flame-retardants, corrosion inhibitors, biocides, and drugs. While the basic clay is non-toxic, derivatization can alter its health and environmental toxicity [32].

The halloysite nanotubes (HNTs) are found in many areas of the world, and due to their non-hazardous nature and easy availability, they are used in various fields of science. To increase the utility and applicability of halloysite nanotubes, they are functionalized by different modification methods such as covalent functionalization and non-covalent functionalization [33]. The HNTs are carriers with attached functional groups. Since the attached/entrapped functional groups determine the toxicity of the composite, this represents a different way by which these compounds become ECC, compared to ILs.

The basic aluminum silicate composition of HNTs is basically inert and non-toxic. Due to its high mechanical strength, thermal stability, biocompatibility, and abundance, HNTs can have many potential applications in polymer nanocomposites. Polymer nanocomposites reinforced with HNTs possess highly increased tensile and flexural strength, elastic moduli, and improved toughness. HNTs-polymer nanocomposites can also exhibit elevated thermal resistance, flame retardancy, and unique crystallization behavior [34]. The presence of the tubular microstructure, the hydroxy groups, and the silicon oxide moieties allow HNTs modification so that they can and function as chemical/drug carriers. HNTs have demonstrated good drug encapsulation and sustained release abilities, allowing them to be used as tissue engineering scaffolds and drug carriers. Finally, the ability to modify HNT and use them as polymer nanocomposites opens potential applications in high performance composites for aircraft/automobile industries, environmental protection, and biomaterials.

4.3.1 How Do HNTs Become ECC?

HNTs are a type of kaolinite clay and are non-toxic. However, the process of modifying them and using the modified materials will affect its environmental toxicity and perhaps transform them into an ECC. A recent article measures how this occurs [35]. Alginate-based nanocomposite beads have been used as adsorbents for heavy metal removal from aqueous solutions. Simply, they are effective and are relatively inexpensive. They can be regenerated after use, so they are also sustainable. Authors incorporated HNTs into blank alginate beads, then measured the environmental impacts, and the impacts of acid regeneration followed by reuse of these beads. Using an LCA on alginate-HNTs nanocomposite beads for the removal of lead(II) from aqueous solutions, they provided a measure of the extent that composites become ECC. The incorporation of HNTs into blank alginate beads improves the mechanical stability of the absorbent system, but had only a small effect on the environment. The real effect was felt in the production of the beads, which required a large quantity of HCl to be used for the extraction of alginate from seaweed, followed by truck transportation of the beads. The regeneration and reuse of the spent beads created even more burdens over the entire life cycle. This is a typical analysis and result for HNTs composites.

4.3.2 Environmental Applications

HNTs can be used in a various industries. As stated above, HNTs are a promising material for drug delivery, as polymer filler, as matrix for photocatalysts and adsorbents for environmental and biomedical applications. Potential halloysite/metal oxides are excellent adsorbents and due to their magnetic properties can be easily separated from the treated solutions by using of external magnetic field. The presence of aluminol and siloxane surface groups on the HNTs allow the encapsulation of the magnetic materials. Such composite systems have the capacity for high adsorption and can be used for effective water purification from inorganic and organic pollutants [36]. The electrostatically charged lumens of HNTs enable the encapsulation and sustained release of active agents into the surrounding environment via diffusion processes. Furthermore, the large surface areas and high aspect ratio of HNTs greatly enhance the possible crosslinking and load-bearing capacities of these HNTs, while also providing accessible sites that facilitate environmental attachment and proliferation [37].

4.3.3 Chemistry

The basic chemistry of HNTs is simple. The aluminum silicate backbone of these clays is useful as a carrier for more reactive functional groups, or as inorganic reinforcing materials for polymers. Loading the HNTs' approximately 15-nm diameter lumens with chemical agents, including bioactive molecules (self-healing, anticorrosion, antimicrobial agents, proteins, DNA, drugs, etc.), and then doping them into polymers allows a controlled sustained release, providing these nanocomposites with new smart properties [38].

The range of chemicals for inclusion in is limited only by size. The lumen and exterior surfaces can be separately modified. Chemically active agents, such as corrosion inhibitors, antifouling agents, are loaded into/onto the HNTs and then doped into plastics producing smart polymeric nanocomposites with improved mechanical strength from the clay as well as added chemical activity [39]. HNTs, as a natural nanotubular material, can also be used as excellent additives for biomaterials. Besides, nanotubular structure, surface charge, and adsorption properties provide favorable conditions for separation and enrichment of metal ions, dyes, organic compounds, and so on [40].

4.3.3.1 Protective Coatings

When added to coating materials and loaded with appropriate chemicals, functionalized HNTs can afford protection from corrosion. The coating efficiency is generally not sufficient for applications in aggressive and extreme environments, but it can provide a low-level, continuous protection. The addition/embedment of corrosion inhibitors within protective coatings enhances the metal protection of these coatings. In this application, HNTs are used to deliver over time corrosion inhibitors. The net result is a technology that can extend the lifetime of the surface [41].

The basis of self-healing materials also involves the release of chemicals on demand. The stimulus for the chemical release is a break or sheer in the system. The possibility to accomplish this has generated enormous scientific attention, as they offer wide range of applications in conjunction with long-lasting performance. As an example, self-healing polymers are desirable in construction materials, automotive parts, electrical encapsulation, adhesives, coatings, and other areas. A simple example of this technology is found in epoxy resins. While epoxy resin coatings are widely used in commercial vehicles and high compact aircraft, their susceptibility toward scratches and microcracks is a major concern. Inhibitor and polymeric healing agents encapsulated in HNTs for the self-healing epoxy coating and contained within halloysite nanotubes have been utilized as reservoirs to store the self-healing agents [42].

4.3.3.2 Excipients

Due to the high aspect ratio of HNTs and their size, HNTs can be used as excellent reinforcement materials for some polymers. The chemical interactions between the HNTs and polymers can be tailored using differing surface modification of the HNT. The resulting mechanical properties of the polymers can be adjusted according to the strength, modulus, stiffness, and impact resistance of polymers [43]. The possibility of accomplishing this with differing polymers opens the possibility for many uses in differing fields. Synthetic polymers, natural polymers, or any material with these components are potential starting materials with either industrial or biocompatibility properties. The fabrication of poly(vinyl alcohol-HNT (PVA-HNT) nanocomposites, as an example, can be a potential way to address some of PVA's limitations. The properties of the resulting PVA-HNT nanocomposite currently has application in medicinal and biomedical areas, such as wound dressings, drug delivery, targeted-tissue transportation systems, soft biomaterial implants, and batteries [44].

A second class of polymers where HNTs are finding use is in rubber. Natural rubber is widely used and may be found in numerous commercial commodities or applications (e.g., tires, dampers, suspension elements). The composition of rubber, however, is flammable, so it is necessary to add excipients that will dampen combustion. By adding flame-retardants to the lumen of HNTs, mixing the HNT composites into the rubber formula and allowing the chemicals to be gradually released, the rubber acquires flame retardancy. As is the case with other polymers, the added HNT composites simultaneously add chemical functionality and enhance the mechanical sturdiness of the rubber [45].

4.3.3.3 Biopolymers

Earlier we discussed the emerging importance of the biorefinery and green chemistry. HNTs find utility in both GC and BF. The use of HNTs to enhance biopolymers shows points of intersection with ILs. More importantly, it provides insight as to where and how HNTs may become ECCs. Biopolymer use is expected to increase within green chemistry and the biorefinery, as they will have reduced human and environmental toxicity. The challenge will become how to stabilize them and disperse them in reactions; this becomes one of the functions of HNTs. Work on the adsorption of biopolymers onto HNTs shows results that are promising [46].

One example of the use of biopolymers adhering to HNTs and reinforcing the strength of the biopolymers is in tissue substitutes. There has been a significant progress in the fabrication of natural and synthetic polymer composites with nanoscale dopants. HNTs as a nanofiller significantly enhance the compressive strength, compressive modulus, and thermal stability of polymers. As they do with rubber, these nanotubular clay minerals reinforce polymer matrixes [47]. In polymer engineering, different research studies utilize HNTs that exhibit a beneficial enhancement in the properties of polymer-based nanocomposites. HNTs can reinforce polysaccharides and provide improved mechanical, thermal, and swelling properties [48]. A secondary but important result is that these HNT composites are biocompatible, showing reduced toxicity.

4.3.3.4 Catalysts

As previously mentioned, HNTs have two distinctly different surfaces that can be chemically modified. The inner lumen (with hydroxy groups) is more electropositive relative to the outer surface (with silicon oxide groups). Both surfaces can be modified, separately and/or simultaneously. If halloysite is used to support metal nanoparticles, the resulting compound can be used for catalytic applications with tunable properties. The morphology of HNTs both allows the dispersion of the catalytic composites and allows access to the catalytic sites. Even more interesting is the ability of the lumen to be used as a confined reaction environment [49].

4.3.4 Examples of Applications

Chemical Adsorbents: Adsorption on to HNTs is a very promising and efficient technology for the removal of hazardous pollutants from water. In order to accomplish this, the HNTs must be modified with the appropriate functional groups. HNTs can be an efficient adsorbent for water treatment, as shown from many adsorption studies using functionalized HNT materials. Properties of the adsorbent can be tuned by surface modification of the HNT for an optimized performance. The modified HNT can be modified in ways to accommodate solution pH, pollutant concentration, contact time, temperature, nature, and dose of adsorbent [50].

HNTs can also be used as efficient removal agents for both aliphatic and aromatic hydrocarbons. Fabrication of sustainable composite HNTs with decontamination capacity toward inorganic and organic pollutants is the way in which HNT can be used in this area. There are hybrid systems based on clay HNTs and organic molecules that are used as efficient removal agents for both aliphatic and aromatic hydrocarbons. Due to their high specific surface area, these modified HNTs can be used as fillers in composite membranes with excellent filtration capacity. The preparation of composite gel beads based on biopolymers (alginate and pectin) and HNTs will also have adsorption capacities for heavy metals and organic dyes. Tubular inorganic micelles can be prepared by the functionalization of the halloysite cavity with anionic surfactants [51].

As an example, L-Dopa-functionalized HNTs were prepared the reaction of L-dopa with HNTs in a weak alkaline solution. Different concentrations of L-dopa coated HNTs were mixed with cellulose acetate to prepare enhanced performance ultrafiltration membranes. The differing membranes exhibited excellent antifouling performance [52].

Water desalination: The increasing global demand for potable and usable water has increased the research on production of water. The development of thin-film nanocomposite membranes utilizing HNT offers a new technology for desalination. The derivatized HNTs are used as effective filters. These nanoparticles are effective in terms of water flux, salt rejection, chlorine resistance, and antifouling properties of HNT membranes. When compared to the typical thin-film composite membranes, they perform better. While still early, it is evident that the treated HNTs possess properties which advance the potential for desalination [52].

Sensors: One of the primary methods for developing functional HNTs is through loading their lumen with desired active molecules. The modified HNTs are able to act as a carrier for "active" nanomaterials. These materials are either dispersed or can function within the HNT. If they function within the HNT, the clay material "distributes" them and then they can be function according to their intended use.

For example, electrochemical sensing applications are possible in HNT due to the high surface area of the HNT. The free lumen in HNTs can be utilized for hosting compounds for obtaining nanostructures with excellent synergistic properties [53].

Energy storage: The development of novel materials and approaches for effective energy consumption and the development of renewable energy systems is also an important subject in modern goals of a twenty-first century sustainable society. Research focused on the effective harvesting and storage of solar energy for various applications has been identified as an area of interest [53]. The possibility of encapsulating phase change materials (PCMs), known to be able to store thermal energy of the sunlight due to adsorption and release of latent heat, is a promising to accomplish this goal. The loading in natural porous or layered clay HNTs appears to be a simple and cost-effective method of encapsulation that significantly improves the shape and cyclic sustainability of PCMs. Further modification of these HNTs would allow their improved physical properties [54].

Ionogels: One last application is the area of ionogels. Applications in flexible electronics require the unique structures of gelatinous ionic media. Ionogels provide mechanical properties, electrical conductivity, and thermal stability. These new ionogels are being developed by combining the fields of ILs and HNTs [55].

4.3.5 Medicine

The use of HNTs in medicine provides two types of information: medical results and information on HNT toxicity. Halloysite is non-toxic and is abundantly available from natural deposit material with minimal processing required [54]. Natural HNTs are like miniature vehicles for carrying and releasing drugs and proteins. The release time required to release trapped molecules may be tuned from 10 to 200 h by coating the tube with surface polymers. These HNTs (pristine or drug-loaded) are mixable with polar and low-polar polymers, allowing for functional biocomposites with enhanced mechanical strength, adhesivity, and slow release of drugs or other chemical agents. The chemistry of the inner and outer halloysite surface allows for selective drug adsorption inside or outside the clay nanotubes. This also allows the design for a time release over 1–2 weeks or allowing for triggered release, which makes them into promising controlled delivery systems [56].

Major challenges still remain in medical applications, and they are associated with toxicity, poor biodistribution, insufficient functionality, ineffective drug incorporation in delivery devices, and subsequent drug release. The chemistry of the HNTs flexible to allow the delay and/or target drug release or even improve drug dissolution. Controlled drug release from the lumen has already lead to proven applications in drug delivery, gene delivery, tissue engineering, cancer and stem cell isolation, and bioimaging [57]. The covalent grafting of suitable organic moieties on the external surface or in the halloysite lumen has been explored as ways to improve the loading and release of several biologically active molecules [49]. The toxicity and bioacceptability of these modified HNTs is a function of the molecules and functional groups that have been added to the native HNT. This is a key conclusion.

Building upon early results, there is work on further development and industrialization in the area of the design of third generation active biomaterials [58]. The backbone of the HNT will remain important in this work. New research on modified halloysite nanotubes and their applications in biological and medical fields will extend their use in medicine [59].

One interesting area of active research involves the incorporation on HNTs with various antimicrobial agents. The HNT delivers these antimicrobial agent to the pathogenic microorganisms in a very interesting and highly efficient processes. The main benefits gained by using HNTs to these nanocomposites appear to be the enhanced thermal and mechanical stability of the antimicrobial nanocomposites, prolonged durability, and release of the antimicrobial agents in a sustained manner. These antimicrobial nanocomposites are synthesized in different forms (powder, film, nanocapsule, and adhesive). Then depending upon the application, they can be used food packaging, water decontamination, waste water management, healing of wounds, antimicrobial agents for surfaces, and for the treatment of microbial infections [60].

The last example involves taking advantage of the natural chemistry of these clays. The unique structure and surface properties of HNTs allow them to assemble into ordered chemical lattices. The applications of these self-assembled HNT structures can be found in biomedical areas such as capture of circulating tumor cells, guiding oriented cell growth, controlling cell germination, and delivery of drugs or nutrients. The fabrication of functional biodevices based on HNTs for tissue engineering, cancer diagnosis/therapy, and personal healthcare products is an ongoing area of research [61].

4.3.6 HNTs and ILs as ECCs

The fields of ILs and HNTs, while presented separately here, are both increasing in importance and will play significand roles in global sustainability in the twenty-first century. However, they will also produce effects on the environment. Currently, GC is providing tools and guidance in the progress of the new era of chemical processes. Sustainability, circularity, and degree of environmental friendliness are becoming how new chemistries (ILs and HNTs) are evaluated. These criteria should be continually evaluated when choices are made regarding the types of ILs/HNTs prepared and their intended uses [55]. While distinctly different, these two fields have some overlap. The confinement of ILs into nanoporous hosts like HNTs opens new realms of possibilities. Reciprocally, the modification of HNTs using ILs will lead to new fields of possible application [62].

4.4 Conclusions

Ionic liquids (ILs) and halloysite nanotubes (HNTs) have been presented as compounds that have extensive value in chemistry, environmental applications, medicine, industry, and research. Their use and utility are expanding. Therefore, they will also have the potential to become emerging chemicals of concern (ECC) over time. This chapter has shown how these chemicals are prepared and modified, which provide insights on how to monitor their effect on the environment. The growth in their use warrants further study of their properties. Unlike the ILs which may be toxic themselves, HNTs may become toxic due to what is loaded in/on them. While the utility of these chemicals is unquestionable, understanding their environmental behavior, fate, and effects is critical.

References

1. Welton T (2015) Solvents and sustainable chemistry. Proc Royal Soc A: Math, Phys Eng Sci 471:2183
2. Isosaari P, Srivastava V, Sillanpää M (2019) Ionic liquid-based water treatment technologies for organic pollutants: current status and future prospects of ionic liquid mediated technologies. Sci Total Environ 690:604–619
3. Heckenbach ME et al (2016) Meta-analysis of ionic liquid literature and toxicology. Chemosphere 150:266–274
4. Costa SPF et al (2017) Environmental impact of ionic liquids: recent advances in (eco)toxicology and (bio)degradability. Chemsuschem 10(11):2321–2347
5. Majid MF et al (2020) Futuristic advance and perspective of deep eutectic solvent for extractive desulfurization of fuel oil: a review. J Mol Liq 306
6. Mudhoo A et al (2020) Fabrication, functionalization and performance of doped photocatalysts for dye degradation and mineralization: a review. Environ Chem Lett 18(6):1825–1903
7. Abdulra'uf LB et al (2020) Review of ionic liquids in microextraction analysis of pesticide residues in fruit and vegetable samples. Chromatograph 83(1):11–33
8. Pandey A et al (2019) A brief review on ionic fluids and its application in machining. In: 9th International Conference of Materials Processing and Characterization, ICMPC. Elsevier Ltd.
9. Banerjee P et al (2020) Recent developments in stimuli-responsive poly(ionic liquid)s. J Poly Res 27(7)
10. Quijada-Maldonado E et al (2020) Possibilities and challenges for ionic liquids in hydrometallurgy. Separation Purification Tech 251
11. Keijer T, Bakker V, Slootweg JC (2019) Circular chemistry to enable a circular economy. Nat Chem 11(3):190–195
12. Rogers RD, Seddon KR (2003) Ionic liquids—solvents of the future? Science 302(5646):792–793
13. Ren F et al (2020) Applications of ionic liquids in starch chemistry: a review. Green Chem 22(7):2162–2183
14. Arumugam V, Redhi G, Gengan RM (2018) The application of ionic liquids in nanotechnology. Fundamentals of nanoparticles: classifications, synthesis methods, properties and characterization. Elsevier, pp 371–400
15. Khandelwal S, Tailor YK, Kumar M (2016) Deep eutectic solvents (DESs) as eco-friendly and sustainable solvent/catalyst systems in organic transformations. J Mol Liq 215:345–386
16. Meksi N, Moussa A (2017) A review of progress in the ecological application of ionic liquids in textile processes. J Clean Prod 161:105–126

17. Singh G, Aggarwal V, Singh S (2020) Critical review on ecological, economical and technological aspects of minimum quantity lubrication towards sustainable machining. J Clean Prod 271
18. Katna R, Suhaib M, Agrawal N (2020) Nonedible vegetable oil-based cutting fluids for machining processes–a review. Mater Manuf Processes 35(1):1–32
19. Cañadas R, et al (2020) Overview of neoteric solvents as extractants in food industry: a focus on phenolic compounds separation from liquid streams. Food Res Int 136
20. Kim SH et al (2019) A perspective on galactose-based fermentative hydrogen production from macroalgal biomass: trends and opportunities. Biores Technol 280:447–458
21. Mahmood H et al (2019) Recent advances in the pretreatment of lignocellulosic biomass for biofuels and value-added products. Current Opinion in Green and Sustainable Chemistry 20:18–24
22. Galbe M, Wallberg O (2019) Pretreatment for biorefineries: a review of common methods for efficient utilisation of lignocellulosic materials. Biotech Biofuels 12(1)
23. Gupta M, Paul S, Gupta R (2010) General aspects of 12 basic principles of green chemistry with applications. Curr Sci 99(10):1341–1360
24. Ohno H, Yoshizawa-Fujita M, Kohno Y (2019) Functional design of ionic liquids: unprecedented liquids that contribute to energy technology, bioscience, and materials sciences. Bull Chem Soc Jpn 92(4):852–868
25. Lopes JM et al (2017) Ionic liquid as reaction media for the production of cellulose-derived polymers from cellulosic biomass. Chem Eng 1(2):1–28
26. Gomes JM, Silva SS, Reis RL (2019) Biocompatible ionic liquids: fundamental behaviours and applications. Chem Soc Rev 48(15):4317–4335
27. Zandu SK, Chopra H, Singh I (2020) Ionic liquids for therapeutic and drug delivery applications. Curr Drug Res Rev 12(1):26–41
28. Florio W et al (2019) Comparative evaluation of antimicrobial activity of different types of ionic liquids. Mat Sci Eng C 104
29. Miskiewicz A et al (2019) The use of liquids ionic fluids as pharmaceutically active substances helpful in combating nosocomial infections induced by Klebsiella Pneumoniae New Delhi strain, Acinetobacter Baumannii and Enterococcus species. Int J Mol Sci 19(9)
30. Du M, Guo B, Jia D (2010) Newly emerging applications of halloysite nanotubes: a review. Polym Int 59(5):574–582
31. Leporatti S (2017) Halloysite clay nanotubes as nano-bazookas for drug delivery. Polym Int 66(8):1111–1118
32. Lazzara G et al (2018) An assembly of organic-inorganic composites using halloysite clay nanotubes. Curr Opin Colloid Interface Sci 35:42–50
33. Saif MJ, Asif HM, Naveed M (2018) Properties and modification methods of halloysite nanotubes: a state-of-The-art review. J Chil Chem Soc 63(3):4109–4125
34. Liu M et al (2014) Recent advance in research on halloysite nanotubes-polymer nanocomposite. Prog Polym Sci 39(8):1498–1525
35. Chiew CSC et al (2020) Life cycle assessment on alginate-based nanocomposite beads for the removal of lead(II) from aqueous solutions. J Water Process Eng 45
36. Danyliuk N, Tomaszewska J, Tatarchuk T (2020) Halloysite nanotubes and halloysite-based composites for environmental and biomedical applications. J Mol Liq 309
37. Wong LW et al (2022) Natural hollow clay nanotubes and their applications as polymer nanocomposites in tissue engineering. J Sci: Adv Mat Dev 7(2)
38. Lvov Y, Abdullayev E (2013) Functional polymer-clay nanotube composites with sustained release of chemical agents. Prog Polym Sci 38(10–11):1690–1719
39. Abdullayev E, Lvov Y (2011) Halloysite clay nanotubes for controlled release of protective agents. J Nanosci Nanotechnol 11(11):10007–10026
40. Zhang H, Li G, Hu Y (2018) Applications of Halloysite nanotubes in separation and enrichment. Prog Chem 30(2–3):198–205
41. Zahidah KA et al (2017) Halloysite nanotubes as nanocontainer for smart coating application: a review. Prog Org Coat 111:175–185

42. Pulikkalparambil H, Siengchin S, Parameswaranpillai J (2018) Corrosion protective self-healing epoxy resin coatings based on inhibitor and polymeric healing agents encapsulated in organic and inorganic micro and nanocontainers. Nano-Struct Nano-Obj 16:381–395
43. Ferrari PC, Araujo FF, Pianaro SA (2017) Halloysite nanotubes-polymeric nanocomposites: characteristics, modifications and controlled drug delivery approaches. Ceramica 63(368):423–431
44. Gaaz TS et al (2015) Properties and applications of polyvinyl alcohol, halloysite nanotubes and their nanocomposites. Molecules 20(12):22833–22847
45. Wan L et al (2020) Flame retardation of natural rubber: strategy and recent progress. Polymers 12(2)
46. Bertolino V et al (2017) Biopolymer-targeted adsorption onto halloysite nanotubes in aqueous media. Langmuir 33(13):3317–3323
47. Naumenko E, Fakhrullin R (2019) Halloysite nanoclay/biopolymers composite materials in tissue engineering. Biotech J 14(12)
48. Wu Y et al (2019) Advances in halloysite nanotubes-polysaccharide nanocomposite preparation and applications. Polym 11(6)
49. Massaro M et al (2018) Chemical modification of halloysite nanotubes for controlled loading and release. J Mat Chem B 6(21):3415–3433
50. Anastopoulos I et al (2018) A review on halloysite-based adsorbents to remove pollutants in water and wastewater. J Mol Liq 269:855–868
51. Cavallaro G et al (2019) Organic-nanoclay composite materials as removal agents for environmental decontamination. RSC Adv 9(69):40553–40564
52. Mu K et al (2017) Enhanced permeability and antifouling performance of cellulose acetate ultrafiltration membrane assisted by L-DOPA functionalized halloysite nanotubes. Carbohyd Polym 174:688–696
53. Zou Y et al (2022) Application of aluminosilicate clay mineral-based composites in photocatalysis. J Environ Sci (China) 115:190–214
54. Lvov Y, Aerov A, Fakhrullin R (2014) Clay nanotube encapsulation for functional biocomposites. Adv Coll Interface Sci 207(1):189–198
55. Hanid NA, Wahit MU, Guo Q (2017) Performance of regenerated cellulose nanocomposites fabricated via ionic liquid based on halloysites and vermiculite. In: Handbook of composites from renewable materials. Wiley, pp 249–273
56. Santos AC et al (2018) Halloysite clay nanotubes for life sciences applications: from drug encapsulation to bioscaffold. Adv Coll Interface Sci 257:58–70
57. Khatoon N, Chu MQ, Zhou CH (2020) Nanoclay-based drug delivery systems and their therapeutic potentials. J Mat Chem B 8(33):7335–7351
58. Voronin D et al (2020) Naturally derived nano- and micro-drug delivery vehicles: Halloysite, vaterite and nanocellulose. New J Chem 44(15):5638–5655
59. Massaro M et al (2017) Covalently modified halloysite clay nanotubes: synthesis, properties, biological and medical applications. J Mat Chem B 5(16):2867–2882
60. Saadat S et al (2020) Nano-interfacial decoration of Halloysite Nanotubes for the development of antimicrobial nanocomposites. Adv Colloid Interf Sci 275
61. Zhao X, Zhou C, Liu M (2020) Self-assembled structures of halloysite nanotubes: towards the development of high-performance biomedical materials. J Mat Chem B 8(5):838–851
62. Zhang S et al (2017) Nanoconfined ionic liquids. Chem Rev 117(10):6755–6833

Chapter 5
New Frontiers for Heterostructured Nanocomposites with Interfacial Functionalities Synthesized via Laser Ablation Synthesis in Solution (LASiS)

Dibyendu Mukherjee

Abstract Emerging materials and manufacturing technology form the quintessential cornerstones to cater to the rapidly-growing demands in water-food-energy (WEF) nexus due to increasing human population and human-technology interfaces that has severely strained our resources and environment globally. To respond to such demands, twenty-first century science and research in nanomaterials have revolutionized new frontiers for materials engineering and development due to the unique properties and functionalities that emerge at nanoscale. Thus, global market size for metal-based composite nanomaterials is projected to reach US$ 620.4 M by 2027—growing at a CAGR of 4.4% over 2020–2027. However, scalable, facile, and chemically clean synthesis of functional heterostructured nanocomposites (HNCs) with low product variability (<10–15%) and desired functionality is still elusive yet, imperative for technological translation of such materials into US's engineering sector. Specifically, enabling science and technology that can tailor these HNCs to tune their structure–property relations for desirable interfacial properties are imperative for their use in catalytic, optoelectronic, and electrochemical applications—all of which, bear impacts for the fast-paced sustainable energy and environmental research. To this end, this book chapter reports the development and deployment of a facile, non-equilibrium and yet, "green" synthesis route—called Laser Ablation Synthesis in Solution—Galvanic Replacement Reaction (LASiS-GRR)—that allows a disruptive merger of high-energy LASiS with chemically reactive GRR in solution-phase for one-pot manufacturing of diverse and complex HNCs comprising metal, metal oxide (M/MO_x), and intermetallic nanocomposites (NCs)/nanoalloys (NAs). The chapter will showcase a few of the authors' formative works in employing LASiS-GRR for the synthesis of advanced HNCs comprising: (1) graphitic shell coated Al NPs as energetic nanomaterials (ENMs), and (2) PtCo NAs in CoO_x matrices as superior bi-functional ORR/OER electrocatalysts. The detailed

D. Mukherjee (✉)
Department of Chemical and Biomolecular Engineering, Nano-Bio Materials Laboratory for Energy, Energetics and Environment (nbml-E3), University of Tennessee, Knoxville, TN, USA
e-mail: dmukherj@utk.edu

M. Shukla et al. (eds.), *Emerging Materials and Environment*,
Challenges and Advances in Computational Chemistry and Physics 37,
https://doi.org/10.1007/978-3-031-39470-6_5

research and discussions presented here is anticipated to provide a broad overview on the future of hierarchically-designed HNCs synthesized via LASiS-GRR route for sustainable electrochemical energy conversion/storage and defense/national security applications. It should be noted here that various sections of this chapter are adopted/reused in part or whole from the principal author, D. Mukherjee's publications, *Appl.Surf. Sci.* 473, 156–163 (2019), *Appl. Catal. B: Environ.* 182, 286–296 (2016), and book chapter in Multifunctional Nanocomposites for Energy & Environmental Applications with copyrights and contents permission via **License Nos.: 5283130233336 (Apr 06, 2022), 5287761493623 (Apr. 14, 2022)** from **Elsevier Publishing Company,** and **License Nos.:** 5340460703005 **(July 01, 2022) from John Wiley and Sons respectively.**

5.1 Introduction

Nanomaterials (NM) have, over the years, generated a great deal of interest from scientists and engineers in nearly all disciplines. This interest has been largely due to a number of desired physico-chemical properties such as magneto-optic properties, specific heats, melting points, and surface reactivity exhibiting unique size-dependent characteristics at the nanoscale. These unique properties are widely believed to be a result of high surface-to-bulk atom ratios as well as, bridging states in which nanophase materials exist between atomic and bulk materials. Furthermore, intermetallic compounds and alloys have garnered attention among solid-state materials in modern science and technology [1]. They possess a surprisingly diverse range of properties, including ferromagnetism [2, 3], shape-memory effects [4], catalytic activity [5–7], hydrogen storage [8] thermoelectric properties [9, 10], and corrosion resistance [11] that find use in various industrial and technological applications. As for all solid-state materials, targeted synthesis of these multimetallic compounds is crucial for controlling their performance in a multitude of applications. To that end, this chapter will be specifically dedicated to the new frontiers in laser-based non-equilibrium synthesis techniques employing the laser ablation synthesis in solution (LASiS) approach to synthesize disparate classes of heterostructured composite nanomaterials, including nanocomposites (NCs) and nanoalloys (NAs), that have increasingly become prominent among the most important solid-state materials in modern science and technology. As representative case studies, two classes of heterostructured nanocomposites (HNCs) synthesized via LASiS-driven techniques, and their detailed characterizations in relation to their engineered interfacial functionalities for high-energy density energetic and electrocatalytic applications will be presented here.

The possibility of integrating well-established properties of intermetallic compounds with emerging nanotechnology research has generated a renewed interest in controlling their synthesis. Tailoring the structure, composition, and morphology of multimetallic nanomaterials becomes particularly of interest in light of their critical role in tuning the functionality and utility of these materials for diverse engineering applications [12, 13]. In nanoscale regime, materials (especially metals and metal oxides) can neither be perceived as atomic species represented by well-defined molecular orbitals, nor as standard bulk materials represented by electronic band structures, but rather by size-dependent broadened energy states [14]. Thus, quantum confinement effects coupled with large surface-to-volume ratios of NPs endow them with unique surface reactivity and interfacial properties. Moreover, fast formation during high-energy manufacturing processes can promote the existence of unbalanced and metastable structures at nanoscale. Thus, the growing industrial importance of IM materials makes it imperative to analyze their properties from small clusters to nanocomposites. Although nanoscale colloidal metals are of interest to scientists of many disciplines, precise control of their physico-chemical properties through directed solution-phase synthesis routes has remained a challenge for material scientists. Till date the biggest scientific hurdle for the nanomaterial community lies in the controlled production of desirable heteronanostructures of intermetallic compositions in the absence of external chemical reagents such that their structural characteristics can be specifically designed to alter their properties (electrical, optical, interfacial, etc.) without causing any surface contamination. Specifically, such chemically clean HNCs become indispensable for the design of functional materials with engineered interfacial activities for catalytic and/or, energetic applications in various energy conversion, storage and/or, dissipation systems. In such application domains, two prominent areas of materials science interests that have burgeoned in our research group—and shall be explicitly presented in this chapter section—are electrocatalytic materials for electrochemical energy conversion in fuel cell applications, and energetic materials for thermal energy dissipation in solid jet propellant/munition applications.

A growing need for high-energy density materials in rocket propulsion for space expeditions and national security missions has resulted in a large volume of research work in the last few decades that have focused on a class of novel metallic materials that demonstrate enhanced energetic properties and enable large thermal energy dissipation under external stimulations, thereby finding applications as propellants, explosives, and pyrotechnics [15–17]. To this end, past studies have investigated mixtures of aluminum (Al) powders and oxidizers as heterogeneous, composite solid propellants that demonstrated high burning rates and enhanced ignition [18–21]. A substantially larger surface area arising from fuel-oxidizer interfaces in the nanoscale regime promotes the kinetically controlled ignition processes. In this regard, the early investigations of energetic nanomaterials (ENMs) primarily focused on the use of metal nanoparticles (NPs), especially nano-Al, in explosives [22–26]. The energetic behavior in these materials is largely driven by the excess interfacial area and small diffusion length scales. In fact, notable works in the past [24, 27] have analyzed the unique combustion properties of various energetic materials at nanoscale as compared to their properties at micro-scale. And yet, the promise of large heat release from these 1st generation energetic metal NPs (e.g., Al NPs) has been often offset by hindered diffusion-limited detonation rates due to excess oxide shell formations and surface area loss from aggregations.

Additionally, heterostructured and intermetallic NMs have found direct engineering applications in recent years in the rapidly expanding areas of efficient and environmentally benign energy conversion/storage technologies to meet the increasing global demand for sustainable energy. To this end, multi-elemental composite NPs including NAs and NCs have exhibited promising potential as superior electrocatalysts in fuel cell energy conversion systems. Herein, the typical catalytic applications involve: (1) cathodic and anodic electrocatalysts for oxygen reduction reactions (ORR) and methanol/ethanol/formic acid oxidation reactions (MOR/EOR/FAOR) in polymer electrolyte membrane fuel cells (PEMFCs); (2) bi-functional ORR/oxygen evolution reaction (OER) electrocatalysts in regenerative fuel cells. In fact, the rate determining step of cathodic ORR is well-known to be energetically expensive that becomes the most critical bottleneck for efficient and clean electrochemical energy conversions in PEMFCs. For the most part, Pt-based nanocatalysts have been extensively used as ORR electrocatalysts in commercial PEMFCs. However, the vulnerability of Pt in terms of its stability and durability under the highly corrosive and acidic conditions of fuel cell operations, added to the high cost of precious metals-based catalysts, have prompted a flurry of recent research in materials science to develop alternate transition metal-based alloys/intermetallic materials with low Pt contents. Thus, recent research efforts have increasingly focused on mixed valence oxides of transition metals with spinel structures (Co_3O_4, Mn_3O_4, NiO, etc.) as electrocatalysts, in spite of their catalytic activities being relatively low [28–32]. In addition, Pt-M NAs with various atomic structures, and morphologies have been shown to manifest similar or even better ORR catalytic performances than Pt [33–37]. On the other hand, OER plays a critical role in water splitting processes during solar fuel production [38–40], or, charge–discharge processes in rechargeable fuel cells and batteries. Hence, electrodes in ideal regenerative electrochemical devices should have high conversion rates for both ORR and OER processes. Unlike the ORR systems, transition metal oxides (e.g., CoO_x, NiO_x, TiO_2, IrO_2, etc.) or perovskites are typically considered as excellent catalytic candidates for OER on account of their high activities and good photo-stability [41–44]. Hence, a key challenge in the design of ideal reversible electrodes is in the low-cost, reproducible synthesis of bi-functional catalysts that can outperform both the ORR and OER activities of commercially available electrocatalysts.

To that end, an important criterion for pushing the boundaries of aforesaid high-impact energy-related materials research is the disruptive development of facile and environment-friendly synthesis techniques that can economically manufacture well-designed and tailored HNCs with tunable interfacial functionalities. Till date, the existing synthesis techniques for nanomaterials can be broadly divided into: (1) "top-down" routes that involve attrition, milling, pyrolysis, etc. to decompose bulk solids into nanoparticles, and (2) "bottom-up" approaches involving liquid/gas-phase nucleation and condensation of atoms or molecules. Specifically, better controls on the environmental and process parameters for the "bottom-up" approaches have resulted in the extensive use of techniques such as chemical reduction/oxidation, solvothermal/hydrothermal, replacement reactions, microemulsions, as well as biomimetic synthesis, and sol–gel methods to synthesize NMs with diverse structures, compositions and morphologies [34, 45–48]. However, an inherent drawback to these techniques is that they inevitably involve unwanted chemicals such as surfactants and leftover reducing agents that can leave behind various chemical by-products on the surfaces of NMs which, in turn, can dramatically affect their interfacial properties and activities [49–51].

5.2 Laser Ablation Synthesis in Solution (LASiS): An Overview

Laser ablation synthesis in solution (LASiS) or (pulsed) laser ablation in liquid (PLAL) has recently seen a resurgence as an emerging green synthesis technique for manufacturing nanoscale materials [52–54]. The general procedure of LASiS involves a high-energy laser beam hitting on a metal target that is immersed in a liquid solution, which in turn creates colloidal nanoparticles in a metastable phase. The specific interest in LASiS arises due to its distinctive advantages of being a simple, elegant and yet surfactant-free technique that requires minimal manual operation and low start-up cost [53, 55, 56]. One of the earliest implementations of laser ablation technique in solutions was reported in 1987 by Patil et al. who used a pulsed Ruby laser on an iron target immersed in water solution [57] to produce a metastable phases of iron oxide. Following this, the late 1990's saw researchers use laser ablation in liquids to synthesize diverse nanomaterial (NMs) that led to more recent works using LASiS to synthesize various types of NPs comprising of heavy metals [58, 59], transition metals [60, 61], intermetallic nano-colloids [62]. Additionally, a few morphologically complex nanostructures such as core–shell [63, 64] and hollow [65] NPs have also been synthesized using LASiS. However, few efforts till date have fully exploited the potentials of LASiS techniques for the synthesis of higher-order and complex multi-elemental composite nanostructures, [66, 67] of which the recent synthesis of a nanosheet water oxidation catalyst is noteworthy [68]. Herein, the unique advantages of charge stabilization and solution-phase chemistry in a liquid-confined plasma in LASiS can offer pivotal technological breakthroughs in the fabrication of metal/intermetallic composite NPs with controlled sizes, shapes, structures, compositions, and interfacial chemistry that can be effectively tuned for diverse engineered functionalities in materials for catalysis, interfacial energetics and optoelectronic applications, to name a few.

The complex physics of laser-matter interactions leading to the nucleation of NPs via LASiS is still not well-understood [56, 69]. Unlike gas-phase laser ablation used for the relatively straight-forward NP synthesis routes, [26] the extremely high temperature and pressure on metal targets resulting from a liquid-confined plasma plume during LASiS leads to explosive boiling in the solution phase due to the thermal vaporization of the metal target within the solvent [69]. It has been extensively discussed in previous studies that the nucleation of seeding NPs ensue inside the oscillating cavitation bubbles emanating from the laser-induced plasma plume.

Furthermore, upon the collapse of the cavitation bubble, the ensuing seeding NPs experience collisional quenching at the bubble–liquid interface [70–73]. Figure 5.1, adapted with permission from Yang et al. [74], schematically illustrates the time sequence for the LASiS mechanism initiated by a nanosecond laser comprise the following processes: (a) The laser ablates the target surface to generate a plasma plume at the interface with the solid target as the front part of the incident laser pulse ablates the surface the target; (b) The plasma expands under the confinement of liquid to transfer energy to the surrounding liquid, thereby creating a vapor layer surrounding the surface of the plasma; (c) The vapor layer gradually expands into a liquid-confined cavitation bubble around the plasma plume, and compresses the plasma to shrink. At this stage, some NPs formed in the plasma will be released into this bubble; (d) The plasma quenches, thereby releasing all the NPs in the plasma into the bubble; and finally, (e) The bubble collapses to release the seeding NPs into the liquid environment and form colloid [75]. Thus, in the first few milliseconds of ablation onset, the LASiS process can be divided into the following stages: liquid penetration of the laser pulse, coupling/absorption of laser energy by target, detachment of materials from ablated target, expansion and quenching of plasma plume, expansion and collapse of cavitation bubble, followed by subsequent aggregation/sintering of NPs [75].

Of these stages, the nucleation process is initiated by the detachment of ablated materials that is typically driven by various ablation processes such as vaporization, normal boiling and explosive boiling [69]. To that end, the laser fluence (laser energy per unit area) and pulse duration time play critical roles in driving the target surface temperature which, in turn, control the onset of each of the aforesaid ablation mechanisms. Specifically, vaporization results in particle emissions from sublimation or evaporation initiated by electron–phonon coupling and can ensue at any laser fluence. Normal boiling requires low laser energy input and initiates heterogeneous nucleation which can be driven by the bubble diffusion. Finally, explosive boiling results in the high-efficiency ablation processes that initiates homogeneous nucleation at superheating conditions requiring higher energy thresholds than normal boiling [53, 69]. A few recent in-situ spatio-temporal analyses of laser-induced plasma emissions and/or, scattering techniques during laser ablation synthesis have indicated that solutes/

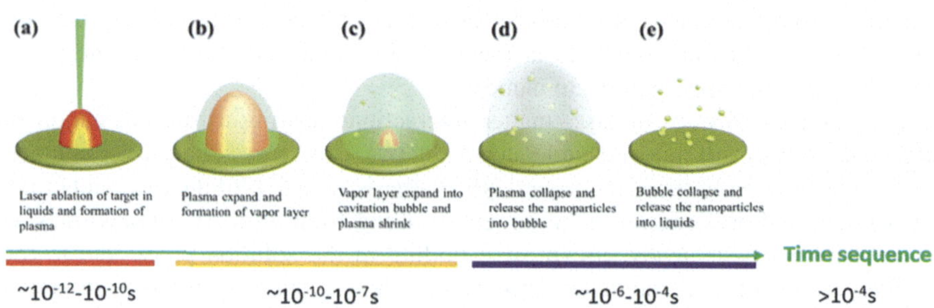

Fig. 5.1 Schematic diagram for the time sequence of LASiS mechanism initiated by nanosecond laser. **Printed with permission from Ref.** [74]. **Copyright (2017) (Science Direct Publishing)**

ions can also be present in the cavitation bubble, thereby initiating the GRR processes during the target plasma quenching and nucleation [72, 76–80]. Besides, Matsumoto et al. had also proposed that it is the phase explosion of water due to the strong interaction with the high temperature plasma that leads to the incorporation of ions in the initial bubbles [73, 81]. On the other hand, an extensive and detailed review article on the laser-based synthesis and processing of colloidal NPs by Zhang et al. had suggested the presence of ionic species in the supercritical vapor surrounding the plasma plume in the early stages of the cavitation bubble evolution, that can take part in galvanic displacement and other chemical reactions in the liquid phase [82]. In such a scenario, one can imagine that the inclusion of secondary metal salts in the solution can enable the reduction of the solution-phase precursor metal atoms/nucleus by the seeding metal NPs that can then subsequently undergo alloy/composite formations with the remaining and newly-quenched target metal atoms to form intermetallic composites. Once the internal pressure reduces to that of the surrounding liquid, the bubble collapses at a time scale of hundreds of microseconds thereby enabling the seeding metal/intermetallic NPs to further grow and agglomerate at bubble–liquid interface. The unique advantages of such a reactive LASiS process stem from the extreme out-of-equilibrium conditions inside the laser-induced plasma plume and cavitation bubble under which the laser irradiation can assist the secondary chemical reactions simultaneously occurring with the NP nucleation. In turn, this assists in the formation of NPs with uniform elemental distributions while the surface charges induced by the plasma environment provide the charge screening effects that hinders particle agglomeration without the need for any surfactant.

The laser parameters and environmental conditions during LASiS dictate the structures, shapes, and properties of the as-synthesized NPs. Laser parameters includes fluence [59, 69], wavelength [83], pulse duration [84], frequency or repetition rates, beam profiles/area, etc. The heat absorbed by the target during laser ablation and in turn, the target surface temperature and corresponding ablation mechanisms are all dictated by both the laser fluence and pulse duration [59]. The laser wavelength choice can significantly affect laser-solvent interactions and re-absorption of laser energy by the solution-phase colloidal NPs during the passage of the laser through the liquid column [83]. The extent to which the already-formed NPs in solution undergo "re-radiation" can significantly modify the final sizes and shapes of the as-synthesized NPs. To that end, it is well-known that the shorter laser wavelengths are more readily re-absorbed by the solution-phase NPs that result in considerable re-structuring of the particles [69]. Finally, LASiS environmental parameters such as dissolved oxygen contents in solvents and solution pH values determine the solution-phase ion concentrations that can affect nanocrystal growth directions, final composition as well as size/morphology of the as-synthesized NPs [85, 86]. To this end, previous works by have shown that the amount of dissolved oxygen in water during LASiS can significantly alter the size and surface oxidation of the ensuing crystalline Si-based NPs [87]. Similarly, injection of solution-phase oxygen during LASiS on Zn was shown to decrease particle sizes and $Zn(OH)_2/ZnO$ ratios while increasing the crystalline contents in the products [88].

5.2.1 *Chemical Reaction-Interfaced LASiS: Laser Ablation Synthesis in Solution—Galvanic Replacement Reaction (LASiS-GRR)*

In spite of the high suitability of the LASiS technique to fabricate controlled nanostructures in the absence of any external chemical reagents and/or, surfactants/ligands, the commonly perceived notions are its disadvantages in regard to relatively low productivity and poor control on shape/size, compositions and structures—especially when it comes to multi-elemental composite NPs [53, 56]. Yet, heterostructured NPs with multi-elemental compositions, uniform morphology and narrow size distributions can function as building blocks for constructing composites, including alloys with higher order superlattices that can exhibit the joint properties of individual nanoparticles comprising homogeneous compositions. To this end, during the inception and development of fundamental laser ablation synthesis techniques in the Mukherjee group, it was hypothesized that the extreme physicochemical conditions of LASiS [89]—when interfaced with secondary solution-phase chemical reaction routes such as galvanic replacement reactions (GRR) or, solvent pyrolysis—can initiate unique non-equilibrium thermodynamics and kinetics at the plasma-liquid interface. Such reaction pathways can generate novel nanostructures or, nanocomposites with tailored morphologies, compositions, and metastable structures or interfaces that have hitherto not been observed, while being potentially promising for catalytic and/or, energetic performances.

It would be pertinent to highlight here that in general, GRR techniques are known to offer effective and versatile approaches for synthesizing various NMs on the basic premise of the redox potentials for oxidant species being higher than those for the reductants. Specifically, the conventional GRR process involves the oxidation of one metal that is usually referred to as a sacrificial template by another metal salt ion with higher reduction potential. Upon contacting each other in a solution phase, the metal template will be oxidized and then dissolute in the solution; meanwhile the other metal salt quickly gets reduced and plated onto the outer shell of the template, in which case the final product typically keeps the original shape of the first metal template [90–93]. In addition, during reactivity-selected GRR processes, the oxidation and dissolution of the template metal atoms preferentially occur at some certain sites/facets due to the difference in kinetic battier and the shielding effect, therefore results in formation of nanostructures with various shapes and structures including hollow, porous, and core–shell structures. But the first-step in such methods requires synthesis of the seeding metal NPs as the sacrificial reductant via chemical reduction routes. In turn, such routes are rarely deemed to be one-step processes or, devoid of various chemical reagents (reducing agents, surfactants, ligands etc.) that are used to nucleate and stabilize the NPs and yet, known to alter interfacial properties and functionalities of the as-synthesized metal/metal oxide NPs [94–96].

To this end, the motivation of combining LASiS and GRR in tandem arises from the superior ability for GRR to control the size, shape, composition, and morphology of the resultant NPs along with the intrinsic charge screening effects provided by

LASiS which prevent NPs from severe aggregations without the need for additional surfactants and stabilizing agents that are potentially harmful for the NC applications. Bearing such a unique high-energy laser-induced chemical synthesis route in mind, Hu and Mukherjee have recently developed and patented the *laser ablation synthesis in solution-galvanic replacement reactions (LASiS-GRR;* ***Patent: US 10,326,146 B2 (2019)****)* [89, 97–104]—as a one-pot non-equilibrium and "green" synthesis technique for multi-elemental NCs with outstanding catalytic, energy storage/dissipation, and energetic activities. The scientific rationale here is that a disruptive merger of "top-down" LASiS-induced plasma synthesis with secondary "bottom-up" solution-phase chemistry (GRR, or other solvent reactions) in desired organic/inorganic solvent environments would allow LASiS-GRR to tailor-make functional intermetallic (IM)/composite NPs heretofore inconceivable via equilibrium synthesis routes. In conjunction with GRR, recent efforts have also been directed toward other solution-phase chemistries including laser-induced organic solvent pyrolysis as an effective way to create graphitic C shell coated metal NPs to tailor their interfacial activities [98]. A fundamental scientific concept here is that the charge-stabilized seeding NPs from rate-limiting LASiS participate in the highly reactive and rate-controlling solvent chemistry (including pyrolysis) to initiate non-equilibrium nucleation pathways at the laser-induced plasma-liquid interface. This facilitates the production of NPs with metastable structures and tailored sizes/shapes, compositions without using any external surfactants, ligands, etc.

5.2.2 Experimental Setup for LASiS/LASiS-GRR Processes

The schematic in Fig. 5.2a shows our in-lab built experimental setup for LASiS/LASiS-GRR. As seen from the schematic, it includes a Q-switched Nd:YAG pulsed laser (Model: Brilliant Easy) loaded with modular 1st and 2nd harmonic generators that can generate 532 nm and 1064 nm wavelengths with 165 and 330 mJ/pulse energy respectively, while operating at 0 Hz repetition rates with a pulse with of 4 ns duration. The surface of the metal target immersed in a desired solvent is ablated by the focused pulsed laser beam inside a sealed stainless-steel reactor cell. Typically, a convex lens is used to focus the laser beam through a high damage threshold-tested cell window. The four side-viewing windows on the cell allow for accurate viewing and adjusting of the laser focal point. The cell can be purged with suitable inert gases by using the provided gas inlet and outlet ports on the cell. A stepper motor on which the target platform is mounted allows for continuous rotation of the target to ensure uniform ablation of the target surface. The solution temperature in the reactor cell can be accurately controlled by the heating coils at the base of the cell, while the temperature is monitored by an immersed thermocouple. The ultrasonic dismembrator is used for continuous in-situ de-agglomeration of the synthesized NPs. Finally, a programmable injection unit allows simultaneous infusion of desired solvents (aqueous/organic) and metal salt precursor solutions during the tandem LASiS-GRR process.

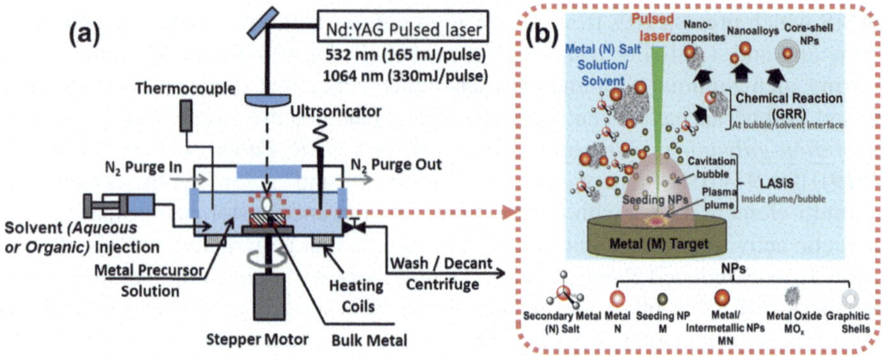

Fig. 5.2 Schematics showing **a** experimental setup and **b** reaction processes for LASiS-GRR technique

As seen from Fig. 5.2b, the transformative concept for LASiS-GRR route lies in the disruptive merger of *rate-limiting LASiS* generating seeding metal NPs inside the plasma-cavitation bubble and *rate-controlling solvent chemistry* (e.g., GRR in aqueous or, pyrolysis in organic solvents) at the liquid-bubble interface to create a facile yet, elegant reaction coupling between the two that drives the formation of composite NPs. LASiS involves thermal vaporization of liquid-confined metal targets with high-energy pulsed laser ($\lambda = 532$, 1064 nm; energies ~165, 330 mJ/pulse) that initiates a micro plasma with excess T ~4000–5000 K and P ~10–15 GPa out of which metal NPs nucleate inside the plasma-induced cavitation bubble. The seeding NPs undergo ultrafast collisional quenching and chemical reactions along with solvent pyrolysis at bubble–liquid interface (Fig. 5.2b) [69].

Here, the scientific rationale is that the thermodynamics of heterogeneous phase formations, although rate-limited by nucleation of seeding NPs, is finally controlled by rate kinetics of the reactions at the bubble/solvent interface [101]. Finally, secondary metals from the chemical reactions and/or, graphitic C from solvent pyrolysis undergo heterogeneous nucleation on the seeding NPs. Thus, the non-equilibrium plasma reaction pathways curved out by the high-energy laser-matter interactions during LASiS, when coupled with the GRR or other solution-phase chemical reactions, can be tuned by laser parameters *[laser fluence, wavelength]*, solvent properties *[pH, type of solvents (aqueous, organic or cryogenic), solution-phase metal salt concentrations]*, and redox conditions *[redox potential gradients between metal targets and aqueous salt solutions]*. Such diverse yet, precise tuning of process parameters allows the thermodynamics and kinetics of the technique [58] to drive desired structure–property relations in complex composite nanostructures by tailoring their sizes/shapes, crystallinity, architecture, and compositions. Additionally, rate controlled chemical kinetics of the process can be easily integrated into continuous flow reactors to ensure high productivity, stability, and low-cost manufacturing. Unique advantages of the technique are: **(1)** facile, yet rapid synthesis devoid of unwanted by-products and post-processing steps; **(2)** "green chemistry" without any surfactants or, chemically harsh reducing agents; **(3)** charge screening effects from densely charged species in plasma prevent NP aggregations [71]; **(4)** low primary energy consumption (<0.5 MJ/kg of materials), with easily controlled process parameters; **(5)** non-equilibrium thermodynamics and kinetics of liquid-confined high-energy plasma promotes unique morphology, metastable structures; **(6)** ability to synthesize functional NPs with interfacial structures otherwise inconceivable via traditional wet chemistry routes [105], and **(7)** easy scale-up for large-scale productions.

5.3 Synthesis of HNCs with Engineered Interfacial Functionalities via LASiS/LASiS-GRR

The aforesaid transformative concepts and unique advantages of the LASiS/LASiS-GRR synthesis technique have prompted past preliminary investigations into structure–property relations (specifically, for catalytic studies) in composite/intermetallic nanomaterials emerging from complex plasma interactions/chemical reactions during LASiS. Ernesto Jimenez et al. reported the manufacturing of core shell SiO_2 capped noble metal NCs by carrying out LASiS on Si targets within various salt solutions [63]. Feng Lin et al. produced Ag/Au embedded metal oxides with enhanced surface plasmon resonance (SPR) and unique magnetic properties [106]. But, a resurgence of growing interests in laser ablation routes for complex nanomaterials manufacturing have led to myriad recent publications reporting on the synthesis of composite/intermetallic NPs via pulsed laser ablation in liquid (PLAL) or, LASiS-based techniques. Specifically, increasing demands for photonic/optoelectronic and catalytic materials in the energy and environmental sectors have inspired LASiS-interfaced solution chemistry methods to synthesize composite NPs comprising precious and transition metals in intermetallic and/or shell-core form. Such efforts include the synthesis of Ag/Pd core–shell NPs via successive PLAL and chemical reduction reactions [107], colloidal Si@Au and Si@Ag NPs via nanosecond IR laser-induced plasma assisted reductions [108], Au/Fe alloyed NPs with SPR effects via nanosecond laser ablation of multi-metal thin films in acetone or water [109], or fundamental studies on the role of dissolved oxygen in organic/aqueous solutions for the formation of metal oxide/ alloyed NPs of Cu and Pt-Cu systems [110]. On the other hand, LASiS-based techniques have been extended toward the synthesis of multi-elemental NCs comprising metal oxide and carbonaceous matrices that include one-step reactive LASiS for Au@CdO core–shell [111] and Au NPs stabilized in SiO_2 for catalytic applications [112], morphology-controlled composite metal/metal oxide NPs comprising different oxide phases of Cu and Zn [113, 114], and more involved NCs indicating the formation of Cu and Ni Phyllosilicates via femtosecond reactive LAL (fs-RLAL) techniques [115, 116].

In our group, the foundational work in LASiS-driven synthesis of metal/metal oxide NPs was established by our earliest works investigating the role of dissolved

O_2, solution-phase pH and solvent temperatures during laser ablation of metal targets (Co, Mn) to synthesize size and morphology-controlled metastable hexagonal β-$Co(OH)_2$ nanocrystals, spherical Co_3O_4 NPs, Co_3O_4 nanorods, and Mn_2O_3 nanowires (Fig. 5.3) [89]. This was successively followed by a series of works that demonstrated the versatility and prowess of LASiS for synthesizing diverse classes of functional (electrochemical, optical, energetic) composite/intermetallic nanomaterials with tailored sizes, shapes/morphology, crystallinity, and compositions (see Fig. 5.3), that included NCs of binary PtCo alloyed NPs in CoO_x matrices as superior bi-functional ORR/OER catalysts [102], in conjunction with binary PtCo and ternary PtCuCo NAs with excellent ORR activities [100, 101], Ag/TiO_x and Ag/ZnO shell-core NPs and NCs with tailored optical and plasmonic properties, and more recently, a one-step LASiS-driven synthesis of nanoporous metal organic frameworks (ZIF-67) [103]. In our continuing efforts to establish LASiS-GRR technique at the forefront of nanomanufacturing, our group has also extended the technique toward the fabrication of HNCs such as of Co_3O_4 NPs encapsulated in N-doped graphene oxide (N-GO) layers for selective ORR and/or supercapacitive activities [99], shell-core structures of graphitic (C) shell coated Al NPs as high-energy density materials [98], and MOF-derived HNCs of C shell coated PtCo NAs embedded in Co_3O_4-decorated carbonaceous matrices (Pt_3Co@C/Co_3O_4 NCs) [104] as ORR electrocatalysts via in-situ or post-processing pyrolytic treatments.

In order to showcase a few representative case studies for the synthesis of HNCs via LASiS, Fig. 5.4 depicts the scope of the following sections in this chapter that specifically discusses the detailed experimental procedures, characterizations and results for two metal-based HNCs made via LASiS-based techniques in our group—namely, (1) Graphitic shell coated Al NPs as energetic nanomaterials (for thermal energy dissipation in solid-state jet propellant and munition applications) and (2) PtCo/CoOx NCs as ORR/OER electrocatalytic nanomaterials (for electrochemical energy conversion in fuel cell applications). Sections 5.3.1 and 5.3.2 are adopted in part or whole from the journal articles, *Appl.Surf. Sci.* **473**, 156–163 (2019) [98] and *Appl. Catal. B: Environ.* **182**, 286–296 (2016) [102] with the principal author, D. Mukherjee's copyrights and contents permission from Elsevier Publishing Co.

5.3.1 Graphitic (C) Shell Coated Al Nanoparticles as Energetic Nanomaterials

In search for high-energy density materials as solid fuels, the combustibility and reactivity of Al NPs have been of considerable research interest due to the large negative enthalpy of combustion for Al (e.g., for bulk Al, $\Delta H_B = -1675$ kJ/mol and for a single Al atom, $\Delta H_1 = -2324$ kJ/mol). The difference in enthalpy between bulk and atomic Al has been attributed to the differences in the kinetics of bulk and nano-sized particles [16]. Although extensive studies have looked into the gas-phase oxidation of micron-sized Al particles in the continuum regime [117], it is obvious from these

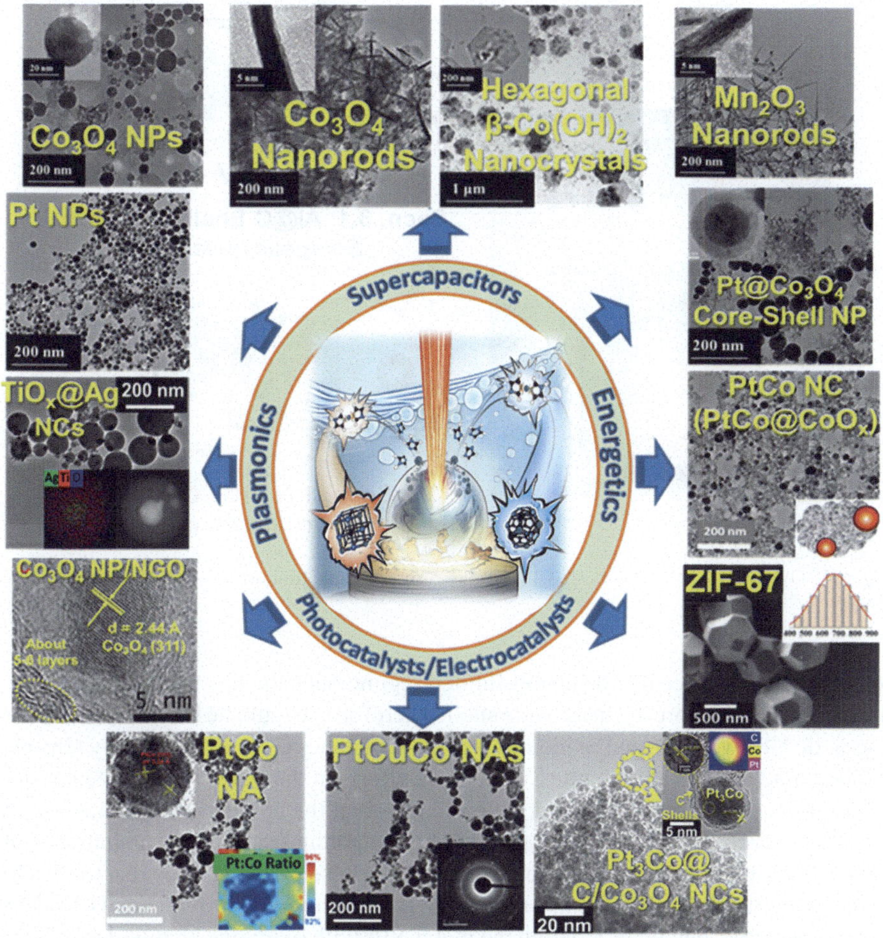

Fig. 5.3 Diverse classes of composite nanomaterials made via LASiS-based techniques

and other theoretical studies that the criteria for continuum burning (where particle size is significantly larger than mean free path of the gas) cannot be extended to ultra-fine metal particles (<100 nm) where their size-dependent properties start exhibiting unique behaviors [17, 26, 118–120]. Al NP oxidation and its energy release can be described by several mechanisms related to the transport of the participating reactive species and surface reactions, namely: **(1)** reaction between the background O_2 and the surface Al atoms; **(2)** formation of oxide layer on the surface Al; **(3)** diffusion of O_2 and Al through the oxide layer for subsequent reactions. It has been shown that the morphology of the particles, especially the oxide layer thicknesses, changes the energetic properties of Al NPs drastically [121]. The formation of the native oxide layer affects the surface reactivity and oxidation kinetics of Al NPs significantly, as the reaction mechanism proceeds from reaction-limited kinetics at the initial stages

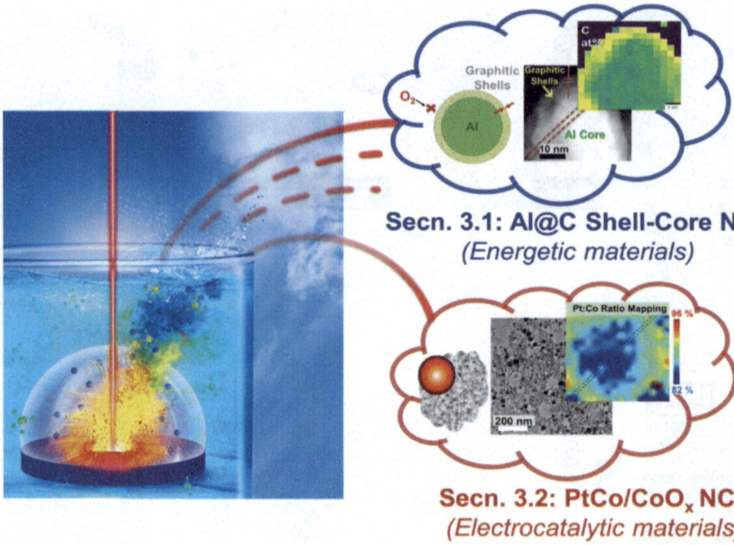

Fig. 5.4 Overall scope of Sects. 5.3.1 and 5.3.2 showcasing functional HNCs synthesized via LASiS

on the bare Al surface to a diffusion-limited regime once the reactive species have to diffuse through the pre-formed oxide shell. Therefore, it could be beneficial to design energetic NPs with optimal oxide shells to passivate the Al surface for safety reasons, while tuning their oxidation rates. Increasing efforts in the design of next-generation ENMs have been directed toward the basic idea that the coatings should not simply retard the reactivity, but promote safety while enabling performance enhancement. Hence, various scenarios have been suggested to preserve the surface reactivities by using noble metals [122], metal oxides [123], and other potent oxidizers [123–125] to coat the active metal NP. Coating the metal NPs with oxidizing agents also reduces the solid fuel-oxidizer diffusion lengths, thereby promoting the burning rates [126]. Carbon coatings have been demonstrated to show similar features [127], with the added advantage that while the coating itself gets oxidized into gaseous products (CO_2, CO etc.) to prevent any solid ash residues on the particle surface, the Al NP aggregation rates are retarded due to the presence of the carbon shell.

Lab-scale and industrial synthesis of metal NPs in gas-phase typically employ rapid condensation of supersaturated metal vapor (monomers) generated from thermal evaporation of the bulk metal, electric arc discharge, laser ablation, flame reactors, or plasma reactors. However, it quickly became apparent that a significant drawback of gas-phase synthesis of metal NPs results from the kinetics of the process, which leads to high particle collision-coalescence rates, rapid aggregation, and loss of interfacial activities. On the other hand, chemical synthesis routes are powerful and have been extensively used to-date. However, they inevitably suffer from harsh residual chemical contaminations and equilibrium growth pathways for the particles, while requiring the use of unwanted surfactants and ligands during the synthesis [82].

To this end, as discussed in details above and developed in our group, LASiS-based techniques emerged as an enabling "green chemistry" route that does not involve the use of surfactants and harsh reducing agents that typically poison the active surface areas. While conventional laser ablation has been employed in the past for the production of bare Al NPs [128], a few studies have also looked into the size and morphology of Al NPs made from Al target ablation in organic solution [129, 130]. Unfortunately, uncertainties in size, morphology, and structure of the Al NPs prevented further development of the technique into reliable and tunable systems that can effectively tailor the surface functionalities and/or coatings on the synthesized NPs. In fact, a vast majority of these studies have largely focused on the mere synthesis of these materials, and are devoid of any comprehensive investigations into tuning the surface functional activities of these materials through systematic coatings to preserve solid Al fuel cores and prevent them from unwanted surface oxidation.

This section of the chapter will discuss the successful implementation of LASiS specifically operated under different organic solvents (acetone, toluene) and laser properties for facile yet, rapid synthesis of Al NPs with carbon/graphitic coatings as the next-generation energetic nanomaterials. The results for the size/morphology, interfacial structures, and compositions of the shell-core C/Al NPs—as adopted from our recent publication [98]—are presented here in light of their energetic performances as compared to various samples made under different experimental conditions, as well as with commercial native oxide shell coated nanometer-sized Al powders.

5.3.1.1 Experimental Methods

Synthesis of Al NPs: The Al pellet was purchased from Kurt J. Lesker (99.99% purity, 1″ diameter and 1/4″ height). Two different organic solvents from Sigma-Aldrich (acetone, 99.95% purity and toluene, 99.9% purity) were used in this study for ablation of Al in solution. The Al target was placed in ~8 ml of the organic solvent and ablated with the laser at room temperature. Following laser ablation, the NP suspensions in the organic solvents were centrifuged, decanted, and washed with methanol prior to deposition on the TEM grids for imaging, while the dried powder stored under an inert environment was used for all other characterization purposes. Commercial Al NPs (80 nm) were purchased (NovaCentrix) for comparison to the LASiS-synthesized particles.

Laser-induced air shock from energetic materials (LASEM): All energetic measurements on the Al/C-based composite NPs were carried out via Laser-induced Air Shock from Energetic Materials (LASEM), a technique recently developed and deployed by scientist at the U.S. Army Research Laboratory (ARL, Aberdeen, MD) facilities for measuring microsecond-timescale energy release of residue materials to estimate the energetic properties of exothermic and explosive materials [131–133]. Briefly, a shock wave is generated in the air above the sample using a focused Nd:YAG laser pulse (1064 nm, 6-ns, 850 mJ) to create a laser-induced plasma on the sample surface. Schlieren imaging with a high-speed camera is subsequently used to track the expansion of the laser-induced shock wave; the measured velocity of the laser-induced shock wave is indicative of the extent of the energy release on the microsecond timescale. For conventional military explosives, a strong correlation between the characteristic laser-induced shock velocity of the material and the measured detonation velocities from large-scale testing was observed. In this work, all LASEM analyses for the as-synthesized and commercial NPs have been carried out at the U.S. Army Research Laboratory facilities. The LASEM samples were prepared by spreading the dry powder in a thin residue affixed to double-sided tape. Excess material not embedded in the tape was ejected away from the sample surface by the laser-induced shock wave and did not chemically react in the laser-induced plasma.

5.3.1.2 Structural Characterization

Figure 5.5 illustrates the STEM images for the NPs generated from the ablation of the Al pellets for 4 min under either acetone or toluene at a laser fluence of 2.6 J/cm^2; HRTEM images for individual particles prepared in acetone or toluene are also shown. For both solvents, the LASiS-synthesized samples consist of Al NPs embedded in a carbon matrix. Figure 5.6 illustrates the Raman spectra for samples prepared in either acetone or toluene. It shows that both samples include the D (defect) and G (graphitic) bands at ~1375 and 1585 cm^{-1}, respectively. The observation of D and G bands suggest the presence of amorphous carbon and graphitic structures in both samples. It is noted that the presence of D-band stems from the incomplete formation of graphitic shells and/or the amorphous carbon matrix reveled in Fig. 5.5. For detailed investigation of the interfacial structures of the as-produced C/Al composite NPs, electron energy loss spectroscopy (EELS) has been used to analyze the morphology of the particles (see Fig. 5.7). Here, it can be observed that for the synthesis in acetone, the Al NP has been covered with a graphitic shell coating. Moreover, the elemental distribution analysis across the dotted line shown in Fig. 5.4d confirms the core–shell structure by revealing a carbon-rich shell and central Al core. EELS measurements (details provided in the Supplementary Information of the original publication) for the samples prepared in toluene also revealed nanostructures consisting of Al cores with graphitic shells, although the cores were more crystalline than the samples produced in acetone. Current investigations are underway in our group to further elucidate the detailed structure-composition analysis of the core–shell nanostructures synthesized under different organic solvents.

By comparing Fig. 5.5a, b, one can immediately observe that the ablation in toluene resulted in larger particles as compared to the samples from the ablation in acetone. To confirm these visual observations, the particle size distributions for both cases were analyzed. Figure 5.8 shows the size distribution of the particles synthesized in acetone and toluene at a laser fluence of 2.6 J/cm^2 for 4 min. Both samples indicate a lognormal distribution. Based on the size distribution, it can be observed that the samples synthesized in acetone show a unimodal distribution with the peak value at 15.7 nm (log-normal standard deviation of $\sigma = 11.0$ nm). On the

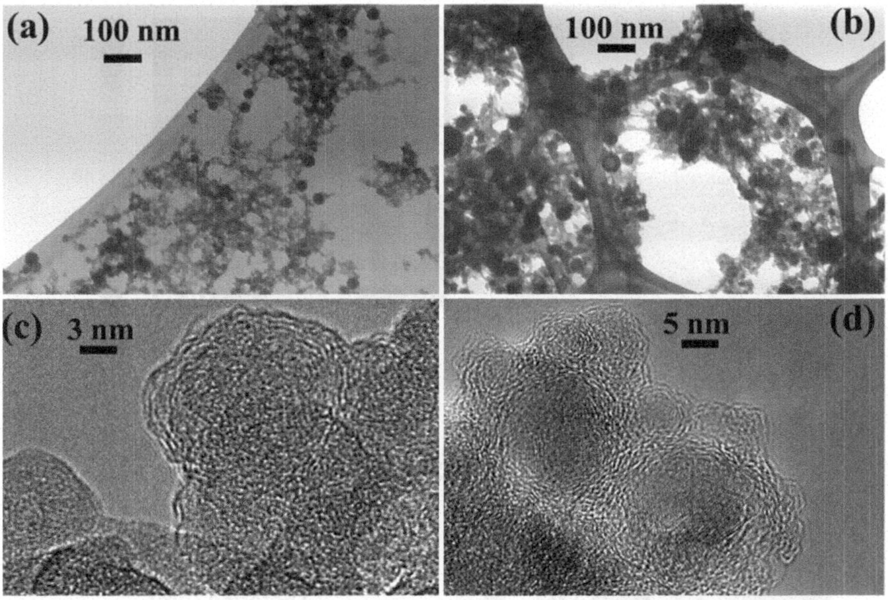

Fig. 5.5 Scanning transmission electron microscope (STEM) images of composite C/Al NP samples prepared via LASiS in **a** acetone and **b** toluene. High-resolution transmission electron microscope (HRTEM) images of composite C/Al NP samples prepared via LASiS in **c** acetone and **d** toluene. **Printed with permission from Ref.** [98]. **Copyright (2019) (Elsevier Publishing)**

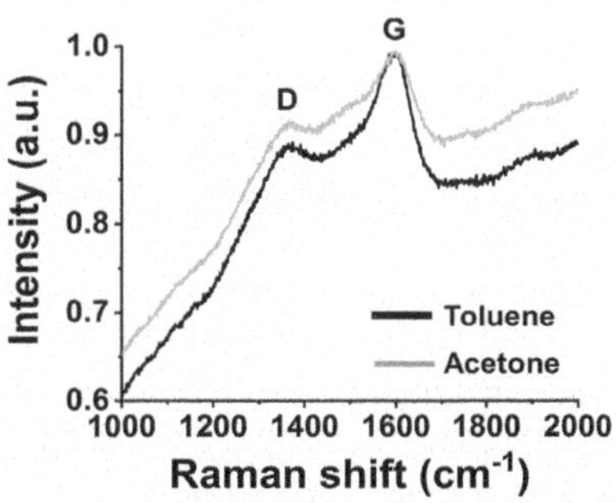

Fig. 5.6 Raman spectra for samples prepared in acetone and toluene indicating the well-known D&G bands. **Printed with permission from Ref.** [98]. **Copyright (2019) (Elsevier Publishing)**

other hand, the samples synthesized in toluene indicate a bimodal distribution with a major peak at 25 nm ($\sigma = 25.3$ nm) and a minor one at 77.2 nm ($\sigma = 17.2$ nm). These results could be explained in relation to the distinct vapor pressure (V.P.) differences between acetone (V.P. ~ 32 kPa) and toluene (V.P. ~ 3.8 kPa). Particle growth is governed by the nucleation of Al monomers in the plasma-cavitation bubble, which leads to the formation of the primary particles. In general, these primary particles are

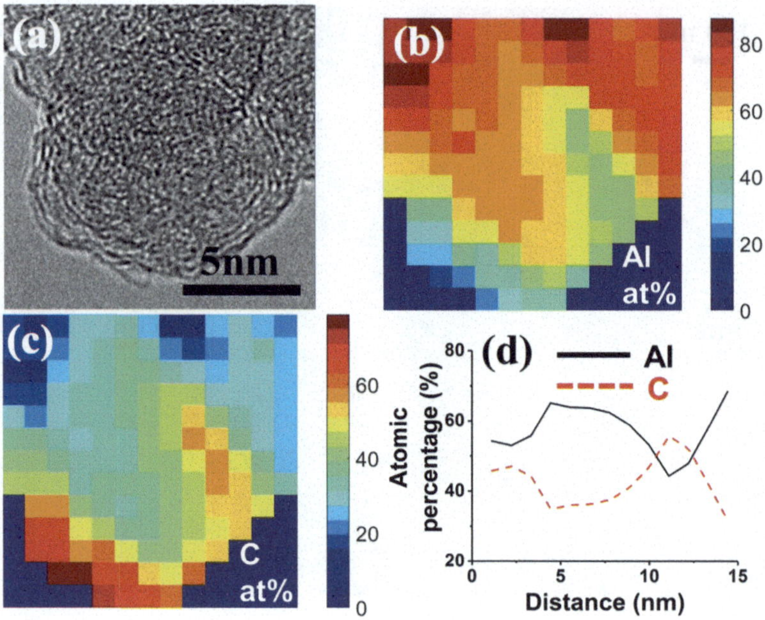

Fig. 5.7 **a** HRTEM image of the Al/C composite prepared in acetone indicating graphitic shell coatings on the Al NP core. EELS elemental mapping for the corresponding **b** C and **c** Al. **d** Al and C atomic ratio (%) distribution across the line scan shown as the red dashed line. **Printed with permission from Ref.** [98]. **Copyright (2019) (Elsevier Publishing)**

released into the solution once the cavitation bubble collapses [82, 134, 135]; they then experience further growth due to coagulation and Al-Al collisions. However, the presence of amorphous carbon produced by the pyrolysis of organic solvents retards the collisions between Al particles and, subsequently, further growth of the Al nanoparticles. Since the vapor pressure of acetone is ~1 order of magnitude greater than toluene, more carbon will be produced due to pyrolysis of organic solvent. Hence, from a simple reaction kinetics analysis where the number of collisions between C and Al particles $N_{\text{carbon}-\text{Al}} = k_{\text{carbon}-\text{Al}} n_c n_{Al}$ such that $k_{\text{carbon-Al}}$ is the free molecular collision kernel for the C and Al monomer concentrations n_c and n_{Al}, respectively, $N_{\text{carbon-Al}}$ will be greater in acetone as compared to toluene (i.e., n_c will be larger for acetone). This proposed route of retardation in Al aggregation due to enhanced pyrolysis is also supported by the smaller particle sizes as well as the lower standard deviation in the particle size distribution obtained from acetone.

The effect of ablation time on the particles size distribution has been investigated for the samples in toluene. Figure 5.9 shows the particle size distributions obtained for the Al pellets ablated in toluene for 2, 4, 6, and 8 min at 2 J/cm^2 laser fluence. Here, the size distribution shows a bimodal behavior. As the laser ablation time increases from 2 to 8 min, the first peak becomes wider ($\sigma = 10.7$ nm for 2 min and $\sigma = 18.1$ nm for 8 min). Moreover, the second peak at 2 min (~77 nm) shifts to

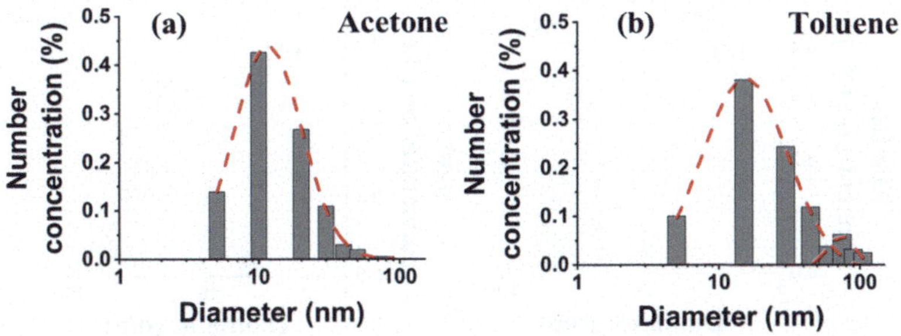

Fig. 5.8 Effect of organic solvents on the size distributions of the C/Al NPs prepared in **a** acetone and **b** toluene for ablation at 2.6 J/cm^2 laser fluence. **Printed with permission from Ref.** [98]. **Copyright (2019) (Elsevier Publishing)**

smaller particles at 8 min (~25 nm) as its intensity correspondingly increases. The observation indicates that as the ablation time increases, the difference between the first and second peaks in terms of the modal size and concentration decreases. Upon considering the free molecular collision kernel for the suspended particles in the solution: $k_{\text{particles}} \sim \sqrt{T_{\text{Solution}}}$, where T_{solution} represents the solution temperature, one can easily observe that the particle size distribution evolution as a function of the ablation time can be explained by the increase in the solvent temperature and thermal stresses which, in turn, promote aggregation and sintering of the particles.

Further investigations have been carried out on the effect of laser fluence on the particles size distributions. Figure 5.10 illustrates the particle size distributions from Al targets ablated in acetone and toluene at 2.6 and 10.5 J/cm^2 laser influences for 4 min. In both the cases, one observes that as the laser fluence increases, the NP modal sizes and the standard deviations become smaller. In the case of acetone, the distribution peak of 15.7 nm ($\sigma = 11.0$ nm) at the laser influence of 2.6 J/cm^2 reduces to 12.3 nm ($\sigma = 6.3$ nm) at 10.5 J/cm^2. The same results can be seen in toluene wherein a bimodal distribution with peak sizes at 77.2 nm ($\sigma = 17.2$ nm) and 25 nm ($\sigma = 25.3$ nm) at the laser fluence of 2.6 J/cm^2 shifts to a unimodal distribution with peak size at 17.0 nm ($\sigma = 10.9$ nm) for the laser fluence of 10.5 J/cm^2. These results are also in agreement with previous studies from our group on cobalt nanostructures [89]. The systematic decrease in the particles size can be explained based on the nucleation and formation of primary Al NPs. As the laser fluence increases more Al is ablated, which results in greater Al monomer concentration (n_{Al}). The saturation ratio (S) of Al monomers that encounter gas-phase synthesis inside the cavitation bubble is defined as: $S = n_{\text{Al}}/n_s$, where n_{Al} is the Al monomer concentration and n_s is the saturated monomer concentration. Therefore, the volume loading and S of Al monomers increase as n_{Al} increases. According to classical nucleation theory (CNT) [136, 137], the size of the nucleated primary particles (d*) is proportional to the inverse of saturation ratio logarithm:

$$d^* = \frac{2\theta}{3\ln(S)}\left(\frac{6v_1}{\pi}\right)^{1/3}$$

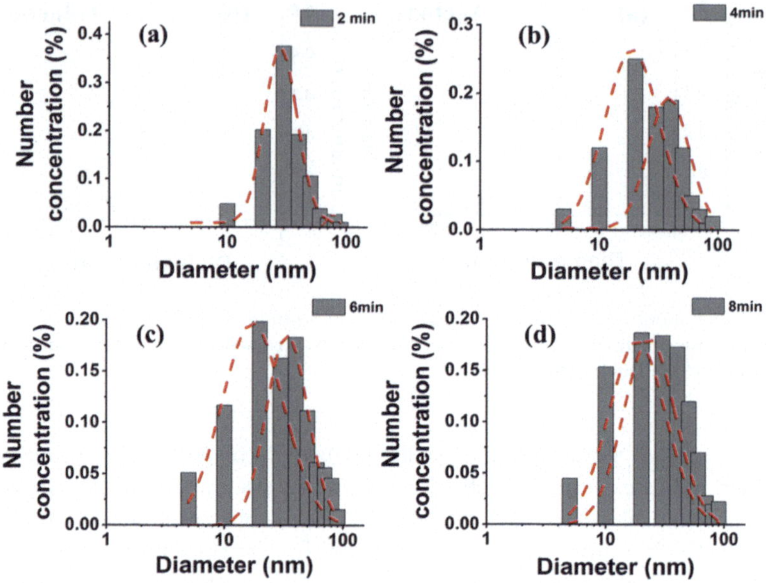

Fig. 5.9 Effect of laser ablation time on size distribution of the C/Al NPs prepared from LASiS in toluene at 2 J/cm^2 laser fluence. **Printed with permission from Ref.** [98]. **Copyright (2019) (Elsevier Publishing)**

where θ is the normalized surface tension of Al particles and v_l is the volume of Al monomer. Hence, the increase in laser fluence results greater saturation ratio and consequently leads to smaller primary particles.

5.3.1.3 Energetic Characterization

For analyzing the energetic properties of the aforementioned C/Al NP composites, four samples were prepared by ablating the Al target for 4 min under various laser fluxes in acetone or toluene and tested using LASEM. Figure 5.11 illustrates the laser-induced shock velocities measured for these samples, as compared to commercial nano-sized energetic Al powders with a native oxide layer. The blank indicates the laser-induced shock velocity for the inert substrate (double-sided tape). Comparing the LASEM results for the laser-induced shock velocities (m/s) from blank substrate (control) to the Al samples, one can immediately notice the increase in the velocities for the Al-containing particles as a result of fast exothermic reactions on the microsecond timescale. The increase in laser-induced plasma temperature due to the exothermic reactions between Al powder and oxygen (either from the sample or air entrained in the plasma) has been previously observed [138]. The

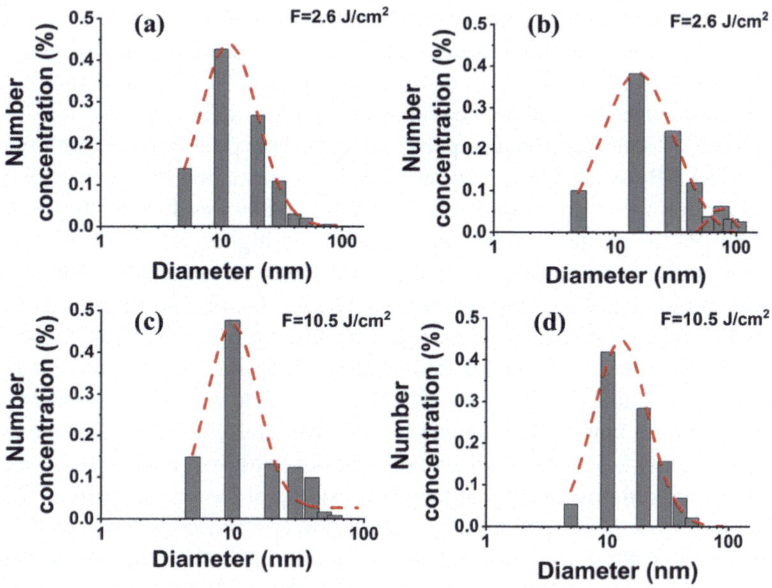

Fig. 5.10 Effect of laser fluence (F, J/cm^2) on the size distribution of C/Al NPs prepared via LASiS in **a** and **c** acetone and **b** and **d** in toluene. **Printed with permission from Ref.** [98]. **Copyright (2019) (Elsevier Publishing)**

current results confirm the ability of the LASEM technique to detect exothermic energetic behavior via laser-induced shock velocity increases from known metal-based energetic material analytes.

Moreover, it can be seen that the C/Al NP composites prepared in toluene or acetone show higher characteristic laser-induced shock velocities (>690 m/s) as

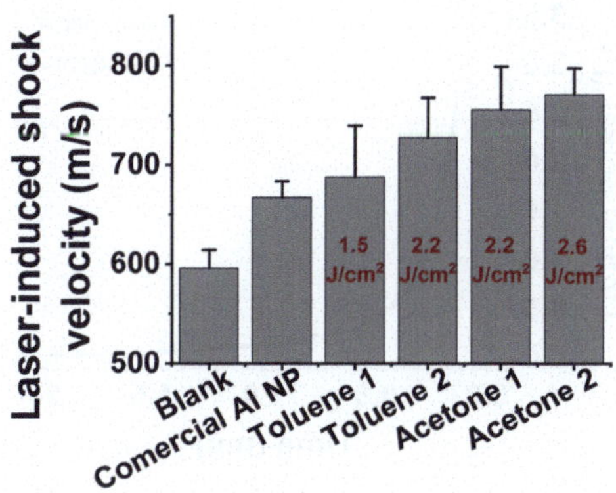

Fig. 5.11 Comparison of laser-induced shock velocities from LASEM measurements on samples prepared in acetone and toluene at various laser fluence values as compared to commercial Al NPs and blank. **Printed with permission from Ref.** [98]. **Copyright (2019) (Elsevier Publishing)**

compared to the commercial nano-sized Al powder (~670 m/s). In fact, the samples prepared in acetone with laser fluences of 2.2 and 2.6 J/cm^2, having peak sizes ~15 nm and relatively narrow size distributions, produce the highest shock velocities (>755 m/s). One can explain the enhanced energetic behavior observed from the synthesized C/Al samples from two perspectives: (1) the protective C/graphitic shells protect the Al NP surfaces from excess oxide layer formation that would hinder the reaction kinetics due to diffusion-limited oxidation pathways through the oxide shell layers; and (2) the earlier onset of pyrolysis in the higher vapor pressure acetone solvent promotes early formation of the amorphous carbon matrix and graphitic structures, dramatically slowing down particle aggregation and thereby generating smaller C/Al NPs (particularly in acetone) which exhibit greater reactivity due to the larger surface area and smaller fuel-oxidizer diffusion length scales. It is well-known from past studies on energetic NPs that as particle sizes decrease, the burning rates and oxidation kinetics become more effective. Here, in the case of the toluene samples, particles show a bimodal size distribution with two modal peaks at 25 nm and 77.2 nm. It is highly likely that the NPs residing at the second peak of 77.2 nm do not undergo complete oxidation on the microsecond-timescale due to the larger sizes and lower surface-to-volume ratios. Therefore, the bulk of the Al core possibly acts as a dead load during the energy release and retards the energetic reactions. Confirmation of this is the more extensive combustion emission on the millisecond timescale from the C/Al NPs synthesized in toluene, as compared to the C/Al NPs from acetone that primarily react on the microsecond timescale (Fig. 5.12). While the initial strong emission signal shown in Fig. 5.12 is a result of the laser-induced plasma emission, combustion of the Al particles following the decay of the plasma appears as a shoulder on this saturated peak. Exothermic reactions that occur later than about 10 μs do not contribute to the laser-induced shock velocity, since at this point in time the shock wave has traveled above the plasma and is only affected by the drag from air.

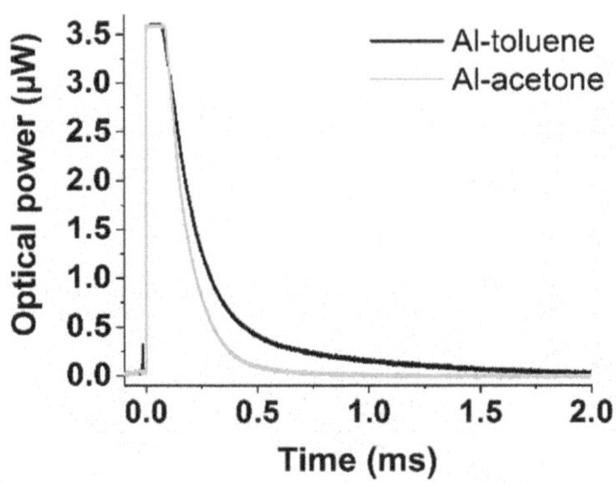

Fig. 5.12 Time-resolved emission from the reaction of laser-excited C/Al NPs prepared in toluene or acetone. **Printed with permission from Ref.** [98]. **Copyright (2019) (Elsevier Publishing)**

In a recent study by Gottfried et al. [131]. LASEM was effectively employed to correlate laser-induced shock velocity to detonation velocities in order to classify various energetic materials such as TNT (trinitrotoluene), PETN (pentaerythritol trinitrate) as exothermic and/or explosive materials [139]. Upon comparing the LASEM recorded shock velocities from our various C/Al NP composites prepared in toluene and acetone with the known energetic data from the aforesaid study as a function of denotation velocities [139], the C/Al NPs synthesized in toluene clearly indicate exothermic rather than explosive behavior, whereas the samples prepared in acetone fall into the explosive range. However, LASEM estimates of detonation performance assume the material is detonable, and it is not clear how the laser-induced shock velocities of pure metallic samples would correlate to the detonation velocities of organic military explosives [140].

5.3.1.4 Oxidative Behavior of C/Al NPs: The Mechanistic Picture

For further investigation into the mechanistic picture behind the excess burning process for the LASiS-synthesized C/Al NPs, Fig. 5.13 shows the thermo-gravimetric analysis (TGA) for the C/Al sample with the highest LASEM activity (i.e., for samples synthesized in acetone at 2.6 J/cm^2 fluence in Fig. 5.11). Figure 5.13 illustrates thermal decomposition/combustion of the sample in air, by plotting weight as a function of temperature (°C). It can be observed that the sample weight decreases continuously up to ~470 °C. This can be readily explained by the fact that as the amorphous carbon matrix/shell burns, it turns to gaseous by-products such as CO and CO_2 and the sample weight decreases. Between ~470 °C up to ~700 °C the weight does not change drastically, possibly due to phase changes from solid to molten Al prior to the onset of Al oxidation. As also discussed and explained in earlier works for oxide coated Al NPs [141], here the core Al NP melts inside the C shell to generate excess internal pressure due to the large volume expansion which, in turn, ruptures the C shell with relatively lower thermal expansion coefficients. This mechanical rupture releases the highly active and hot molten Al to the oxidizing environment resulting in instantaneous burning/oxidation. Here, one needs to bear in mind that the low young's modulus (~4–30 GPa), high melting point (~3500 °C), and thermal conductivity (~10–114 W/m–K) of graphitic shells play a critical role in heating the encapsulated Al core to its melting point, producing large internal stresses without buckling the shell. Upon reaching a critical activation T and P, such scaffolds typically collapse via melt dispersion or nanoexplosions [142, 143] to initiate instantaneous kinetically-driven oxidation of the as-released core nanoclusters. It should be noted that size-dependent Laplace pressure (~surface tens ion/particle radius) and stress inside such NPs can exceed ambient pressure by many orders of magnitude (~10–20 GPa) [144]. Thus, in the TGA data in Fig. 5.10, we observe a sudden increase in the sample weight at around ~700 °C. This behavior can be explained by rupture of the remaining graphitic shells between ~470 and 700 °C and Al oxidation initiation ~700 °C, which consequently increases the sample weight. Previous mass spectrometry studies have also recorded the onset of Al NP oxidation at ~600–700 °C for

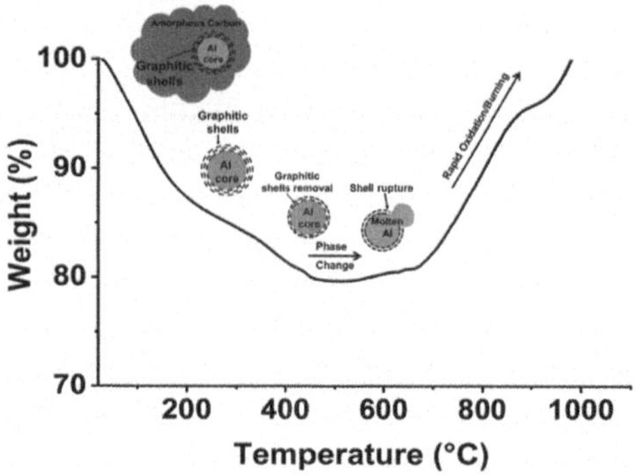

Fig. 5.13 TGA analysis for the sample with highest LASEM activity (acetone, 2.6 J/cm^2). **Printed with permission from Ref.** [98]. **Copyright (2019) (Elsevier Publishing)**

particle sizes < 50 nm [26]. Added to this, one also needs to bear in mind that while internal pressures inside the shell can rise dramatically for smaller NPs (Laplace pressure effects~surface tension/radius) [145], melting points can drop drastically (Tolman effects) [146]—accelerating the phase transition of the Al. In summary, these results are in good agreement with previous studies that have captured similar behavior and temperatures for the oxidation of coated Al NPs [127].

5.3.2 *PtCo/CoO$_x$ NCs as Bi-Functional ORR/OER Electrocatalystic Nanomaterials*

Turning our attention to the search for high-efficacy and stable ORR/OER elecytrocatalysts for robust, reliable, and cost-effective PEMFC operations for sustainable electrochemical energy conversions, alloyed nanocatalysts have gained tremendous research interest in the past decade due to their unique geometric and/or electronic characteristics that dramatically enhance their catalytic activities, while reducing the net Pt content [147–149]. Herein, alloying Pt with 3-d transition metals, such as Co, Ni, Fe, have been found to effectively shrink the lattice constant (geometric effect) and downshift the d-band center (electronic effect), resulting in a reduced oxygen binding energy (eV) and consequently improved specific and mass activities for electrocatalyitc ORR processes. On another hand, regenerative alkaline fuel cells require electrocatalysts to have high activities for both ORR and OER. In such a case NCs composed of a noble metal such as Pt/Pd embedded in metal oxide matrices (mainly the first or second row transition metals, e.g., Co_3O_4, NiO) or perovskite materials through physical/chemical adsorption have been reported to show outstanding bi-functional ORR/OER catalytic activities due to the synergetic "spill-over" effect that benefit the adsorption of reactant molecules on the desired surface as well as the charge transport process within the composites [150–152]. To this end, our group has specifically carried out some foundational works in the early days of developing LASiS-based techniques for the synthesis of tailored IM NPs such as PtCo and PtCuCo NAs with excellent ORR activities [100, 101] as well as, HNCs such as Co_3O_4 NPs/nanorods (NRs) embedded N-doped/reduced GO with selective ORR/supercapacitive activities [99] and more recently, MOF-derived C shell coated PtCo NAs embedded in Co_3O_4-decorated carbonaceous matrices (Pt_3Co@C/Co_3O_4 NCs) [104] as ORR electrocatalysts.

In light of our success at the synthesis of diverse classes of intermetallic HNCs for electrochemical energy conversion/storage applications, this section is dedicated to the results and discussions from one of our seminal works that showcased the use of LASiS-GRR technique for the synthesis of HNCs of binary PtCo NAs in CoO_x matrices as superior bi-functional ORR/OER catalysts [102].

5.3.2.1 Experimental Methods

All LASiS-GRR experiments were carried out in an in-house built cell equipped with facilities for simultaneous injection of metal salt solutions, temperature control as well as ultrasonication, as depicted in Fig. 5.2a earlier. For the synthesis of PtCo/CoO_x NCs via tandem LASiS-GRR technique, different amounts of K_2PtCl_4 salts were dissolved in 40 ml of DI-water to produce four different concentrations of K_2PtCl_4 solutions, namely 25, 60, 120, and 250 mg/l. Thereafter, each of the freshly prepared K_2PtCl_4 solution was transferred into the LASiS cell through the injection unit. The Co pellet was then ablated in those K_2PtCl_4 solutions for 20 min. The obtained colloidal solutions were aged for three days under ambient temperature and pressure conditions. Then, the samples were centrifuged at 4700 rpm for 15 min and decanted after washing with DI-water for two times. Both LASiS and tandem LASiS-GRR experiments were carried out at room temperature with simultaneous ultrasonication. Catalytic activity tests for both pure Co_3O_4 NPs and PtCo/CoO_x NCs were carried out via rotating disk electrode (RDE) measurements in 1 M KOH solution. For ORR tests, polarization curves were recorded at a rotating speed of 1600 rpm in O_2-saturated solution. In order to overcome the electronic conductivity limitations and increase the catalytic surface area for ORR experiments, all the synthesized NCs were dispersed in CB (Vulcan XC-72R, weight ratio NP:CB = 1:4). While for OER tests, products from LASiS-GRR are directly deposited on the working electrode for measurement. Further details for the experimental set-ups for LASiS-GRR synthesis and direct electrochemical measurements can be found elsewhere [102].

5.3.2.2 Mechanistic Picture of the LASiS-GRR Process During Synthesis of PtCo/CoO_x NCs

The galvanic replacement reaction (GRR) between K_2PtCl_4 and Co occurs based on the respective redox potentials for Co/Co^{2+} (−0.28 V vs. SHE) and $PtCl_4^{2-}$/Pt (0.76 V vs. SHE) as: $PtCl_4^{2-} + Co \rightarrow Pt + Co^{2+}$. Theoretically, the displacement

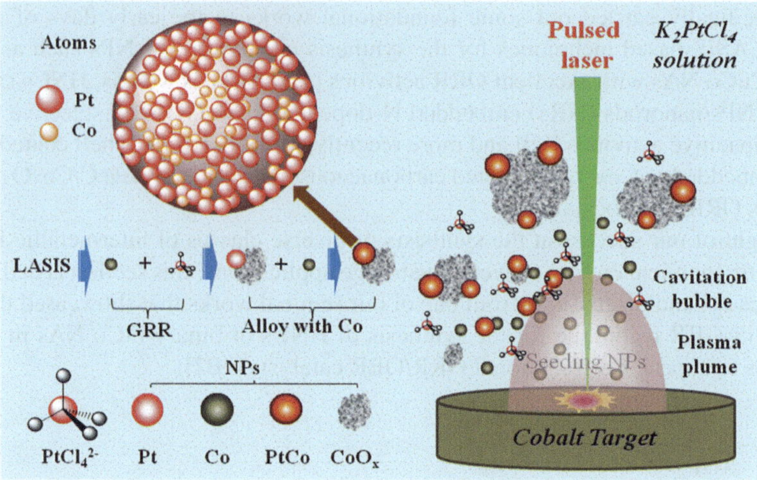

Fig. 5.14 Schematic representation of the tandem LASiS-GRR technique that synthesize PtCo/CoO_x nanocomposites (NCs). **Printed with permission from Ref.** [102]. **Copyright (2016) (Elsevier Publishing)**

reaction can occur between $PtCl_4^{2-}$ and any metal with redox potential < 0.76 V vs. SHE (for $PtCl_4^{2-}$/Pt). Figure 5.14 shows the schematic for the reaction pathways during LASiS-GRR that lead to the NC formation. Pulsed laser beam creates a liquid-confined plasma plume with extremely high temperature and pressure on the Co surface that results in thermal vaporization of the metal target. Large number of seeding NPs then start to form inside an oscillating cavitation bubble resulting from the expanding plasma plume. Thereafter, the seeding NPs undergo collisional quenching and redox reactions with the solution-phase ionic species at the bubble–liquid interface upon the bubbles collapse. However, only part of those seeding Co NPs can go through GRR with K_2PtCl_4 due to the high local concentration but limited reaction rate, thereby leaving the rest of them either coalesce (alloy) with reduced Pt or get oxidized by the solution-phase O_2/H^+. The newly formed intermetallic PtCo NPs then continue to coalesce with each other into large spherical NPs that are partially embedded into the CoO_x matrices.

5.3.2.3 Structure and Composition Characterizations

In this section, LASiS on cobalt with four different initial K_2PtCl_4 concentrations (*i.e.*, 25 mg/l, 60 mg/l, 120 mg/l, and 250 mg/l) produces four different NP colloidal samples labeled here onwards as PtCo-1, PtCo-2, PtCo-3, and PtCo-4, respectively. The corresponding Pt and Co NP concentrations in each of the samples, as revealed from ICP-OES measurements, were summarized in Table 5.1. The results indicate that ~60–70% of the precursor Pt salt is transformed into Pt NPs upon ablation at

60 J/cm^2 fluence for 30 min. The Pt:Co molar ratios was calculated to be 1:9, 1:4, 1:2, and 1:1 for PtCo-1 to PtCo-4, respectively. The unreacted residual K_2PtCl_4 as well as any excess KCl formed are washed away during the centrifugation step. This is also confirmed by the TEM images, as shown in Fig. 5.15a–d, indicating that mainly two types of nanostructures are found in all four samples, namely individual spherical NPs (darker contrast) embedded in a large amount of highly porous "sponge-shaped" nanostructures (lighter contrast). These structures are expected to be CoO_x (lighter) and Pt and/or PtCo intermetallic (darker) NPs, respectively, where the different contrasts in the transmitted electron intensities are due to their different mass to charge ratios. The mean sizes increased from 8.5 nm to 17.7 nm with the increase of the Pt:Co ratios. This is due to the increased concentrations of Pt NPs promoting enhanced coalescence. The SAED patterns (insets in Fig. 5.15a–d) and the calculated d-spacing values (Table 5.2) confirm the existence of Pt and CoO_x NCs from their respective crystalline structures.

The elemental composition and mapping were verified from EDX tests carried out in the STEM mode images. Specifically, for the PtCo-3 NCs, the high-angle annular dark-field (HAADF) images (Fig. 5.16a) show the detailed elemental mappings for Pt, Co, and O, (Fig. 5.16b–d). The clear observation of Pt distribution in the four bright and large spherical NPs at the center can be confirmed from the HAADF image, while the Co is mainly identified in the e background gray areas although it is also strongly detected inside the aforesaid brighter NPs. In contrast, O is uniformly distributed in the whole image. Figure 5.16e, f also shows the corresponding EDX spectra for the marked *e* and *f* areas in Fig. 5.16a. Strong Pt and Co peaks and a relatively weak O peak are noted inside the marked particles. In comparison, Fig. 5.16f indicates negligible Pt peak along with a relatively strong O peak in the background areas. It is noted here that the Cu and C signals in the spectra are from the TEM carbon film with copper grids. The absence of any other detectable elements in the EDX data indicates that all the residual chemicals such as K_2PtCl_4 and KCl have been washed away by the centrifuging process. Aforementioned results for the elemental mappings clearly suggest the formation of PtCo NA inside the bright spherical NPs. This also validates the earlier assumption that the NCs are made of PtCo NA (brighter NPs) embedded in the CoO/Co_3O_4 matrices.

Table 5.1 Pt and Co concentrations from ICP-OES measurements for various $PtCo/CoO_x$ NCs synthesized by LASiS-GRR at different initial K_2PtCl_4 salt concentrations

	Initial K_2PtCl_4 Conc. (mg/l)	Pt Conc. (mg/l)	Co Conc. (mg/l)	Molar Pt/Co ratio
PtCo-1	25	7.3	20	1:9
PtCo-2	60	19	23	1:4
PtCo-3	120	42	25	1:2
PtCo-4	250	81	25	1:1

Printed with permission from Ref. [102]. **Copyright (2016) (Elsevier Publishing)**

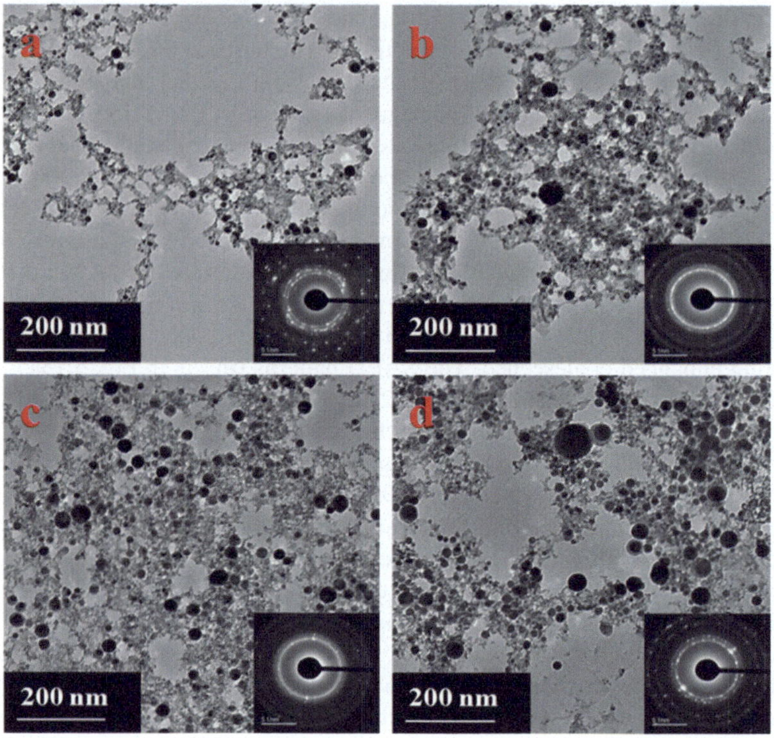

Fig. 5.15 **a–d** TEM images along with corresponding SAED patterns (inset) for PtCo/CoO_x NCs synthesized by LASiS-GRR with different Pt:Co atomic ratios of: **a** 1:9 (PtCo-1); **b** 1:4 (PtCo-2); **c** 1:2 (PtCo-3); **d** 1:1 (PtCo-4). Scale bar in for SAED patterns is 5 (1/nm); **(e–h)** Corresponding size distributions for dark colored PtCo alloyed NPs in **a–d**. **Printed with permission from Ref.** [102]. **Copyright (2016) (Elsevier Publishing)**

Table 5.2 Experimental d-spacing values calculated from the SAED results in Fig. 5.15

Experimental d-spacing (Å)							
PtCo-1		PtCo-2		PtCo-3		PtCo-4	
2.86	Co_3O_4 (220)	2.47	CoO (111)	2.26	Pt (111)	2.27	Pt (111)
2.43	Co_3O_4(311)	2.27	Pt (111)	1.95	Pt (200)	1.96	Pt (200)
2.1	CoO (200)	1.95	Pt (200)	1.39	Pt (220)	1.39	Pt (220)
1.95	Pt (200)	1.5	CoO (220)				
1.55	Co_3O_4 (333)	1.39	Pt (220)				
1.29	CoO (311)						

Printed with permission from Ref. [102]**. Copyright (2016) (Elsevier Publishing)**

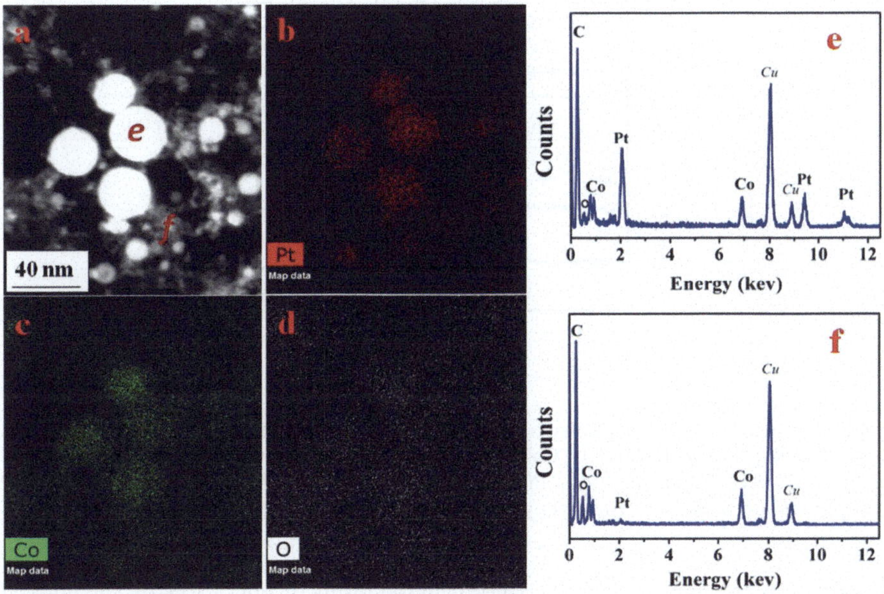

Fig. 5.16 EDX results for PtCo-3 sample along with: **a** HAADF image, and elemental mappings for **b** Pt, **c** Co, and **d** O, respectively; **e, f** corresponding EDX spectra at the positions marked in **a** (Cu and C signals are from the TEM carbon film with copper grids). **Printed with permission from Ref.** [102]. **Copyright (2016) (Elsevier Publishing)**

To corroborate the abovementioned results, Fig. 5.17 summarizes the XRD data for all the PtCo/CoO_x NCs dispersed in carbon black, where the sample from LASiS on Co (without the K_2PtCl_4 salt) displays the distinct Co_3O_4 (311) peak (PDF-#42–1467) at $2\theta = 36.9°$ but does not show any characteristic peak for CoO (PDF-#43–1004). In contrast, the PtCo-1 sample indicates a CoO (200) peak at $2\theta = 42.4°$ along with a minor peak at $2\theta \sim 36.6°$ which is assigned as an overlapped peak of CoO (111) ($2\theta = 36.5°$) and Co_3O_4 (311) ($2\theta = 36.9°$). These results further confirm the previous SAED data in Fig. 5.15 (d) that the CoO gets partially conserved in the PtCo/CoO_x NCs that do not undergo further oxidation during the ageing process. Meanwhile, two distinct characteristic peaks are also observed for this sample at $2\theta = 40.2°$ and $2\theta = 46.7°$ that are assigned to the (111) and (200) peaks for PtCo alloy. Here, one needs to note that the 2θ values for standard Pt are 39.8° (111) and 46.2° (200) (PDF-#04–0802). The shift of these peaks to higher angles is attributed to the Pt alloying with Co that result in a shrunken lattice constant calculated to be ~3.88 Å as compared to 3.92 Å for standard Pt. Besides, Fig. 5.17 also reveals that the alloyed PtCo peaks quickly become dominant with the increase of Pt:Co ratio. Specifically, for the case of PtCo-3 and PtCo-4 (PtCo ratio equals to 1:2 and 1:1), the CoO_x peaks are barely discernable, which is ascribed to the much higher crystallinity and hence, the diffraction pattern intensity for Pt alloyed with Co as compared to that for CoO_x. It needs to be noted here that this phenomenon is also

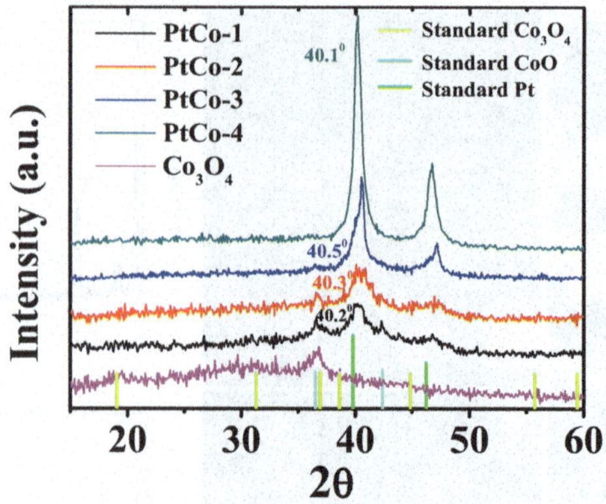

Fig. 5.17 Comparison of XRD patterns for PtCo/CoO_x NCs with different Pt:Co ratios along with standard peak positions for Co_3O_4, CoO, and Pt indicated by yellow, purple, and green vertical lines, respectively. Values marked in the figure indicate the PtCo (111) peak positions. **Printed with permission from Ref.** [102]. **Copyright (2016) (Elsevier Publishing)**

supported by the previous SAED patterns in the insets of Fig. 5.15a–d. The alloyed Pt:Co ratios from PtCo-1 to PtCo-4 are estimated to be 8.6:1, 6.7:1, and 4.6:1 and 11.5:1, respectively according to the Vegard's law [153]. As also described earlier in Fig. 5.14, the seeding Co NPs from the cavitation bubble go through two competing reactions, namely, either GRR with Pt^{2+} or oxidation by solution-phase O_2 and H^+. When the initial K_2PtCl_4 concentration is low, a large portion of Co will react with solution-phase O_2/H^+, thereby leaving few available seeding Co NPs to alloy with Pt formed via GRR mediated reduction by Co. Hence, the Co:Pt ratio in the PtCo alloys rises at first with increasing initial K_2PtCl_4 concentrations in the solution. However, beyond a critical value, further increase of Pt^{2+} ion concentrations lead to oxidation of Co atoms in the initially formed PtCo alloy. Such a de-alloying process reduces the amount of alloyed Co in the PtCo NAs, and has already been reported by some other works [154].

5.3.2.4 Electrochemical Characterizations for ORR/OER Electrocatalytic Activities

Direct electrochemistry measurements shown in Fig. 5.18a compares the ORR catalytic performance for each of the catalyst samples under study through the linear sweep voltammograms (LSV) tests. The results indicate a remarkable improvement for the ORR activities with the increase of Pt to Co ratio from 1:9 (PtCo-1) to 1:2 (PtCo-3). Specifically, the half-wave potential for the PtCo-3 sample is improved to 860 mV, which is almost comparable to the corresponding values for standard Pt/C samples (870 mV). The ORR overpotential for this sample is calculated to be 370 mV based on the standard potential for reduction of oxygen to water being

1.23 V. However, as the Pt:Co ratio further increases to 1:1 for PtCo-4, the half-wave potential slightly reduces to 850 mV, with the overpotential calculated to be 380 mV. Here, the LSV plots on pure Co_3O_4 NPs and commercial Pt NPs are used for comparing against the respective OER and ORR performances of the PtCo/CoO_x samples synthesized via LASiS. Our earlier studies on LASiS-driven synthesis of Co_3O_4 NPs [89] had indicated that half-wave potentials and diffusion limiting currents measured for Co_3O_4 samples made under high laser fluence (~60 J/cm^2) were considerably better than those made under low laser fluence conditions (~1 J/cm^2). Hence, all catalytic studies presented in this section are for the Co_3O_4 and PtCo/CoO_x samples synthesized under high fluence laser conditions. It needs to be pointed out here that the ORR catalytic activities for the Co_3O_4 NPs itself faired extremely well; previous studies have reported moderately high ORR activities for Co_3O_4 NPs [29, 30, 155]. But, we believe that the clean, one-step synthesis process for LASiS plays a significant role here in generating such high activities from even the Co_3O_4 NPs since the LASiS technique does not involve any additional chemicals (surfactants, reducing agents, etc.) that can retard or poison their catalytic activities. Besides, Fig. 5.18b compares the Tafel plots for the abovementioned samples generated from Fig. 5.18a over the low overpotential regions. The measured Tafel slope values are 53.0, 42.2, 40.3, 39.2, and 54.6 mV/dec, respectively, as compared to ~66.2 mV/dec for standard Pt/C samples, thereby indicating higher charge transfer coefficients for the as-synthesized catalysts. Figure 5.18c further compares the mass activities per unit Pt loading amount at 0.85 V vs. RHE for the PtCo/CoO_x NCs studied. The results indicate a much higher mass activity for PtCo-3 sample (0.73 mA μg^{-1} Pt) than the commercial Pt/C (0.54 mA μg^{-1} Pt). The other three samples (PtCo-1, PtCo-2, and PtCo-4) demonstrate lower but comparable mass activities, although the PtCo-4 sample with an increased amount of Pt indicates a marked drop in the ORR activity.

The dynamics of the electron transfer process during ORR activities of the aforementioned catalysts are analyzed using the Koutecky-Levich (KL) equation for rotating disk voltammetry (RDV) measurements carried out at different speeds. Figure 5.18d shows the slopes for the KL plots generated from rotation-rate dependent current–potential curves (inset in Fig. 5.18d) for the PtCo-3 NCs in the range of 0.70–0.83 V. The slopes estimate the number of electrons transferred (n) to be 3.9–4.0, thereby indicating an ideal four-electron transport process for the ORR activity. (Detailed RDV data and KL plots for PtCo-1, PtCo-2, and PtCo-4 NCs are provided in the supplementary information with the original article [102] that indicate the corresponding electron transfer numbers to be 3.8–4.0 for all of them).

In addition to the good ORR catalytic activities, chronoamperometric (CA) measurements also reveal excellent stabilities for the PtCo-3 NC samples. Figure 5.18e compares the normalized current density (%) at the corresponding half-wave potentials for the Co_3O_4, PtCo-3, and standard Pt/C samples. As can be seen from the results, the ORR current densities decay by less than 15% over 12000 s of continuous operation for both the Co_3O_4 and PtCo-3 samples. In contrast, the standard Pt/C sample indicates a ca. 28% current density reduction over the same period of time. The decay in the catalytic activities of standard Pt/C samples is attributed to

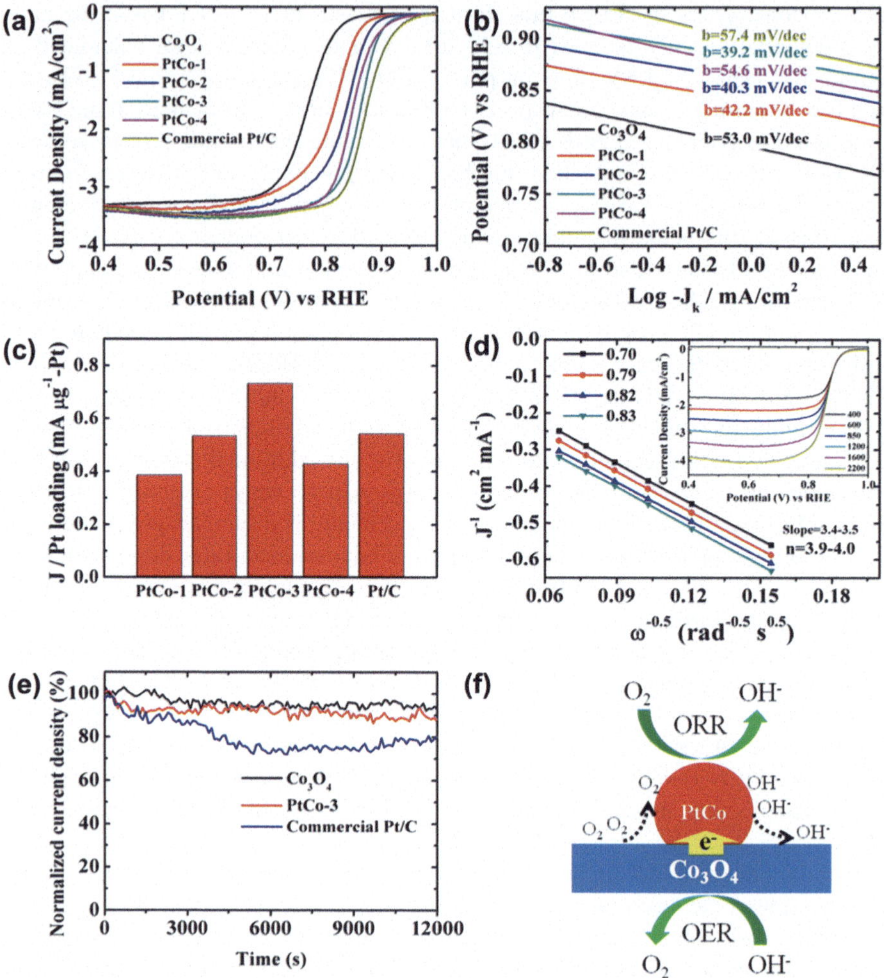

Fig. 5.18 Electrochemistry data for as-synthesized Co_3O_4 NPs and PtCo/CoO_x NCs compared to standard Pt/C samples using: **a** LSV data for ORR analysis in 1 M KOH electrolyte saturated with dissolved O_2 at 1600 rpm and scan rate of 5 mV/s; **b** Tafel plots from **a**; **c** comparison of mass activities per unit Pt mass loading at 0.85 V vs. RHE; **d** Koutecky-Levich plots from rotating disk voltammogram (RDV) data (shown in inset) for PtCo-3 at different potentials (0.70–0.83 V); **e** stability comparison for Co_3O_4, PtCo-3, and standard Pt/C through chronoamperometry (CA) tests; **f** schematic illustrating the *"spill-over"* effect. **Printed with permission from Ref.** [102]. **Copyright (2016) (Elsevier Publishing)**

its surface oxidation as well as particle dissolution and aggregation in the alkaline electrolyte [156, 157], which hardly occur for the metal oxides (Co_3O_4/CoO). In fact, in the case of the PtCo/CoO_x NCs, the existence of the sponge-shaped CoO_x serve as a matrix material that protects the PtCo alloyed NPs from aggregation or dissolution. Meanwhile, the alloyed PtCo structure also helps to slow down the oxidation of Pt due to the much higher electron negativity of Co.

Based on these results, we posit that the excellent ORR activities for these materials are mainly attributed to the following reasons. Firstly, the formation of alloyed PtCo nanostructures shrinks the Pt lattice constants and decreases the effective sites for OH^- adsorption. Added to that, it enhances the Pt-O bonding due to higher 5d orbital vacancies in its electronic structure that promotes the donation of π electrons from O_2 to Pt. Hence, the sites on the PtCo alloy NPs (known for their ORR activities) preferentially promote both O_2 adsorption and OH^- desorption, both benefit the ORR efficiency. Secondly, the NC of the metal-transition metal oxide (NM-TMO) system further benefits the ORR process by providing a synergic "spill-over" effect [150], as depicted schematically in Fig. 5.18f. Specifically, the OH^- reduced from O_2 is readily desorbed from the active PtCo sites and transferred to the sponge-shaped CoO_x sites which are less active according to their respective M–O bonding strength and intermolecular affinities. Thus, the best ORR activity is observed for PtCo-3 catalyst with optimal Pt contents (33.3 molar %, 62.3 wt. %) that leads to higher degree of Pt–Co alloy formation with more lattice shrinkage (confirmed by Fig. 5.18c and Table 5.2) and appropriate NA sizes (mean size = 11.7 nm), combined with the balanced PtCo to CoO_x ratio that potentially maximizes the aforesaid synergic "spill-over" effects.

Finally, the OER catalytic activities for the PtCo NC samples are investigated by comparing with the results for pure Co_3O_4 and standard Pt/C samples, as shown in Fig. 5.19. For comparing the OER overpotentials, all potential measurements were carried out with 10 mA/cm^2 as the metrics due to its well-known relevance to typical solar fuel characterizations. The results reveal that, with an overpotential of 385 mV (Fig. 5.19a) and a measured Tafel slope of 62.4 mV/dec, the LASiS generated Co_3O_4 NPs themselves provide exceedingly well OOR activities. As expected, the standard Pt/C sample shows the worst OER activities (with an overpotential of ~640 mV). As also mentioned earlier, the high OER activity for our Co_3O_4 NPs is attributed to the clean synthesis route of LASiS that produces the sponge-like nanostructures with enhanced surface area that do not have any surface contamination from unwanted chemicals during the preparation. But, the most significant observation from Fig. 5.19a is that the PtCo-1 sample exhibits an even smaller OER overpotential (380 mV) than the Co_3O_4 sample that is typically well-known for OER activities. In fact, the overpotentials for PtCo-2 and PtCo-3 are also comparable (~385 and 386 mV respectively), indicating that these PtCo/CoO_x NCs exhibit good OER catalytic activities as well. Here we also note that the PtCo-4 sample displays a relatively poorer activity (overpotential of ~435 mV) that could be possibly due to the scarcity of CoO_x in the NC catalyst. It is also noted here though that Tafel slopes for the PtCo NCs are slightly larger (~71–72 mV/dec for PtCo-1 and PtCo-4, whereas ~66–67 mV/dec for PtCo-2 and 3) as compared to those for pure Co_3O_4 (~62.4 mV/dec) as seen from Fig. 5.19b. The enhanced OER performance can also be attributed to

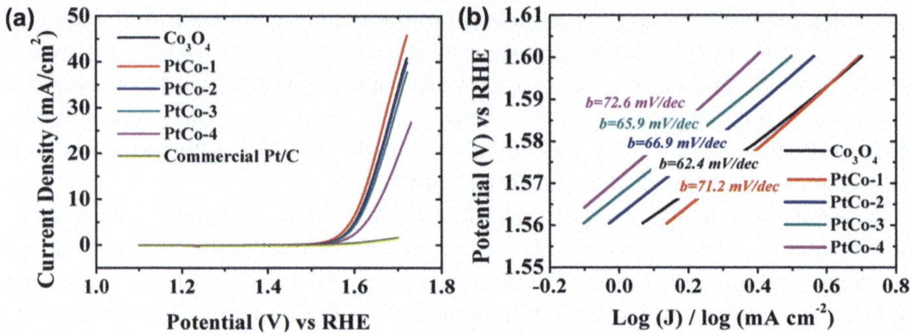

Fig. 5.19 **a** Linear sweep voltammograms for OER analysis on PtCo/CoO_x NCs compared to those for Co_3O_4 NPs and standard Pt/C samples; **b** corresponding Tafel plots in the potential range of 1.55–1.60 V. **Printed with permission from Ref.** [102]. **Copyright (2016) (Elsevier Publishing)**

the aforementioned synergic "spillover" effect for specific NM-TMO systems, where the produced O_2 is desorbed from active CoO_x surfaces and transferred to relatively inert PtCo NA sites with higher O_2 affinity. For better understanding, the schematic shown earlier in Fig. 5.18f illustrates the detailed mechanistic picture behind the synergic "spill-over" effects responsible for the site-specific adsorption/desorption of the desired species to promote the bi-functional catalytic performances in the aforementioned NCs. For both ORR and OER, the PtCo sites and CoO_x matrices provide synergic support for each other wherein each of the sites provides refuge for the undesirable species from the other sites, thereby promoting both the reactions.

The combined overpotentials for PtCo-1 to PtCo-4 are calculated to be 795, 775, 756, and 815 mV respectively, which remarkably outperform the overpotentials for either Co_3O_4 or the commercial Pt/C (i.e., 845 mV for Co_3O_4 and 1000 mV for commercial Pt/C), as shown in Fig. 5.20a, b as well as charted out in details in Table 5.3. These results indicate that the synthesized PtCo NCs exhibit superior bi-functional catalytic performances for both ORR and OER processes, especially for the PtCo-3 sample with an optimal Pt amount (33.3 molar %, 62.3 wt. %). Overall, we attribute the excellent bi-functional catalytic properties of the PtCo/CoO_x NCs to the unique heteronanostructuring of alloyed PtCo NPs embedded in the sponge-shaped CoO_x matrices which, while contributing to the enhanced ORR and OER behaviors due to the synergic "spillover" effects, prevent the PtCo NPs from aggregation and dissolution in the alkaline media.

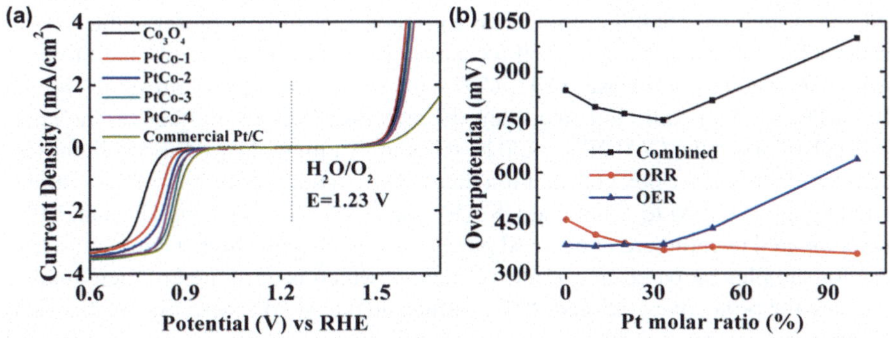

Fig. 5.20 **a** Comparison of bi-functional catalytic activities for both OER and ORR via LSV plots; **b** evolution of ORR, OER and combined overpotentials for catalytic samples understudy with the increase of Pt molar ratio. **Printed with permission from Ref.** [102]. **Copyright (2016) (Elsevier Publishing)**

Table 5.3 Comparison of OER and ORR overpotentials for PtCo/CoO_x NCs, Co_3O_4 and the commercial Pt/C samples along with the respective combined overpotentials

Sample	ORR OP (mV)	OER OP (mV)	Combined OP (mV)
Co_3O_4	460	385	845
PtCo-1	415	380	795
PtCo-2	390	385	775
PtCo-3	370	386	756
PtCo-4	380	435	815
Pt/C	360	640	1000

Printed with permission from Ref. [102]. **Copyright (2016) (Elsevier Publishing)**

5.4 Conclusion

This chapter discussed the detailed experimental procedures, characterizations, and results for the advanced design of two metal-based functional HNCs with engineered interfacial activities synthesized via novel LASiS-based techniques. Specifically, the HNC studies presented in the aforesaid sections—namely, (1) Graphitic shell coated Al NPs and (2) PtCo/CoOx NCs—pave the path for new frontiers in the development of advanced nanocomposites for energetic applications in solid-state jet propulsion and munition studies, and ORR/OER electrocatalytic applications for electrochemical energy conversions.

Section 5.3.1 discussed the LASiS-driven synthesis of energetic NCs comprising graphitic shell coated Al NPs embedded in carbonaceous matrix via laser ablation in acetone or toluene under various laser fluences. The design of these shell-core Al/C NCs provided a creative approach to bypass the diffusion limitations of the combustion of the 1st generation bare Al NPs used as energetic materials. To this end, the graphitic C shell is employed as a protective coating to preserve the Al core surfaces from unwanted oxide shell formation while promoting the C shell to burn with spontaneous gas release (CO2, CO) without; leaving any ash residues on the NP surfaces. The results indicated that the size distributions and hence, the energetic behaviors of the C/Al NPs could be tailored by tuning the LASiS medium (i.e., organic solvent), laser flux and ablation times. The energetic activities of the samples tested by the LASEM technique at US ARL labs (Aberdeen Proving Ground, MD), and compared to commercial nano-Al powders, indicated that higher energetic activities for samples prepared in acetone or toluene which is attributed to their smaller sizes and the protective graphitic shell coatings on the Al NPs. Specifically, the C/Al NPs prepared in acetone exhibited higher laser-induced shock velocities as compared to the samples synthesized in toluene. Although the C/Al NPs made via LASiS in toluene behave as exothermic materials when compared to their counterparts grown in acetone that exhibit more explosive properties (i.e., react more rapidly), a definitive conclusion on their exothermic/explosive behaviors would demand further investigation into their structure–property characteristics and reaction propagation mechanisms. Current ongoing research efforts in our group are focused on the refinement of LASiS process protocols to: (1) tune the energetic behaviors through structure–property relations of the HNCs, and (2) produce more homogeneously distributed Al NPs with C shells, that could provide additional enhancements in the microsecond-timescale energy release. In summary, this work would pave the path for the future design of energetic HNCs comprising metal/intermetallic NPs encapsulated in nanostructured graphitic shell coatings and/or fullerene-like layers that can be tailored via LASiS-based techniques to tune their reactive behaviors for a myriad of defense and solid-state propellant applications.

On the other hand, Sect. 5.3.2 presented the innovative tandem LASiS-GRR technique as a unique one-pot one-step, "green" synthesis route to manufacture complex PtCo/CoO_x NCs with outstanding bi-functional ORR/OER electrocatalytic activities. The transformative concept here is the ability to synthesize such complex HNCs in one step without the need for any harsh chemical reagents and/or, surfactants. This is made possible by the out-of-equilibrium thermodynamic conditions emerging from extreme physicochemical conditions of high-energy LASiS operating in tandem with chemical reduction processes (GRR in this case). The synthesized PtCo NCs exhibited excellent yet, stable catalytic activities for both oxygen reduction reactions (ORR) and oxygen evolution reactions (OER) in alkaline media. Specifically, the results reported a combined overpotential of 756 mV vs. RHE for the PtCo-3 NC (33.3 molar %, 62.3 wt. %), which is the highest value ever reported using carbon black as the supporting material while reducing the active precious Pt contents in the catalyst compositions. This excellent bi-functional activity is attributed to the structural and electronic properties of PtCo NAs that promote oxygen adsorption, as well as to the sponge-shaped CoO_x matrices in which the PtCo NPs are embedded that bring about the synergic "spillover" effects to facilitate both ORR and OER activities. Added to this, the CoO_x matrix also prevents the PtCo NPs from aggregation and dissolution in alkaline electrolytes. The encouraging results from these studies can provide a paradigm shift for clean, facile, and non-equilibrium synthesis routes for the future design and large-scale synthesis of complex intermetallic NCs/NAs in a rapid and cost-effective fashion.

References

1. Ferrando R, Jellinek J, Johnston RL (2008) Nanoalloys: from theory to applications of alloy clusters and nanoparticles. Chem Rev 108:845–910
2. Sun SH, Murray CB, Weller D, Folks L, Moser A (2000) Monodisperse FePt nanoparticles and ferromagnetic FePt nanocrystal superlattices. Science 287:1989–1992
3. Goll D, Kronmuller H (2000) High-performance permanent magnets. Naturwissenschaften 87:423–438
4. Stern RA et al (2003) Fe3Pd ferromagnetic shape memory alloys. J Appl Phys 93:8644–8646
5. Roucoux A, Schulz J, Patin H (2002) Reduced transition metal colloids: a novel family of reusable catalysts? Chem Rev 102:3757–3778
6. Casado-Rivera E et al (2004) Electrocatalytic activity of ordered intermetallic phases for fuel cell applications. J Am Chem Soc 126:4043–4049
7. Zhang CJ, Baxter RJ, Hu P, Alavi A, Lee MH (2001) A density functional theory study of carbon monoxide oxidation on the Cu3Pt(111) alloy surface: Comparison with the reactions on Pt(111) and Cu(111). J Chem Phys 115:5272–5277
8. Kirchheim R et al (1988) Hydrogen in amorphous and nanocrystalline metals. Mater Sci Eng 99:457–462
9. Snyder GJ, Toberer ES (2008) Complex thermoelectric materials. Nat Mater 7:105–114
10. Pei YZ, Heinz NA, Snyder GJ (2011) Alloying to increase the band gap for improving thermoelectric properties of Ag2Te. J Mater Chem 21:18256–18260
11. Lopez MF, Escudero ML (1998) Corrosion behaviour of FeAl-type intermetallic compounds. Electrochim Acta 43:671–678
12. Burda C, Chen XB, Narayanan R, El-Sayed MA (2005) Chemistry and properties of nanocrystals of different shapes. Chem Rev 105:1025–1102
13. Yin YD et al (2004) Formation of hollow nanocrystals through the nanoscale Kirkendall Effect. Science 304:711–714
14. Richards R, Bonnemann H (2005) Nanofabrication towards biomedical applications, Challa JC, Kumar SSR, Leuschner C. Wiley-VCH
15. Mench M, Kuo K, Yeh C, Lu Y (1998) Comparison of thermal behavior of regular and ultra-fine aluminum powders (Alex) made from plasma explosion process. Combust Sci Technol 135:269–292
16. Rai A, Park K, Zhou L, Zachariah MR (2006) Understanding the mechanism of aluminium nanoparticle oxidation. Combust Theor Model 10:843–859
17. Mukherjee D, Davari SA (2017) Energetic materials: from cradle to grave, Shukla MK et al (eds). Springer International Publishing, pp 271–341
18. Chiaverini MJ, Serin N, Johnson DK, Lu YC (1997) Instantaneous regression behavior of HTPB solid fuels burning with GOX in a simulated hybrid rocket motor. Int J Ener Mat Chem Prop 4
19. Renie J, Lilley J, Frederick J, Osborn J, 18th Joint Propulsion Conference, 1110
20. Bakhman N, Belyaev A, Kondrashkov YA (1970) Influence of the metal additives onto the burning rate of the model solid rocket propellants. Phys Combust Explos 6:93
21. Ritter H, Braun S (2001) High explosives containing ultrafine aluminum ALEX. Propellants, Explos, Pyrotech 26:311–314
22. Ilyin A et al (2022) Characterization of aluminum powders I. parameters of reactivity of aluminum powders. Propel, Expl, Pyrotech 27:361–364
23. Brousseau P, Anderson CJ (2002) Nanometric aluminum in explosives. Propellants, Explos, Pyrotech 27:300–306

24. Weiser V, Kelzenberg S, Eisenreich N (2001) Influence of the metal particle size on the ignition of energetic materials. Propellants, Explos, Pyrotech 26:284–289
25. Mukherjee D, Rai A, Zachariah MR (2006) Quantitative laser-induced breakdown spectroscopy for aerosols via internal calibration: application to the oxidative coating of aluminum nanoparticles. J Aerosol Sci 37:677–695. https://doi.org/10.1016/j.jaerosci.2005.05.005
26. Park K, Lee D, Rai A, Mukherjee D, Zachariah MR (2005) Size-resolved kinetic measurements of aluminum nanoparticle oxidation with single particle mass spectrometry. J Phys Chem B 109:7290–7299. https://doi.org/10.1021/jp048041v
27. Pantoya ML, Granier JJ (2005) Combustion behavior of highly energetic thermites: nano versus micron composites. Propellants, Explos, Pyrotech 30:53–62
28. Liang Y et al (2011) Co3O4 nanocrystals on graphene as a synergistic catalyst for oxygen reduction reaction. Nat Mater 10:780–786
29. Wang Y et al (2013) Carbon supported MnOx–Co3O4 as cathode catalyst for oxygen reduction reaction in alkaline media. Int J Hydrogen Energy 38:13611–13616
30. Xu J, Gao P, Zhao TS (2012) Non-precious Co3O4 nano-rod electrocatalyst for oxygen reduction reaction in anion-exchange membrane fuel cells. Energy Environ Sci 5:5333–5339
31. Hong WT et al (2015) Toward the rational design of non-precious transition metal oxides for oxygen electrocatalysis. Energy Environ Sci 8:1404–1427
32. Yuan C, Wu HB, Xie Y, Lou XW (2014) Mixed transition-metal oxides: design, synthesis, and energy-related applications. Angew Chem Int Ed 53:1488–1504
33. Wang D et al (2013) Structurally ordered intermetallic platinum-cobalt core-shell nanoparticles with enhanced activity and stability as oxygen reduction electrocatalysts. Nat Mater 12:81–87
34. Hong JW et al (2012) Controlled synthesis of Pd-Pt alloy hollow nanostructures with enhanced catalytic activities for oxygen reduction. ACS Nano 6:2410–2419
35. Loukrakpam R et al (2013) Atomic ordering enhanced electrocatalytic activity of nanoalloys for oxygen reduction reaction. J Phys Chem C 117:20715–20721
36. Shan S, Luo J, Yang L, Zhong C-J (2014) Nanoalloy catalysts: structural and catalytic properties. Catal Sci Technol 4:3570–3588
37. Godinez-Salomon F, Hallen-Lopez M, Solorza-Feria O (2012) Enhanced electroactivity for the oxygen reduction on Ni@Pt core-shell nanocatalysts. Int J Hydrogen Energ 37:14902–14910
38. Yin Q et al (2010) A fast soluble carbon-free molecular water oxidation catalyst based on abundant metals. Science 328:342–345
39. Kanan MW, Nocera DG (2008) In situ formation of an oxygen-evolving catalyst in neutral water containing phosphate and Co^{2+}. Science 321:1072–1075
40. Kanan MW, Surendranath Y, Nocera DG (2009) Cobalt-phosphate oxygen-evolving compound. Chem Soc Rev 38:109–114
41. Higashi M, Domen K, Abe R (2013) Fabrication of an efficient $BaTaO_2N$ photoanode harvesting a wide range of visible light for water splitting. J Am Chem Soc 135:10238–10241
42. Ohno T, Bai L, Hisatomi T, Maeda K, Domen K (2012) Photocatalytic water splitting using modified GaN:ZnO solid solution under visible light: long-time operation and regeneration of activity. J Am Chem Soc 134:8254–8259
43. Grimaud A et al (2013) Double perovskites as a family of highly active catalysts for oxygen evolution in alkaline solution. Nat Commun 4:2439
44. Suntivich J, May KJ, Gasteiger HA, Goodenough JB, Shao-Horn Y (2011) A perovskite oxide optimized for oxygen evolution catalysis from molecular orbital principles. Science 334:1383–1385
45. Liao M et al (2012) Co_3O_4 nanoparticles as robust water oxidation catalysts towards remarkably enhanced photostability of a Ta_3N_5 photoanode. Adv Funct Mater 22:3066–3074
46. Thangavelu K, Parameswari K, Kuppusamy K, Haldorai Y (2011) A simple and facile method to synthesize Co_3O_4 nanoparticles from metal benzoate dihydrazinate complex as a precursor. Mater Lett 65:1482–1484

47. Mazumder V, Chi M, More KL, Sun S (2010) Synthesis and characterization of multimetallic Pd/Au and Pd/Au/FePt core/shell nanoparticles. Angewandte Chemie-Int Edition 49:9368–9372
48. Le HA et al (2011) Chemical vapor synthesis and characterization of manganese oxides. Chem Vap Deposition 17:228–234
49. Zhang Y et al (2008) Near-monodisperse Ni–Cu bimetallic nanocrystals of variable composition: controlled synthesis and catalytic activity for H2 generation. J Phys Chem C 112:12092–12095
50. Wiechen M, Zaharieva I, Dau H, Kurz P (2012) Layered manganese oxides for water-oxidation: alkaline earth cations influence catalytic activity in a photosystem II-like fashion. Chem Sci 3:2330–2339
51. Armbruester M, Wowsnick G, Friedrich M, Heggen M, Cardoso-Gil R (2011) Synthesis and catalytic properties of nanoparticulate intermetallic Ga-Pd compounds. J Am Chem Soc 133:9112–9118
52. Barcikowski S, Compagnini G (2013) Advanced nanoparticle generation and excitation by lasers in liquids. Phys Chem Chem Phys 15:3022–3026
53. Amendola V, Meneghetti M (2013) What controls the composition and the structure of nanomaterials generated by laser ablation in liquid solution? Phys Chem Chem Phys 15:3027–3046
54. Amans D et al (2011) Synthesis of oxide nanoparticles by pulsed laser ablation in liquids containing a complexing molecule: impact on size distributions and prepared phases. J Phys Chem C 115:5131–5139
55. Tsuji T, Hamagami T, Kawamura T, Yamaki J, Tsuji M (2005) Laser ablation of cobalt and cobalt oxides in liquids: influence of solvent on composition of prepared nanoparticles. Appl Surf Sci 243:214–219
56. Itina TE (2010) On nanoparticle formation by laser ablation in liquids. J Phys Chem C 115:5044–5048
57. Patil P et al (1987) Pulsed-laser–induced reactive quenching at liquid-solid interface: aqueous oxidation of iron. Phys Rev Lett 58:238
58. Amendola V, Meneghetti M (2009) Laser ablation synthesis in solution and size manipulation of noble metal nanoparticles. Phys Chem Chem Phys 11:3805–3821
59. Cristoforetti G, Pitzalis E, Spiniello R, Ishak R, Muniz-Miranda M (2010) Production of palladium nanoparticles by pulsed laser ablation in water and their characterization. J Phys Chem C 115:5073–5083
60. Amendola V, Riello P, Meneghetti M (2011) Magnetic nanoparticles of iron carbide, iron oxide, iron@iron oxide, and metal iron synthesized by laser ablation in organic solvents. J Phys Chem C 115:5140–5146
61. Blakemore JD, Gray HB, Winkler JR, Müller AM (2013) Co_3O_4 nanoparticle water-oxidation catalysts made by pulsed-laser ablation in liquids. ACS Catal 3:2497–2500
62. Hui T-B, Phuoc TX, Chyu MK (2009) Synthesis of Mg-Al and Zn-Al layered double hydroxide nanocrystals using laser ablation in water. Opt Lasers Eng 47:695–700
63. Jiménez E, Abderrafi K, Abargues R, Valdés JL, Martínez-Pastor JP (2010) Laser-ablation-induced synthesis of SiO_2-capped noble metal nanoparticles in a single step. Langmuir 26:7458–7463
64. Zeng HB, et al (2006) Violet photoluminescence from shell layer of Zn/ZnO core-shell nanoparticles induced by laser ablation. Appl Phys Lett 88
65. Niu KY, Yang J, Kulinich SA, Sun J, Du XW (2010) Hollow nanoparticles of metal oxides and sulfides: fast preparation via laser ablation in liquid. Langmuir 26:16652–16657
66. Zhang JM et al (2012) Preparation of PtAu alloy colloids by laser ablation in solution and their characterization. J Phys Chem C 116:13413–13420
67. Serbezov V (2013) Pulsed laser deposition: the road to hybrid nanocomposites coatings and novel pulsed laser adaptive technique. Recent Pat Nanotechnol 7:26–40
68. Hunter BM et al (2014) Highly active mixed-metal nanosheet water oxidation catalysts made by pulsed-laser ablation in liquids. J Am Chem Soc 136:13118–13121

69. Miotello A, Kelly R (1999) Laser-induced phase explosion: new physical problems when a condensed phase approaches the thermodynamic critical temperature. Appl Phys A-Mat Sci Process 69:S67–S73
70. De Giacomo A, Dell'Aglio M, Colao F, Fantoni R (2004) Double pulse laser produced plasma on metallic target in seawater: basic aspects and analytical approach. Spectrochimica Acta Part B 59:1431–1438
71. Wagener P, Ibrahimkutty S, Menzel A, Plech A, Barcikowski S (2013) Dynamics of silver nanoparticle formation and agglomeration inside the cavitation bubble after pulsed laser ablation in liquid. Phys Chem Chem Phys 15:3068–3074
72. Lam J et al (2016) Dynamical study of bubble expansion following laser ablation in liquids. Appl Phys Lett 108:074104
73. Matsumoto A et al (2015) Transfer of the species dissolved in a liquid into laser ablation plasma: an approach using emission spectroscopy. J Phys Chem C 119:26506–26511
74. Xiao J, Liu P, Wang CX, Yang GW (2017) External field-assisted laser ablation in liquid: An efficient strategy for nanocrystal synthesis and nanostructure assembly. Prog Mater Sci 87:140–220
75. Yang GW (2007) Laser ablation in liquids: applications in the synthesis of nanocrystals. Prog Mater Sci 52:648–698
76. Ibrahimkutty S, et al (2015) A hierarchical view on material formation during pulsed-laser synthesis of nanoparticles in liquid. Scientific Rep 5
77. Tanabe R, Nguyen TTP, Sugiura T, Ito Y (2015) Bubble dynamics in metal nanoparticle formation by laser ablation in liquid studied through high-speed laser stroboscopic videography. Appl Surf Sci 351:327–331
78. Ibrahimkutty S, Wagener P, Menzel A, Plech A, Barcikowski S (2012) Nanoparticle formation in a cavitation bubble after pulsed laser ablation in liquid studied with high time resolution small angle x-ray scattering. Appl Phys Lett 101
79. Soliman W, Takada N, Sasaki K (2010) Growth processes of nanoparticles in liquid-phase laser ablation studied by laser-light scattering. Appl Phys Express 3:035201
80. Tamura A, Matsumoto A, Fukami K, Nishi N, Sakka T (2015) Simultaneous observation of nascent plasma and bubble induced by laser ablation in water with various pulse durations. J Appl Phys 117:173304
81. Matsumoto A, Tamura A, Fukami K, Ogata YH, Sakka T (2013) Two-dimensional space-resolved emission spectroscopy of laser ablation plasma in water. J Appl Phys 11
82. Zhang D, Gökce B, Barcikowski S (2017) Laser synthesis and processing of colloids: fundamentals and applications. Chem Rev 117:3990–4103
83. Mortazavi SZ, Parvin P, Reyhani A, Golikand AN, Mirershadi S (2011) Effect of laser wavelength at IR (1064 nm) and UV (193 nm) on the structural formation of palladium nanoparticles in deionized water. J Phys Chem C 115:5049–5057
84. Intartaglia R et al (2011) Optical properties of femtosecond laser-synthesized silicon nanoparticles in deionized water. J Phys Chem C 115:5102–5107
85. Alrehaily LM, Joseph JM, Biesinger MC, Guzonas DA, Wren JC (2013) Gamma-radiolysis-assisted cobalt oxide nanoparticle formation. Phys Chem Chem Phys 15:1014–1024
86. Compagnini G et al (2012) Monitoring the formation of inorganic fullerene-like MoS2 nanostructures by laser ablation in liquid environments. Appl Surf Sci 258:5672–5676
87. Blandin P et al (2013) Femtosecond laser fragmentation from water-dispersed microcolloids: toward fast controllable growth of ultrapure Si-based nanomaterials for biological applications. J Mat Chem B 1:2489–2495
88. Singh SC (2011) Effect of oxygen injection on the size and compositional evolution of ZnO/$Zn(OH)_2$ nanocomposite synthesized by pulsed laser ablation in distilled water. J Nanopart Res 13:4143–4152
89. Hu S, Melton C, Mukherjee D (2014) A facile route for the synthesis of nanostructured oxides and hydroxides of cobalt using laser ablation synthesis in solution (LASIS). Phys Chem Chem Phys 16:24034–24044

90. Sun Y, Wang Y (2011) Monitoring of galvanic replacement reaction between silver nanowires and $HAuCl_4$ by in situ transmission X-ray microscopy. Nano Lett 11:4386–4392
91. Zhang H et al (2011) Synthesis of Pd−Pt bimetallic nanocrystals with a concave structure through a bromide-induced galvanic replacement reaction. J Am Chem Soc 133:6078–6089
92. Ortiz N, Skrabalak SE (2011) Controlling the growth kinetics of nanocrystals via galvanic replacement: synthesis of Au tetrapods and star-shaped decahedra. Cryst Growth Des 11:3545–3550
93. Xia X, Wang Y, Ruditskiy A, Xia Y (2013) 25th anniversary article: galvanic replacement: a simple and versatile route to hollow nanostructures with tunable and well-controlled properties. Adv Materials 25:6313–6333
94. Hussain I et al (2005) Size-controlled synthesis of near-monodisperse gold nanoparticles in the 1–4 nm range using polymeric stabilizers. J Am Chem Soc 127:16398–16399
95. Lee H et al (2006) Morphological control of catalytically active platinum nanocrystals. Angewandte Chemie-Int Edition 45:7824–7828
96. Palanisamy P, Raichur AM (2009) Synthesis of spherical NiO nanoparticles through a novel biosurfactant mediated emulsion technique. Mat Sci Eng C-Biomimetic Supramol Syst 29:199–204
97. Mukherjee D, Hu S (2019) Compositions, systems and methods for producing nanoalloys, and/or nanocomposites using tandem laser ablation synthesis in solution-galvanic replacement reaction. USA patent US 10,326,146 B2
98. Davari SA et al (2019) Graphitic coated Al nanoparticles manufactured as superior energetic materials via laser ablation synthesis in organic solvents. Appl Surf Sci 473:156–163
99. Hu S et al (2017) Hybrid nanocomposites of nanostructured Co3O4 interfaced with reduced/nitrogen-doped graphene oxides for selective improvements in electrocatalytic and/or supercapacitive properties. RSC Adv 7:33166–33176
100. Hu S et al (2017) A facile and surfactant-free route for nanomanufacturing of tailored ternary nanoalloys as superior oxygen reduction reaction electrocatalysts. Catal Sci Technol 7:2074–2086
101. Hu S, Tian M, Ribeiro EL, Duscher G, Mukherjee D (2016) Tandem laser ablation synthesis in solution-galvanic replacement reaction (LASiS-GRR) for the production of PtCo nanoalloys as oxygen reduction electrocatalysts. J Power Sources 306:413–423
102. Hu S, Goenaga G, Melton C, Zawodzinski TA, Mukherjee D (2016) PtCo/CoOx nanocomposites: bifunctional electrocatalysts for oxygen reduction and evolution reactions synthesized via tandem laser ablation synthesis in solution-galvanic replacement reactions. Appl Catal B 182:286–296
103. Ribeiro EL, Davari SA, Hu S, Mukherjee D, Khomami B (2019) Laser-induced synthesis of ZIF-67: a facile approach for the fabrication of crystalline MOFs with tailored size and geometry. Mat. Chemistry Frontiers 3:1302–1309
104. Ribeiro EL et al (2021) MOF-derived PtCo/Co_3O_4 nanocomposites in carbonaceous matrices as high-performance ORR electrocatalysts synthesized via laser ablation techniques. Catalysis Sci & Tech. https://doi.org/10.1039/D0CY02099K
105. Richards R, Bönnemann H (2005) Nanofabrication towards biomedical applications 1–32. Wiley-VCH Verlag GmbH & Co, KGaA
106. Lin F et al (2010) Laser synthesis of gold/oxide nanocomposites. J Mater Chem 20:1103–1106
107. Mottaghi N et al (2014) Ag/Pd core-shell nanoparticles by a successive method: Pulsed laser ablation of Ag in water and reduction reaction of PdCl2. Appl Surf Sci 292:892–897
108. Saraeva IN et al (2018) Laser synthesis of colloidal Si@Au and Si@Ag nanoparticles in water via plasma-assisted reduction. J Photochem Photobiol, A 360:125–131
109. Amendola V, Scaramuzza S, Carraro F, Cattaruzza E (2017) Formation of alloy nanoparticles by laser ablation of Au/Fe multilayer films in liquid environment. J Colloid Interface Sci 489:18–27
110. Marzun G et al (2017) Role of dissolved and molecular oxygen on Cu and PtCu alloy particle structure during laser ablation synthesis in liquids. ChemPhysChem 18:1175–1184

111. Mostafa AM, Yousef SA, Eisa WH, Ewaida MA, Al-Ashkar EA (2017) Au@CdO core/shell nanoparticles synthesized by pulsed laser ablation in Au precursor solution. Appl Phys A 123
112. John MG, Tibbetts KM (2019) One-step femtosecond laser ablation synthesis of sub-3 nm gold nanoparticles stabilized by silica. Appl Surf Sci 475:1048–1057
113. Desarkar HS, Kumbhakar P, Mitra AK (2013) One-step synthesis of Zn/ZnO hollow nanoparticles by the laser ablation in liquid technique. Laser Phys Lett 10:055903
114. Amendola V et al (2020) Room-temperature laser synthesis in liquid of oxide, metal-oxide core-shells, and doped oxide nanoparticles. Chem—A Europ J 26:9206–9242
115. John MG, Tibbetts KM (2020) Mechanism of nickel phyllosilicate formation by laser ablation in liquid. J Phys Chem C 124:13273–13282
116. John MG, Tibbetts KM (2020) Controlling the morphology of copper-silica nanocomposites from laser ablation in liquid. Appl Surf Sci 510:145037
117. Yetter RA, Risha GA, Son SF (2009) Metal particle combustion and nanotechnology. P Combust Inst 32:1819–1838
118. Mukherjee D, Prakash A, Zachariah MR (2006) Implementation of a discrete nodal model to probe the effect of size-dependent surface tension on nanoparticle formation and growth. J Aerosol Sci 37:1388–1399
119. Mukherjee D, Sonwane CG, Zachariah MR (2003) Kinetic Monte Carlo simulation of the effect of coalescence energy release on the size and shape evolution of nanoparticles grown as an aerosol. J. Chemical Physics 119:3391–3404
120. Davari SA, Mukherjee D (2018) Kinetic Monte Carlo simulation for homogeneous nucleation of metal nanoparticles during vapor phase synthesis. AIChE J 64:18–28
121. Mukherjee D, Wang M, Khomami B (2012) Impact of particle morphology on surface oxidation of nanoparticles: a kinetic Monte Carlo based study. AIChE J 58:3341–3353
122. Dreizin EL (2009) Metal-based reactive nanomaterials. Prog Energ Combust 35:141–167
123. Prakash A, McCormick AV, Zachariah MR (2005) Tuning the reactivity of energetic nanoparticles by creation of a core-shell nanostructure. Nano Lett 5:1357–1360
124. Sullivan K, Young G, Zachariah MR (2009) Enhanced reactivity of nano-B/Al/CuO MIC's. Combust Flame 156:302–309
125. Prakash A, McCormick AV, Zachariah MR (2005) Synthesis and reactivity of a super-reactive metastable intermolecular composite formulation of Al/$KMnO_4$. Adv Mater 17:900
126. Jouet RJ et al (2005) Surface passivation of bare aluminum nanoparticles using perfluoroalkyl carboxylic acids. Chem Mater 17:2987–2996
127. Park K, Rai A, Zachariah M (2006) Characterizing the coating and size-resolved oxidative stability of carbon-coated aluminum nanoparticles by single-particle mass-spectrometry. J Nanopart Res 8:455–464
128. Baladi A, Mamoory RS (2010) Investigation of different liquid media and ablation times on pulsed laser ablation synthesis of aluminum nanoparticles. Appl Surf Sci 256:7559–7564
129. Kuladeep R et al (2013) Investigation of optical limiting properties of aluminium nanoparticles prepared by pulsed laser ablation in different carrier media. J Appl Phys 114:243101
130. Viau G, Collière V, Lacroix L-M, Shafeev G (2011) Internal structure of Al hollow nanoparticles generated by laser ablation in liquid ethanol. Chem Phys Lett 501:419–422
131. Gottfried JL (2015) Laboratory-scale method for estimating explosive performance from laser-induced shock waves. Propellants, Explos, Pyrotech 40:674–681
132. Gottfried JL, Klapötke TM, Witkowski TG (2017) Estimated detonation velocities for TKX-50, MAD-X1, BDNAPM, BTNPM, TKX-55, and DAAF using the laser–induced air shock from energetic materials technique. Propellants, Explos, Pyrotech 42:353–359
133. Gottfried JL, Bukowski EJ (2017) Laser-shocked energetic materials with metal additives: evaluation of chemistry and detonation performance. Appl Optics 56:B47–B57
134. Ibrahimkutty S, Wagener P, Menzel A, Plech A, Barcikowski S (2012) Nanoparticle formation in a cavitation bubble after pulsed laser ablation in liquid studied with high time resolution small angle x-ray scattering. Appl Phys Lett 101:103104
135. Lam J et al (2014) γ-Al 2 O 3 nanoparticles synthesised by pulsed laser ablation in liquids: a plasma analysis. Phys Chem Chem Phys 16:963–973

136. Girshick SL, Chiu CP (1989) Homogeneous nucleation of particles from the vapor-phase in thermal plasma synthesis. Plasma Chem Plasma P 9:355–369
137. Girshick SL, Agarwal P, Truhlar DG (2009) Homogeneous nucleation with magic numbers: Aluminum. J Chem Phys 131
138. Gottfried JL (2012) Laser-induced plasma chemistry of the explosive RDX with various metallic nanoparticles. Appl Optics 51:B13–B21
139. Gottfried JL (2014) Influence of exothermic chemical reactions on laser-induced shock waves. Phys Chem Chem Phys 16:21452–21466
140. Gottfried JL, Dean SW, Collins ES, Wu C-C (2018) Estimating the relative energy content of reactive materials using nanosecond-pulsed laser ablation. MRS Adv, 1–12
141. Rai A, Lee D, Park KH, Zachariah MR (2004) Importance of phase change of aluminum in oxidation of aluminum nanoparticles. J Phys Chem B 108:14793–14795
142. Sun L, Krasheninnikov AV, Ahlgren T, Nordlund K, Banhart F (2008) Plastic deformation of single nanometer-sized crystals. Phys Rev Lett 101:156101
143. Son SF et al (2016) Energetic nanomaterials: synthesis, characterizations, and application, Zarko VE, Gromov AA (eds). Elsevier
144. Wang Z-Q, Wang Z-C, Zhan S, Ye J-S (2015) A water-soluble iron electrocatalyst for water oxidation with high TOF. Applied Catalysis a-General 490:128–132
145. Hawa T, Zachariah MR (2004) Internal pressure and surface tension of bare and hydrogen coated silicon nanoparticles. J Chem Phys 121:9043–9049
146. Lu HM, Jiang Q (2005) Size-dependent surface tension and Tolman's length of droplets. Langmuir 21:779–781
147. Chen B, Cheng D, Zhu J (2014) Synthesis of PtCu nanowires in nonaqueous solvent with enhanced activity and stability for oxygen reduction reaction. J Power Sources 267:380–387
148. Huang X et al (2015) High-performance transition metal–doped Pt3Ni octahedra for oxygen reduction reaction. Science 348:1230–1234
149. Zhu Z, Zhai Y, Dong S (2014) Facial synthesis of PtM (M = Fe Co, Cu, Ni) bimetallic alloy nanosponges and their enhanced catalysis for oxygen reduction reaction. ACS Appl Mater Interfaces 6:16721–16726
150. Jaksic JM, Labou D, Papakonstantinou GD, Siokou A, Jaksic MM (2010) Novel spillover interrelating reversible electrocatalysts for oxygen and hydrogen electrode reactions. J. Phys Chem C 114:18298–18312
151. Van TTH, Pan CJ, Rick J, Su WN, Hwang BJ (2011) Nanostructured Ti0.7Mo0.3O2 support enhances electron transfer to Pt: high-performance catalyst for oxygen reduction reaction. J Am Chem Soc 133:11716–11724
152. Zhu YL et al (2014) A universal and facile way for the development of superior bifunctional electrocatalysts for oxygen reduction and evolution reactions utilizing the synergistic effect. Chem—A Euro J 20:15533–15542
153. Kristian N, Yu Y, Lee J-M, Liu X, Wang X (2010) Synthesis and characterization of Cocore-Ptshell electrocatalyst prepared by spontaneous replacement reaction for oxygen reduction reaction. Electrochim Acta 56:1000–1007
154. Sun Y, Xia Y (2004) Mechanistic study on the replacement reaction between silver nanostructures and chloroauric acid in aqueous medium. J Am Chem Soc 126:3892–3901
155. Sa YJ, Kwon K, Cheon JY, Kleitz F, Joo SH (2013) Ordered mesoporous Co3O4 spinels as stable, bifunctional, noble metal-free oxygen electrocatalysts. J Mat Chem A 1:9992–10001
156. Jin W, Du H, Zheng S, Xu H, Zhang Y (2010) Comparison of the oxygen reduction reaction between NaOH and KOH solutions on a Pt electrode: the electrolyte-dependent effect. J. Phys Chem B 114:6542–6548
157. Hu S, Mukherjee D (2018) Colloidal synthesis of advanced functional nanostructured composites and alloys via laser ablation based techniques. In: Guo Z, Chen Y, Lu NL (eds) Multifunctional nanocomposites for energy and environmental applications. Wiley, USA, pp 135–172

Chapter 6
Recent Mechanistic Insights into Some Enzyme Mimetic Functions of Ceria

Suman Bhasker-Ranganath and Ye Xu

Abstract Ceria, a prototypical reducible oxide, has been shown to possess enzyme-like catalytic activities in recent years, which makes it an attractive enzyme mimic with possible applications in medicine, biochemistry, and environment remediation. This chapter reviews the current literature for mechanistic understanding of the phosphatase, superoxide dismutase, and catalase mimetic actions of ceria from experimental and theoretical approaches. Some inconsistencies stemming from common views of the role of the Ce^{3+}/Ce^{4+} redox couple are identified and discussed. Alternative interpretations of experimental observations are suggested in hopes of unlocking the technological potentials of ceria.

6.1 Introduction

Natural enzymes stand out among all types of catalyst for being able to carry out highly specific chemical transformations of substrates ranging from small molecules such as N_2, CO_2, and CH_4 to proteins and DNA. An enzymatic reaction usually consists of a series of electron transfer steps targeting specific bonds that interact with single-atom or few-atom cofactors, with attendant ligands creating the necessary micro-environments for orienting or stabilizing the intended substrates. Limitations on the application of natural enzymes, including difficulty of mass production due to separation and purification requirements, and limited stability in harsh environments, long-term storage, and recycling, have prompted researchers to look

S. Bhasker-Ranganath
SUNCAT Center for Interface Science and Catalysis, SLAC National Accelerator Laboratory, Menlo Park, CA 94025, USA

SUNCAT Center for Interface Science and Catalysis, Stanford University, Stanford, CA 94305, USA

Y. Xu (✉)
Cain Department of Chemical Engineering, Louisiana State University, Baton Rouge, LA 70803, USA
e-mail: yexu@lsu.edu

M. Shukla et al. (eds.), *Emerging Materials and Environment*,
Challenges and Advances in Computational Chemistry and Physics 37,
https://doi.org/10.1007/978-3-031-39470-6_6

for potential inorganic mimics of the active centers or cofactors of natural enzymes [1–6] among nanosized, nanostructured, or nanocomposite materials. Such synthetic enzyme mimics have come to be termed nanozymes, with applications being actively sought in medical diagnostics, therapeutics, and environmental remediation [7, 8]. High catalytic efficiency and substrate specificity under mild conditions are also desirable characteristics of catalysts to chemical manufacturers. A great deal of research effort has been put into understanding the action of natural enzymes in hopes of designing novel catalysts for production of hydrogen [9] and ammonia [10] that is more energy-efficient than currently practiced industrial processes.

Among different substances, cerium dioxide, or ceria, has attracted strong attention. Ce complexes can catalyze the hydrolysis of DNA phosphate esters, with Ce^{4+} being much more active than Ce^{3+} [11–13]. Nanoparticles (NPs) of ceria have been shown to have therapeutic effects in cells [14–18]. More recently, ceria NPs have been reported to exhibit catalytic activity that mimics the action of several types of enzyme, including superoxide dismutase, catalase, peroxidase, oxidase, and phosphatase [19–23] at ambient conditions. Ceria is the prototypical example of a reducible oxide, in which the Ce 4*f* states covalently interact with O 2*p* states [24] and the Ce cation can switch between the +3 and +4 oxidation states without significant energy penalty or structural change. As a result, oxygen is absorbed into or released from the lattice relatively easily. In heterogeneous catalysis research, ceria is widely used as a catalyst support and promoter, or directly as a catalyst [25]. It enhances the redox activity of metal catalysts for, or directly catalyzes, many reactions including the automotive three-way catalysis [26], water–gas shift [27, 28], reforming [29, 30], oxidation [31, 32], and ketonization [33, 34], all of which play important roles in the modern chemical industry.

Facile switch between the +3/+4 oxidation states has come to be viewed as an ability for facile electron transfer that underlies the enzyme mimetic properties of ceria. It is commonly accepted that Ce^{3+} occurs with oxygen vacancy (V_O) formation, as the removal of each lattice O atom (O_{latt}) reduces two Ce^{4+} to Ce^{3+}. This favors the consideration of ceria nanoparticles (NPs) as enzyme mimics, because compared to larger-sized particles, NPs not only have larger surface-to-volume ratios but also have higher concentrations of defects on the surface [19, 35]. As a stable, inorganic compound that is insoluble in water and common acids and bases, ceria would be an ideal candidate for use in harsh applications such as degradation of organophosphates found in industrial and municipal wastes, agricultural runoff, and chemical warfare agents.

Ceria is commercially available in a polycrystalline powder format with particle size down to 10–20 nm. To facilitate novel materials and chemistry applications, a procedure for synthesizing single-crystalline ceria NPs was devised by Feng et al. [36]. Synthesis of size- and shape-controlled ceria NPs was reported around the same time [37–42]. A hydrothermal process using simple chemicals (cerium(III) nitrate hexahydrate ($Ce(NO_3)_3{\cdot}6H_2O$) as Ce source, and a precipitant such as $Na_3PO_4{\cdot}6H_2O$, NaOH, or H_2O_2) was developed by Yan et al. [43] to enable the synthesis of single-crystalline ceria NPs in different shapes by adjusting variables such as temperature, solution pH, and reaction time. Among various shapes, nano-octahedra and nanocubes primarily exposed {111} and {100} facets, respectively, whereas nanorods and nanowires can exhibit combinations of {100}, {110}, and {111} facets (Fig. 6.1) [43–46]. This approach has opened the door to the design of ceria nanomaterials and gained popularity rapidly in catalysis research [44–48].

Although a wide range of enzyme mimetic activity of ceria NPs has been reported by recent studies, fundamental understanding of how ceria catalyzes these reactions remains incomplete. Details crucial to further development of nanoceria as enzyme mimics for advanced biological and biochemical applications, including the nature of the active site and specifics of reaction mechanisms, remain speculative. In this chapter, we review the current mechanistic understanding of the mimetic functions of ceria for phosphatase, superoxide dismutase, and catalase, through a survey of recent literature. As has become clear over the past two decades, a critical asset in the toolbox of materials and catalysis researchers for elucidating reaction mechanisms and connecting material properties to catalytic reactions is atomistic computational simulation based on first principles calculations, primarily density functional theory (DFT) [49–51]. Our review will therefore include relevant recent theoretical work that can enhance our understanding of the enzyme mimetic function of ceria.

6.2 Ceria as Phosphatase Mimic

6.2.1 Biological Phosphatases

A phosphatase is an enzyme that catalyzes the dephosphorylation of phosphate monoesters (denoted $ROPO_3H_2$) [52]. The enzyme cleaves the P–OR ester bond of a substrate and displaces the alkoxide leaving group (RO) by attaching a different nucleophile such as water or alcohol to the phosphoryl group, to form an inorganic phosphate or a different monoester. Phosphatases are essential to biological functions because phosphorylation (by kinases) and dephosphorylation (by phosphatases) are fundamental in cellular energy transfer, stability of DNA and RNA, and regulation of protein activity. They are found in all life forms on Earth. In vivo, a phosphatase acts on a specific substrate through interaction with multiple amino acid residues, and correspondingly, there exist a large number of phosphatases.

The *E. coli* alkaline phosphatase is perhaps the best-studied phosphatase [53–55], which reaches its optimal activity in alkaline environments. It is a zinc-containing dimeric enzyme of 86 kDa. Each monomer contains an active center consisting of two Zn^{2+} ions occupying sites A and B, and one Mg^{2+} ion occupying site C, so the full enzyme is referred to as $(Zn_AZn_BMg_C)_2$. The substrate is coordinated via one phosphate O atom to each Zn ion, with the two Zn, two O, and P atoms being nearly coplanar. The other two O atoms in the phosphate are hydrogen-bonded to a nearby guanidinium group. The enzyme can retain partial functionality even if it is missing one Zn ion or the Mg ion from the active center. Substituting Zn with

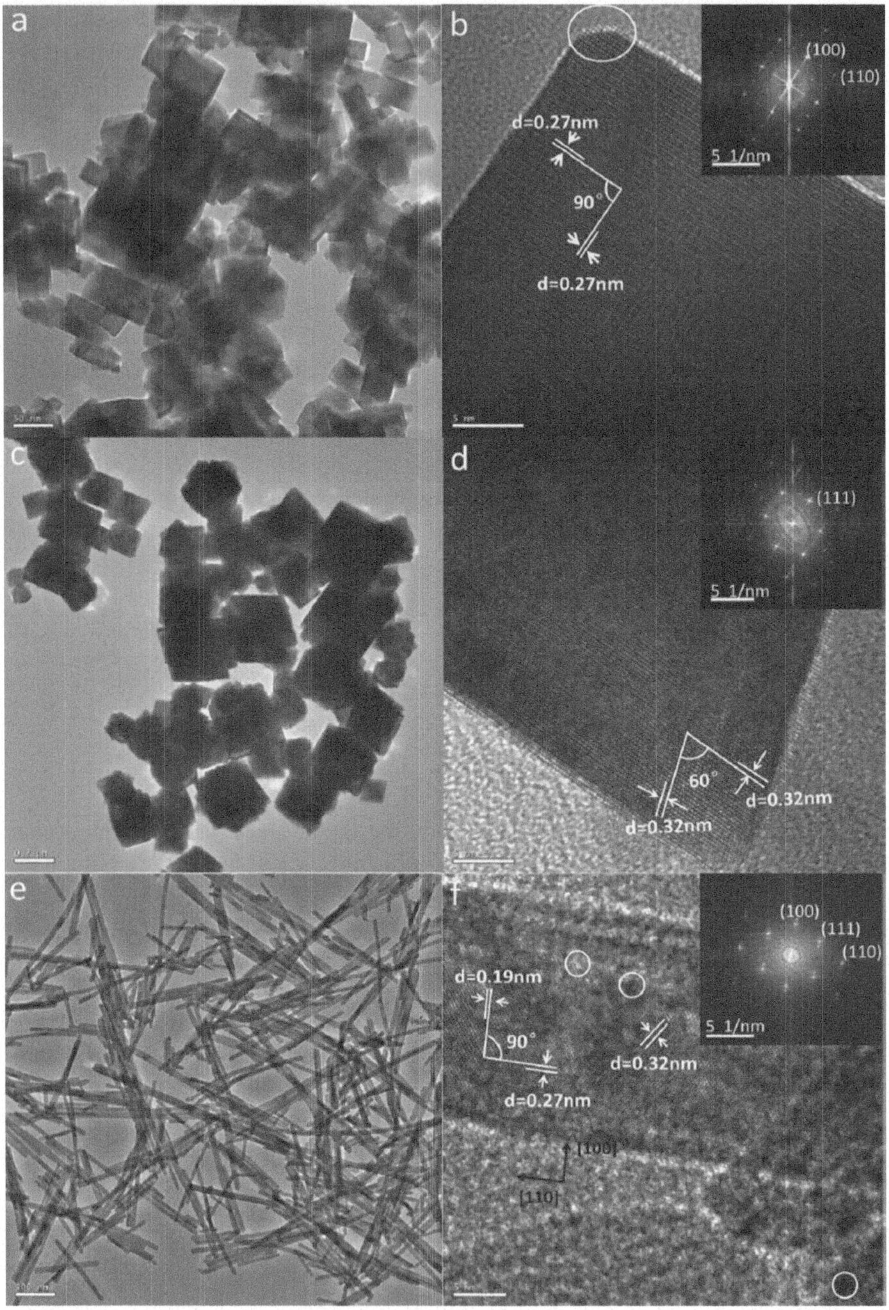

Fig. 6.1 TEM and HRTEM images and corresponding FFT patterns (inset) of shape-controlled ceria NPs: **a**, **b** nanocubes; **c**, **d** nano-octahedra; **e**, **f** nanorods. Adapted from Ref. [44] with permission. Copyright © 2015 Wiley-VCH Verlag GmbH & Co. KGaA, Weinheim

another divalent cation such as Cd^{2+} or Mn^{2+} diminishes activity significantly. A combination of electrostatic and hydrogen bonding interactions that originate from structural pre-organization has been proposed to assist in the optimum positioning of the phosphoryl group, nucleophile, and leaving group in the active site to catalyze the reaction [56–58]. NMR studies have suggested P-OR ester bond scission to be rate-limiting at acidic pH and phosphate desorption to be rate-limiting at alkaline pH [59].

The un-catalyzed aqueous phase hydrolysis reaction serves as a reference for evaluating the catalytic efficiency of phosphatases. The hydrolysis of phosphate monoesters has two common features: The rate of hydrolysis exhibiting a maximum around pH $= 4$ with weak dependence on the leaving group in the hydrolysis of the monoanions; and the high sensitivity of the rate to the pKa of the leaving group for the dianions [60–62]. Multiple mechanisms have been proposed for the reaction, based on whether the reaction is substrate-assisted or solvent-assisted (i.e., whether the attacking nucleophile is deprotonated by the substrate in the rate-limiting step), and whether it occurs via a step-wise or concerted pathway (i.e., whether the reaction proceeds through a stable pentacoordinate intermediate state, or a loose transition state (TS) in which the central P atom is weakly bonded to both the incoming nucleophile and the leaving group) [63]. Computational modeling based on theoretical calculations has been applied to investigate the molecular-level details and energetic profiles of the hydrolysis reaction both un-catalyzed and as catalyzed by an enzyme [63–70].

6.2.2 Experimental Evidence for Ceria as Phosphatase Mimic

Tan et al. [71] and Kuchma et al. [72] were among the first groups to extend the investigation to whether ceria could catalyze this reaction, thus acting as a phosphatase mimetic. Tan et al. experimented with commercial ceria powder that had a grain size of 1–10 μm and found it to be effective at catalyzing the dephosphorylation of phosphopeptides between 0 and 37 °C. Kuchma et al. synthesized 3–5 nm ceria NPs via a micro-emulsion process by reacting $Ce(NO_3)_3 \cdot 6H_2O$ with H_2O_2. The NPs had up to 40% of Ce in the +3 oxidation state, the concentration of which decreased with increasing particle size. The large amounts of Ce^{3+} were attributed to the presence of V_O in the lattice. Kuchma et al. showed that p-NPP was readily hydrolyzed in the presence of ceria NPs, releasing p-nitrophenol and inorganic phosphate at room temperature and neutral pH. Lowering pH yielded a higher rate of reaction. Further treatment of the NPs with H_2O_2, which was considered to convert Ce^{3+} to Ce^{4+} (Fig. 6.2), diminished the dephosphorylation activity, which led the authors to conclude that Ce^{3+}, not Ce^{4+}, was the active site for this reaction. Those ceria NPs also showed some, albeit lower, dephosphorylation activity toward ATP and a phosphorylated amino acid, o-phospho-L-tyrosine, while they were inactive toward plasmid DNA.

Fig. 6.2 Appearance of ceria NPs treated with H_2O_2 at different time intervals. Adapted from Ref. [17] with permission. Copyright © 2006 Elsevier Ltd.

To support their hypothesis, Kuchma et al. [72] performed theoretical calculations using Gaussian 2003 to determine the activation energy (E_a) of P-OR ester bond scission in p-NPP coordinated to a di-cerium-hydroxide complex, used as a model for the active site. Geometry optimization was done at the Hartree–Fock level with single-point energy refinement at unrestricted DFT level using the Boese-Martin functional for Kinetics and 6-31G* basis set with the implicit polarizable continuum model (PCM). The p-NPP molecule in its dianionic form was bonded in a μ-bridging configuration to the two Ce atoms simultaneously via two of the O atoms in the phosphate group (Fig. 6.3a), which echoes how phosphate coordinates to the active center of an alkaline phosphatase. Using this model, they determined the E_a for the P-OR ester bond scission to be 0.99 eV on Ce^{4+}–Ce^{4+}, 0.69 eV on Ce^{4+}–Ce^{3+}, and 0.59 eV on Ce^{3+}–Ce^{3+}. The decrease in E_a with increasing extent of reduction of the Ce centers appeared to support the hypothesis that Ce^{3+} was required for high dephosphorylation activity. The authors further concluded that the origin of the superior catalytic activity of Ce^{3+} was purely electronic and not geometric. The reaction energy profile for the full reaction path was not reported.

Since Kuchma et al., several groups took advantage of the hydrothermal synthesis approach and tested the dephosphorylation of p-NPP on shaped-controlled ceria NPs [44, 46, 48]. Consistent with the original report by Yan et al. [43] nano-octahedra and nanocubes primarily exposed {111} and {100} facets, respectively, whereas nanorods or nanowires exposed a mixture of {110} and {100} facets. Just like polycrystalline ceria, the shape-controlled ceria NPs were catalytically active toward dephosphorylation at ambient temperature, but different shapes showed different activities: On the basis of apparent mass or surface area-normalized catalytic rate constant (k_{cat}), nano-octahedra were more active than nanorods, which were in turn more active than nanocubes [44, 46, 48]. Different authors proposed different explanations for the observed differences in catalytic activity toward dephosphorylation. Wang et al. and Tan et al. attributed it to strong interaction between Ce and phosphoryl oxygen (P=O) [44, 48]. Wang et al. inferred that p-NPP adsorbed at a higher surface density on {111} and {110} facets than on {100} facets, consistent with the higher activity of nano-octahedra than nanocube. The authors further rationalized the adsorption capacity in terms of Lewis acidity of Ce cations, which they asserted to be lower on {100} than on {111} or {110} facets due to greater coordination with

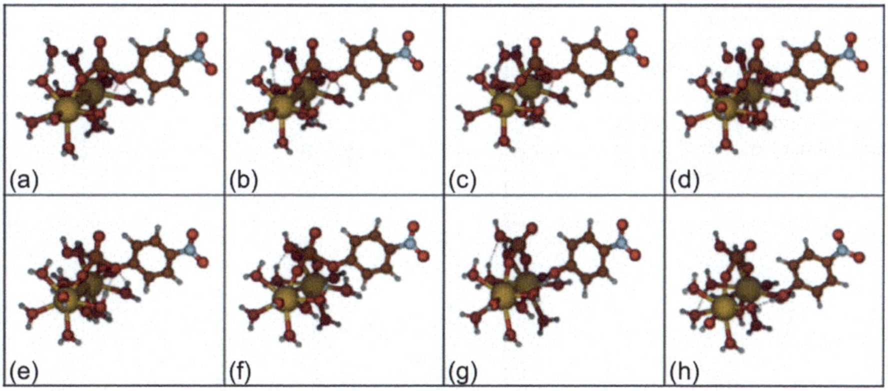

Fig. 6.3 Snapshots of the reaction pathway for p-NPP hydrolysis on a di-cerium-hydroxide complex calculated by Kuchma et al. [72]. **a** Reactant complex. **b** Hydrogen bonding switching to the bridging hydroxide anion. **c** Proton transfer along the newly formed hydrogen bond. **d** Newly formed hydroxide attacking p-NPP. **e** Transition state. **f** Departure of p-nitrophenolate anion. **g** Deprotonation of one of the water molecules in the first coordination sphere. **h** Hydrogen-bonded product complex. Adapted from Ref. [72] with permission. Copyright © 2010 Elsevier Inc.

O_{latt} atoms and thus higher steric hindrance of Ce cations on {100} facets. Later on, Tan et al. used probe-assisted NMR to measure the chemical state of surface Ce cations and reported that they were indeed most acidic in nano-octahedra followed by nanorods and were least so in nanocubes. It was corroborated by X-ray photoelectron spectroscopy (XPS), which indicated that Ce cations in the nano-octahedra were predominantly in the +4 oxidation state, whereas a significantly higher portion of Ce cations was reported to be in the +3 oxidation state in the nanocubes. Dhall et al. [73] synthesized ceria NPs using a wet chemical method by reacting cerium nitrate hexahydrate with either H_2O_2 or NH_4OH as the precipitant, which produced ceria NPs with a high (1.28) or low (0.37) Ce^{3+}/Ce^{4+} ratio, respectively. The Ce^{3+}-rich NPs were inactive toward 4-methylumbelliferyl phosphate, whereas the Ce^{4+}-rich NPs were highly active.

Others reached a different conclusion. Besides nano-octahedra, wires, and cubes, Manto et al. [46] also synthesized sub-10 nm nanospheres and tested commercial ceria nanopowder for comparison. They performed O_2 temperature programmed desorption (TPD) to estimate the surface density of V_O to follow the order of nanospheres > commercial nanopowder > nano-octahedra ~ nanorods > nanocubes. The apparent E_a of p-NPP dephosphorylation thus exhibited a linear trend with respect to the surface density of V_O (Fig. 6.4). Thus, the authors concluded that the active site for dephosphorylation was V_O on the surfaces of the NPs. Yao et al. [74] synthesized porous nanorods of ceria and calcined them in air at 300 and 400 °C. They performed XPS to determine the concentration of Ce^{3+} in the samples and concluded that higher calcination temperature reduced the Ce^{3+}/Ce^{4+} ratio. The activity for p-NPP dephosphorylation was found to be lower the higher the calcination temperature was. These authors, like Manto et al. and Kuchma et al., associated Ce^{3+} with the

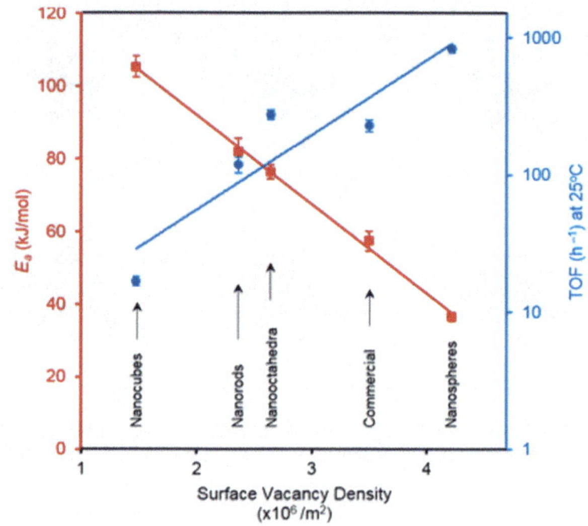

Fig. 6.4 Correlations of E_a and TOF with the surface density of oxygen vacancies estimated from O_2 TPD. Adapted from Ref. [46] with permission. Copyright © 2017 American Chemical Society

active site instead of Ce^{4+}. Yao et al. [74] also reported that their ceria nanorods were more active in higher pH, in contrast to Kuchma et al. [72].

6.2.3 *Proposed Mechanism from Theory*

The basic features of the catalytic dephosphorylation reaction seem straightforward. In analogy to enzymes, it is reasonable to hypothesize that p-NPP adsorbs on ceria first. Since p-nitrophenol and inorganic phosphate groups are both detected in the solution as p-NPP disappears [44, 46, 72], the role of ceria must be to catalyze the cleavage of the P–OR ester bond, followed by product desorption. The details, however, remain unclear, including how the P–OR ester bond is activated by ceria, and as a type of hydrolysis reaction, how water participates in the reaction.

The commonly proposed mechanism of this catalytic reaction on ceria has been in one way or another influenced by the earlier work by Tan et al. [71] who suggested, based on the consideration that ceria is a Lewis acid, that the reaction is mediated by coordination of the phosphate O atoms to Ce^{4+} cations. As Ce^{4+} draws electron away from the O atoms, the central P atom becomes more electropositive and thus more susceptible to nucleophilic attack by water or hydroxide, which causes the P–OR ester bond to break. The problem with this view is that, while ceria nano-octahedra are quite active, Ce^{4+} cations are occluded by a layer of O_{latt} atoms in the {111} facets (Fig. 6.5a) and cannot closely coordinate to phosphate O atoms. On the other hand, while surface V_O expose Ce cations, reduce some of them to Ce^{3+}, and make them more accessible to adsorbates, they should not be expected to remain physically open when exposed to the atmosphere or an aqueous phase, when the adsorption of

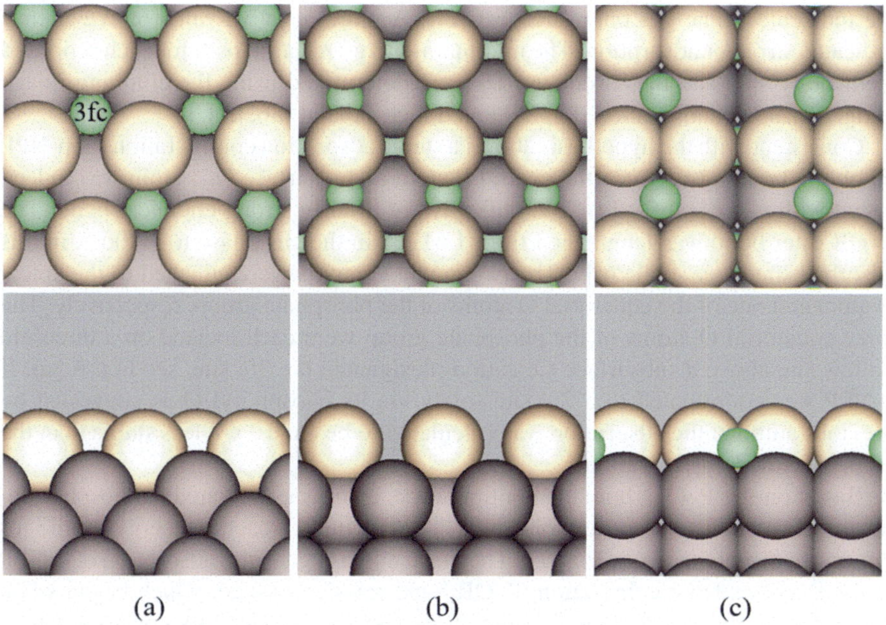

Fig. 6.5 Top (top panels) and side (bottom panels) views of the **a** (111), **b** reconstructed (100), and **c** (110) facet of CeO_2. Color code: green = Ce, light brown = surface O_{latt}, and dark brown = subsurface O_{latt}. Molecular images herein are created using VESTA [77]

O_2 ($\Delta E_{ads} = -1.72$ eV) [75] or dissociative adsorption of water ($\Delta E_{ads} = -1.3$ eV) [76] is strongly exothermic in V_O [75, 76] and irreversible at ambient temperature.

There is another possibility for how a phosphate compound interacts with ceria. Given the fact that {111} facets of ceria expose only O_{latt} atoms, which are negatively charged and therefore nucleophilic, they should be in a position to interact directly with the positively charged P atom in a phosphate compound. The competition by O_{latt} for bonding with P, instead of the coordination of P=O to Ce, would also weaken the existing P–O bonds in the compound. Based on this idea, Zhao et al. carried out DFT calculations to explore an alternative mechanism for the adsorption and dephosphorylation of p-NPP on CeO_2(111) [78].

The calculation parameters are detailed in Ref. [78]. Briefly, the periodic DFT calculations were performed in the generalized gradient approximation (GGA-PW91) [79] using the Vienna Ab initio Simulation Package (VASP) [80]. The DFT + U approach of Dudarev et al. [81] was employed to rectify the delocalization of Ce 4f states resulting from self-interaction error. A U_{eff} value of 2 eV was used, which yielded an equilibrium lattice constant of 5.476 Å for ceria, in good agreement with the experimental value of 5.41 Å [82, 83]. CeO_2(111) was modeled with a $p(3 \times 3)$ slab of three O–Ce–O trilayers. The adsorption energy, ΔE_{ads}, was evaluated as $\Delta E_{ads} = E_{total} - E_{slab} - E_{gas}$, where E_{total}, E_{slab}, and E_{gas} refer to the total energies of the slab with a molecule adsorbed on it, a clean slab without any adsorbate, and the molecule in the gas phase, respectively. A more negative value of ΔE_{ads} thus indicated stronger adsorption. The minimum-energy reaction path for each proposed elementary step and the associated TS were determined using the climbing-image nudged elastic band method [84, 85] and dimer method [86, 87].

Consistent with the hypothesis above, a stable adsorption configuration for p-NPP involving bonding between the central P atom and a surface O_{latt} atom was indeed found with a ΔE_{ads} of -1.04 eV, indicative of a fairly strong chemical bond. The P–OR ester bond was lengthened from 1.619 Å in the gas phase to 1.706 Å in this adsorbed state (see Table 6.1). The two acidic H atoms were attached to the apical O atom and one of the equatorial O atoms of the phosphate group, respectively. The three equatorial O atoms of the phosphate group were each located on a threefold hollow site above a subsurface Ce cation (designated the 3fc site, see Fig. 6.5a). If p-NPP was coordinated to a 3fc site solely via its phosphoryl O as suggested by previous authors, the adsorption was calculated to be ca. 0.7 eV less stable than the P–O_{latt}-bonded configuration.

Based on the P–O_{latt} bonding configuration, a surface-mediated dephosphorylation mechanism was proposed for p-NPP. The calculated reaction energy profile for this mechanism is shown in Fig. 6.6. The mechanism began with adsorption of p-NPP in the P–O-bonded configuration. P–OR bond scission ensued, which began when the proton on the apical O atom of the phosphate group rotated to a position where it could form a hydrogen bond with the O atom of the p-nitrophenolate group and stabilize it once the latter dissociated from the phosphate group. The P–OR ester bond was further lengthened to 1.858 Å in the TS (Fig. 6.6iii). The E_a relative to the adsorbed p-NPP was a mere 0.13 eV. In fact, the E_a ($= \Delta E_{TS} - \Delta E_{ads}$) for P–OR

Table 6.1 Adsorption energy (ΔE_{ads}, in eV) and key bond lengths (in Å) of several generalized acyl compounds adsorbed via X–O_{latt} bonding on $CeO_2(111)$

Compound	ΔE_{ads}	d(X–Z) Gas phase	d(X–Z) Adsorbed	d(X=Y) Gas phase	d(X=Y) Adsorbed	d(X–O_{latt})
Methyl phosphate	−1.01	1.59	1.67	1.48	1.55	1.70
p-nitrophenyl phosphate	−1.04	1.63	1.71	1.47	1.55	1.69
Chloromethyl phosphate	−1.03	1.64	1.71	1.48	1.55	1.69
Phenyl phosphate	−1.03	1.61	1.70	1.47	1.55	1.69
p-chlorophenyl phosphate	−1.00	1.61	1.70	1.48	1.55	1.69
2-pyridyl phosphate	−0.66	1.67	1.70	1.47	1.55	1.69
Methyl formate	−0.49	1.35	1.49	1.21	1.33	1.41
Methyl acetate	−0.36	1.36	1.51	1.22	1.34	1.42
Acetamide	−0.32	1.37	1.50	1.23	1.36	1.45
Benzamide	−0.24	1.37	1.50	1.23	1.35	1.45
Methyl methanesulfonate	+0.32	1.63	1.76	1.45/1.45	1.51/1.52	1.80

Adsorbate coverage is 1/9 ML. Data for phosphate monoesters taken from Ref. [78]

ester bond scission was calculated to be small or mild for a range of organic groups including methyl (Fig. 6.7).

Once the P-OR ester bond was cleaved, the apical H atom on the remaining phosphoryl (H_2PO_3) group was transferred to p-nitrophenolate to form p-nitrophenol (state iv, Fig. 6.6). P-nitrophenol easily desorbed due to weak adsorption ($\Delta E_{ads} = -0.42$ eV) like other small alcohol species. The phosphoryl group (now HPO_3) then re-arranged itself to adopt the minimum-energy configuration that occupied two 3fc sites (state v, Fig. 6.6), prior to undergoing hydration to form phosphoric acid. Both hydrogen transfer and HPO_3 re-arrangement were essentially barrier-less. Hydration could be subdivided into another hydrogen transfer step and a hydroxide attack steps. The overall barrier was ca. 1.0 eV regardless of the sequence of these two substeps. The ΔE_{ads} of phosphoric acid was calculated to be -1.15 eV, the negative of which can be taken to be approximately the desorption barrier for phosphoric acid to gas phase. Thus, while the P–OR ester bond scission was facile, the hydration of the residual phosphoryl group to form phosphoric acid or its desorption was rate-limiting in gas phase. Upon inclusion of solvation energy [88] for phosphoric acid, the barrier for phosphoric acid desorption per se was significantly reduced (state xi, Fig. 6.6). Nonetheless, the reaction energy profile suggests that the surface phosphoryl group constitutes the deepest well on the reaction energy surface. The energetic cost for removing it from the surface as H_3PO_4 is ca. 0.75 eV regardless of the organic group, which is larger than the E_a for P–OR ester bond scission even in methyl phosphate (Fig. 6.7). This predicted invariance of activity with respect to the organic group parallels certain characteristics of alkaline phosphatase. The beneficial effect of water on the reaction rate compared to alcoholic solvents has been noted by Dhall et al. [73].

It should be noted that the above mechanism does not involve any Ce^{3+} or V_O. It could be operative on stoichiometric {111} facets of ceria (as exhibited by nano-octahedra) without direct coordination of p-NPP to any Ce cation. Its fundamental characteristics are consistent with experimental evidence that the dephosphorylation activity of ceria was inhibited through site blocking instead of change in surface oxidation state of Ce [73, 89], but inconsistent with those studies that have attributed catalytic activity for dephosphorylation to Ce^{3+}.

6.2.4 Adsorption of Generalized Ester Compounds on Ceria

The P–O_{latt}-bonded configuration for adsorbed p-NPP can be contrasted to the adsorption of another organic compound with an oxo group, acetaldehyde. For adsorption of acetaldehyde on $CeO_2(111)$, a C–O_{latt}-bonded μ state has been reported to be more stable than adsorption through the carbonyl O (the η state) [90]. For a phosphate compound (e.g., phosphoric acid), the $3p$ orbitals of the P atom allow it to re-hybridize and coordinate to five oxygen atoms simultaneously in a trigonal bipyramidal configuration (Fig. 6.8a), while the P=O bond gives up one bond order as lattice Ce cations partially stabilize the phosphoryl O. This allows the formation

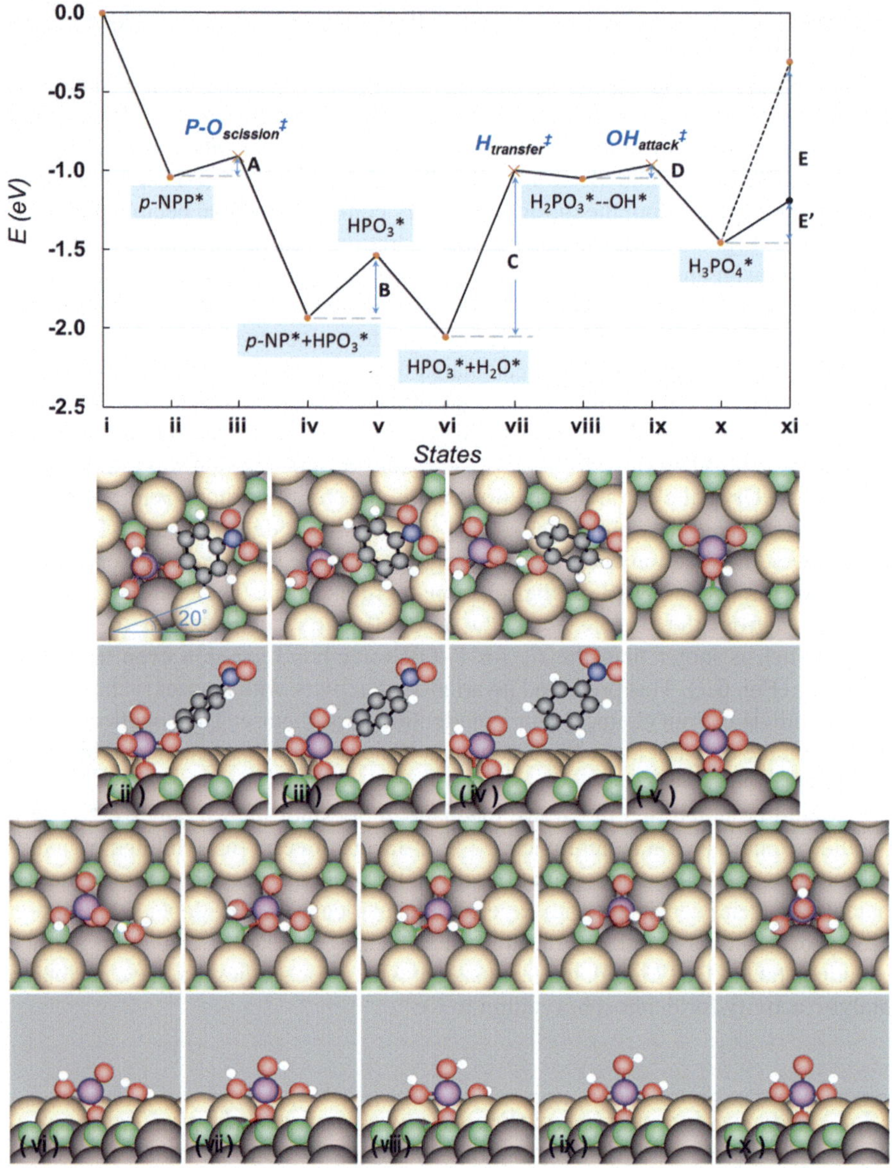

Fig. 6.6 DFT-calculated minimum-energy reaction profile for the surface-mediated dephosphorylation of p-NPP on stoichiometric $CeO_2(111)$ in vacuum. * represents surface adsorbed states and ‡ represents transition states. A-D correspond to activation energies for respective reaction steps, E and E′ correspond to desorption barrier to gas and aqueous phases, respectively. Top and side views of reaction states are included below the energy profile. Color code: green = Ce, light brown = surface O_{latt}, dark brown = subsurface O_{latt}, red = O in molecule, violet = P, black = C, blue = N, and white = H. O_{latt} directly bonded to P is treated as part of the molecule. Adapted from Ref. [78] with permission. Copyright © 2018 Elsevier B.V

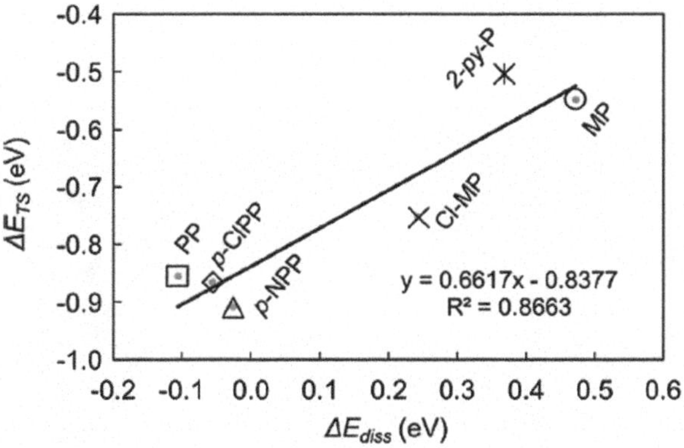

Fig. 6.7 Transition state scaling relation between the adsorption energies of the transition states (ΔE_{TS}) for P–OR ester bond scission in several phosphate monoesters, and the adsorption energies of dissociation products (ΔE_{diss}, for alkoxide and H_2PO_3 at infinite separation) on $CeO_2(111)$. The organic group R = phenyl (in PP); p-chlorophenyl (in p-ClPP); p-nitrophenyl (in p-NPP); chloromethyl (in Cl-MP); 2-pyridyl (in 2-py-P); and methyl (in MP). Adapted from Ref. [78] with permission. Copyright © 2018 Elsevier B.V.

of a strong P–O_{latt} bond (ΔE_{ads} ~ −1 eV; cf. Table 6.1). The same pattern is seen in the adsorption of acetaldehyde, with the main difference being that the central C atom transition from planar (*sp*2 hybridization) to tetrahedral (*sp*3 hybridization) coordination to allow the formation of a C–O_{latt} bond with the surface (Fig. 6.8b). The tetrahedral geometry would, however, inexorably bring some part of the molecule (i.e., the carbonyl O or the methyl group) into close proximity to the surface. As a result, the μ state has a ΔE_{ads} of −0.53 eV, making the C–O_{latt} bond less favorable than the P–O_{latt} bond in the phosphate monoesters.

For an enzyme to act on a substrate, the first step is for it to complex with the latter. In the context of a solid particle acting as an enzyme mimic, this means surface adsorption by the substrate. The possibility of a heteroatom in an oxo group forming a stable bond with a surface O_{latt} atom suggests a universal pattern for the adsorption and activation of a broad class of organic compounds on ceria, in which a heteroatom is involved in the general bonding arrangement of $R_nX(=Y)ZR'_n$. Here X could be, e.g., C, P, or S, and Y and Z are heteroatoms that are more electronegative than X. We term these *generalized ester compounds* (henceforth abbreviated as GECs). We postulate that, depending on the ability of the central heteroatom X to re-hybridize and accommodate additional bonds, and depending on how polarized the X=Y bond is, adsorption via an X–O_{latt} bond on ceria may be energetically favorable, in which case the X–ZR'_n ester bond will be weakened because O_{latt} competes for bonding with the central atom X, thus facilitating the scission of the ester bond. GECs include a wide variety of compounds such as: carboxylate esters (X:C, Y:O, Z:O); amides (X:C, Y:O, Z:N); imamides (X:C, Y:N, Z:N); nitrate esters (X:N, Y:O, Z:O); phosphate and phosphonic esters (X:P, Y:O, Z:O); and sulfate and sulfonate esters (X:S, Y:O, Z:O).

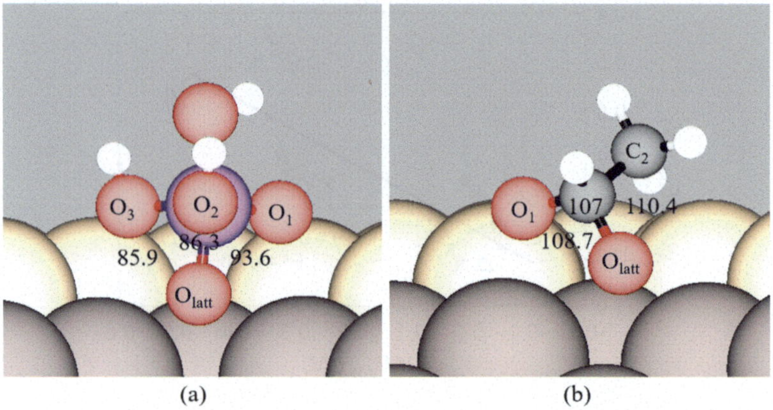

Fig. 6.8 Side views of the DFT-calculated minimum-energy geometries of **a** phosphoric acid and **b** acetaldehyde adsorbed to a surface O_{latt} atom through P/C on $CeO_2(111)$. Angles (in °) formed by the atoms in each adsorbate with the P/C and surface O_{latt} atoms are indicated. Color code: green = Ce, light brown = surface O_{latt}, dark brown = subsurface O_{latt}, red = O in molecule, violet = P, black = C, and white = H. The O_{latt} directly bonded to the central P/C atom is treated as part of the molecule

Many GACs play significant roles in diverse areas including biochemistry, medicine, chemical manufacture, biomass conversion, and environmental protection.

As a step toward understanding trends in the reactivity of ceria toward GECs, we have modeled and calculated the adsorption of several such compounds on $CeO_2(111)$. In the minimum-energy adsorption geometries (representative compounds illustrated in Fig. 6.9), they are each bonded to a surface O_{latt} site through the X atom, with the =Y group located over a 3fc site above a lattice Ce cation. The adsorption energies, ΔE_{ads}, are summarized in Table 6.1 along with several key bond distances.

Phosphoric acid and several phosphate monoesters, including methyl, chloromethyl, phenyl, p-chlorophenyl, p-nitrophenyl, and 2-pyridyl phosphates, are considered. The central P atom is coordinated to four oxygen atoms in a tetrahedral configuration in the gas phase. As mentioned above, upon adsorption P becomes pentacoordinate and forms a bond with an O_{latt} atom as it adopts a trigonal bipyramidal configuration. Existing P–O bonds are lengthened in the adsorbed state (compare, e.g., the lengths of the P–O ester bond and P=O bond in gas phase vs. as adsorbed, Table 6.1). ΔE_{ads} is ca. −1 eV and the P–O_{latt} bond is ca. 1.69 Å, nearly independent of the organic group. ΔE_{ads} of 2-pyridyl phosphate is noticeably weaker (−0.66 eV) due to the loss of a hydrogen bond between N and one of the acidic protons upon adsorption.

Additionally, four compounds with X:C and Y:O are considered: methyl formate, methyl acetate, acetamide, and benzamide, with Z:O for carboxylate esters and Z:N for amides. The ΔE_{ads} of the carboxylate esters is slightly more exothermic (ca. −0.4 eV) than that of the amides (ca. −0.3 eV). Methyl methanesulfonate, a simple

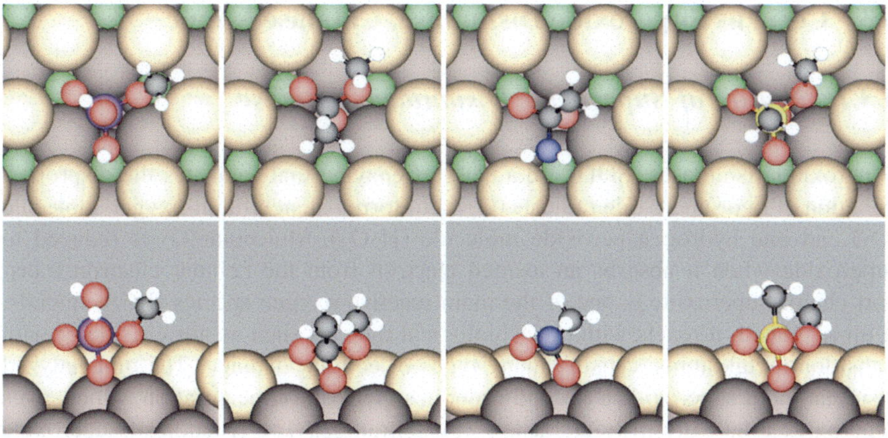

Fig. 6.9 Top (top panels) and side (bottom panels) views of the DFT-calculated minimum-energy geometries of **a** methyl phosphate, **b** methyl acetate, **c** acetamide, and **d** methyl methanesulfonate adsorbed on $CeO_2(111)$. Color code: green = Ce, light brown = surface O_{latt}, dark brown = subsurface O_{latt}, red = O in molecule, violet = P, black = C, blue = N, gold = S, and white = H. The O_{latt} directly bonded to X is treated as part of the molecule

sulfonate ester (X:S), exhibits an endothermic ΔE_{ads}, meaning that adsorption via S–O_{latt} bonding is electronically feasible but energetically unstable.

Table 6.1 suggests distinctly different ΔE_{ads} for different X atoms. For instance, ΔE_{ads} is ca. -1 eV for phosphate monoesters but is -0.3 to -0.4 eV for carboxylate esters and amides. This parallels the experimental observation that inorganic phosphate anions bind ceria strongly while sulfate and carbonate do not [73, 89]. This may be viewed as a level of substrate specificity possessed by ceria. That is, among the GACs, ceria has the highest affinity for X:P, less for X:C, and none for X:S. What is common for all the GECs considered, however, is how both the X=Y double bond and the X–Z ester bond lengthen upon adsorption on $CeO_2(111)$, which supports our view of a universal bonding mechanism for the GECs on ceria, and suggests potential catalytic activity of ceria toward the decomposition of the GECs.

An intriguing possibility therefore exists for amides and imamides on ceria. These groups appear frequently in N-containing heterocycles, important examples of which include the four nucleobases that constitute DNA, i.e., adenine, guanine, cytosine, and thymine. The imamide group appears in adenine, guanine, and cytosine where the central C atom has a $-NH_2$ (amine) substituent group. The imamide group is also a direct part of the five-membered heterocycle imidazole. Many biologically active compounds contain the imidazole ring. Furthermore, imidazole and pyrimidine fuse to form purine, which is the most widely occurring nitrogen-containing heterocycle in nature, appearing in molecules including adenine and guanine. Further research will demonstrate whether ceria is able to catalyze, e.g., the detachment of the amine group in the nucleobases or the opening of imidazole rings.

6.3 Ceria as Superoxide Dismutase, Catalase Mimic

6.3.1 Nature of Superoxide Dismutase, Catalase

Superoxide dismutases (SOD) are a class of enzyme that catalyzes the disproportionation of two superoxide radicals ($O_2^{\cdot-}$) to form one neutral di-oxygen molecule (O_2) and one hydrogen peroxide molecule (H_2O_2). Molecular O_2 is reduced to superoxide when it absorbs an excited electron from the cellular electron transport chain. Superoxide is one of the main reactive oxygen species (ROS) in cells. If un-regulated, it reacts with other biological radicals such as nitric oxide or with transition metal atoms, causing many types of cellular damage. The un-catalyzed dismutation of superoxide is second-order in superoxide concentration, which gives superoxide a half-life of hours at low concentrations. The reaction of superoxide with a SOD, on the other hand, is first-order in superoxide concentration and is essentially diffusion-limited [91], which protects cells from superoxide toxicity by out-competing reactions of superoxide with cellular components.

SOD is classified according to its metal cofactor, which consists of either one or two metal cations: CuZn, which is present in all multicellular organisms; Fe or Mn, which is used by bacteria, with the Mn type also present in mitochondria; Ni, which is present in prokaryotic cells. CuZn-SOD and Mn-SOD are present in all mammals, including humans. The commercially available CuZn-SOD is purified from bovine liver and is the most commonly used SOD in biomedical research [92]. It is a dimeric enzyme of 32 kDa. Each subunit has a Cu^{2+} and a Zn^{2+} cation surrounded mainly by histidine residues within a tertiary structure consisting of eight β strands [93, 94]. The Cu is accessible to substrates whereas the Zn is not. A two-step mechanism is commonly accepted for the CuZn-SOD catalyzed dismutation reaction [95–97], in which one $O_2^{\cdot-}$ reduces the Cu^{2+} to Cu^+ and is converted to O_2, and a second $O_2^{\cdot-}$ oxidizes the Cu^+ back to Cu^{2+} while it itself is reduced to H_2O_2. Essentially the same mechanism has been proposed for SOD with a Fe, Mn, or Ni cofactor [98–101].

$$Cu^{2+} - SOD + O_2^{\cdot-} \rightarrow Cu^{+} - SOD + O_2$$
$$Cu^{+} - SOD + O_2^{\cdot-} + 2H^{+} \rightarrow Cu^{2+} - SOD + H_2O_2$$

Catalases have a similar function to SOD in protecting the cell from oxidative damage. A catalase catalyzes the decomposition of another ROS, H_2O_2, to water and molecular O_2. The vast majority of catalases are heme-containing mono-functional catalases, while some are bi-functional (catalases/peroxidases) or non-heme-containing. Like SOD, the commercially available catalase is isolated from bovine liver. Heme-containing mono-functional catalases is a tetramer with each subunit weighing between 55 and 84 kDa [102]. The active site in each subunit contains Fe bound to a protoporphyrin IX group (heme pocket), forming a core that is deeply buried within a shell of amino acid residues. A two-step mechanism [103,

104] is commonly accepted for the heme-containing catalase (CAT):

$$Fe^{3+} - CAT + H_2O_2 \rightarrow O = Fe^{4+} - CAT + H_2O$$
$$O = Fe^{4+} - CAT + H_2O_2 \rightarrow Fe^{3+} - CAT + H_2O + O_2$$

Details regarding how H_2O_2 interacts with the active site components are not completely understood, which has been the subject of recent theoretical studies [105, 106]. For instance, given that the heme pocket is buried deep within the enzyme, possible channels for the diffusion of the substrate (H_2O_2) to and the products (H_2O and O_2) from the active site have been postulated using molecular dynamics simulation [107, 108]. Besides the degradation of hydrogen peroxide, catalases can also react with alkyl hydrogen peroxides such as methylperoxide and ethylperoxide, and the second H_2O_2 molecule can be replaced by an alcohol as a hydrogen donor.

Catalases and SOD have some of the highest turnover numbers of all known enzymes. They both turn a dismutation reaction that is second-order in reactant concentration into a sequential reaction that is first-order in reactant concentration, rendering it diffusion-limited at ambient temperature. Their combined actions are crucial to the protection of cells from oxidative stress by regulating the concentration of ROS. Common and essential to the function of both enzymes is the oscillation of the oxidation state of the metal ion in the active site (i.e., in SOD, between +1 and +2 for Cu, +2 and +3 for Mn, Fe, and Ni; in CAT, +3 and +4 for Fe).

6.3.2 *Experimental Evidence for Ceria as SOD, Catalase Mimics*

The facile Ce^{3+}/Ce^{4+} redox couple has prompted studies on whether ceria has SOD or catalase functions. Korsvik et al. [92] were the first to report SOD mimetic properties for ceria NPs. They used a wet chemical process to synthesize two polycrystalline ceria NP samples: one which was 3–5 nm in size and contained 40% of Ce in the + 3 state according to XPS; the other which was 5–8 nm in size and contained 22% Ce^{3+}. Both samples led to more H_2O_2 being produced from superoxide than if uncatalyzed, with the smaller NPs producing significantly more H_2O_2 than the larger NPs. The smaller NPs were further shown to effectively out-compete cytochrome C for reduction of superoxide, whereas the larger NPs were less efficient at doing so. The k_{cat} for the smaller NPs was calculated to exceed the most efficient CuZn-SOD by ca. 30%. Heckert et al. [109] further treated ceria NPs with H_2O_2 and confirmed with XPS and UV–vis analysis that the treatment lowered the Ce^{3+}/Ce^{4+} ratio in the samples. The H_2O_2-treated ceria NP samples showed partial to complete loss of superoxide dismutation activity depending on the concentration of the H_2O_2 solution. This was interpreted as conversion of active Ce^{3+} sites to inactive Ce^{4+} sites, akin to Kuchma et al.'s work that linked dephosphorylation activity to Ce^{3+} [72]. Subsequently over the course of two weeks, the NP samples gradually regained most of their pretreatment activity. The authors attributed both the apparent oxidation of Ce^{3+} to Ce^{4+} and the regeneration of Ce^{3+} to H_2O_2.

Ceria NPs have also been reported, initially by Pirmohamed et al. [110], to possess catalase mimetic activity. Two similarly prepared ceria NP samples, one with a Ce^{3+}/Ce^{4+} ratio of 29% and the other 7%, behaved in opposite manners when tested: Over a period of nearly an hour, the Ce^{3+}-rich sample showed no detectable catalase activity whereas the other sample, which was Ce^{4+}-rich, showed significant activity. Incubation of ceria NPs with a phosphate buffer was later reported to have an enhancing effect on the catalase mimetic activity of Ce^{3+}-rich ceria NPs (but an inhibiting effect on SOD mimetic activity) [89]. Later, Singh et al. showed that incubation with phosphate had no effect on the catalase mimetic activity of Ce^{4+}-rich ceria NPs [111].

The catalase mimetic activity of ceria NPs with low Ce^{3+}/Ce^{4+} ratios (and the lack thereof by ceria NPs with high Ce^{3+}/Ce^{4+} ratios) stands in conflict with the conventional wisdom that post-synthesis H_2O_2 treatment oxidizes Ce^{3+} to Ce^{4+} in ceria NPs. As Fig. 6.2 shows, H_2O_2 treatment turns the color of a ceria NP suspension from nearly colorless to orange or even darker [17, 72]. The interpretation is based on the fact that aqueous Ce^{3+} ions are typically colorless, whereas aqueous Ce^{4+} ions are intensely yellow. However, the color of solvated ions should not be equated to the color of solid particles. The conventional wisdom ignores the fact that aqueous Ce^{4+} is yellow and not dark orange, nor does it explain why aging causes the coloring to subside, which is interpreted as Ce^{4+} being reduced back to Ce^{3+} despite a lack of any apparent reductant. The questionable regeneration mechanism has been noted by Celardo et al. [18]. Instead, we suggest that the observed color change upon H_2O_2 treatment may be due to defect formation in ceria NPs, as defective ceria is known to take on a blue to black color, and due to concomitant dissolution of Ce ions that end up in the +4 oxidation state. After H_2O_2 is spent, the defective ceria NPs are gradually re-oxidized while dissolved Ce^{4+} ions precipitate back on to existing NPs, resulting in a (partial) clearing of the solution. More careful and critical experimental work is needed to resolve this conflict.

6.3.3 Mechanistic Insight for Ceria as SOD, Catalase Mimic

In analogy to the mechanism of natural SOD, Korsvik et al. [92] proposed the following mechanism for superoxide dismutation as catalyzed by ceria NPs:

$$Ce^{4+} + O_2^{\cdot -} \rightarrow Ce^{3+} + O_2$$

$$Ce^{3+} + O_2^{\cdot -} + 2H^+ \rightarrow Ce^{4+} + H_2O_2$$

The authors took the common view that ceria NPs with higher Ce^{3+}/Ce^{4+} ratio should catalyze superoxide dismutation, although as proposed, the mechanism should be operative regardless of whether Ce is in the +3 or +4 oxidation state.

Celardo et al. [18] took inspiration from the fact that ceria can exhibit both SOD and catalase mimetic activity and proposed catalytic mechanisms for superoxide and H_2O_2 dismutation on ceria based on redox cycles between the +3 and +4 oxidation states of Ce (Fig. 6.10). An open V_O with two associated Ce^{3+} cations was proposed to be the active site. For superoxide dismutation, the two Ce^{3+} cations are the proposed site for the adsorption of $O_2^{\cdot-}$ (adsorbed without electron transfer). The two Ce^{3+} sites are sequentially oxidized to Ce^{4+} in the reduction/protonation of the first and second $O_2^{\cdot-}$. After two superoxide anions are converted to two H_2O_2, the active site (now as two Ce^{4+} cations) is reduced back to two Ce^{3+} cations by oxidizing one H_2O_2 molecule to O_2. For H_2O_2 dismutation, the cycle begins with the active site as two Ce^{4+}. The last step in the superoxide dismutation cycle now occurs first to generate molecular O_2 and two Ce^{3+} cations by consuming the first H_2O_2. The 2nd H_2O_2 is then reduced/protonated twice by the two Ce^{3+} cations forming two water molecules, thus regenerating the two Ce^{4+} sites.

While considerably more detailed than the previous mechanisms, several areas can be identified with the above mechanism that warrant refinement: (1) As proposed, the mechanism for H_2O_2 dismutation would work whether starting with two Ce^{4+} or two Ce^{3+} sites, which means the reactions ought to be insensitive to the Ce^{3+}/Ce^{4+} ratio in the starting material; (2) as noted above, an open V_O associated with two Ce^{3+} cations is unlikely to persist under ambient conditions; this does not exclude the possibility that an adjacent pair of Ce^{3+} cations can be found on the surface of a ceria NP; (3) at the pH that the SOD and catalase activity assays are carried out (7–7.5), free protons are not readily available as the mechanism requires; (4) the mechanisms are not supported by energetics, making it impossible to determine whether they are consistent with appreciable k_{cat} at ambient temperature.

Recently, Wang et al. [112] pointed out that the redox potential of $O_2/O_2^{\cdot-}$ is 0.34 eV below the energy of the conduction [sic] band of ceria, meaning that direct electron transfer from $O_2^{\cdot-}$ to Ce^{4+} reducing it to Ce^{3+} is infeasible, contrary to common assumption (cf. Fig. 6.10a). This view echoes the findings of previous surface science and theoretical studies reporting that adsorption of gas phase O_2 on Ce^{3+} cations associated with low-coordination sites such as corners, step edges, and dislocations on CeO_2 NPs result in the formation of stable superoxide states (Fig. 6.11) [113–116]. Formation of a superoxide state has also been reported for O_2 adsorption in the presence of other surface defects [117, 118], on different facets of ceria [119], and on doped ceria [120–123], confirmed by the O–O bond length and vibrational frequency. If Ce^{3+} spontaneously reduces O_2 to $O_2^{\cdot-}$, it stands to reason that the reverse process cannot occur spontaneously, especially in aqueous phase where $O_2^{\cdot-}$ is solvated strongly by water and is thus much more stable than in gas phase.

As a result, Wang et al. [112] performed periodic DFT calculations to investigate alternative mechanisms that do not involve direct electron transfer, on the (111) facet of CeO_2. For superoxide dismutation, the authors represented superoxide ($O_2^{\cdot-}$) as $HO_2^{\cdot}$ on the grounds that $O_2^{\cdot-}$ is converted to $HO_2^{\cdot}$ in water. The authors reported intermolecular H transfer between two $HO_2^{\cdot}$ adsorbed on adjacent Ce sites to be a plausible pathway to H_2O_2 formation, which has an E_a of 0.78 eV when occurring

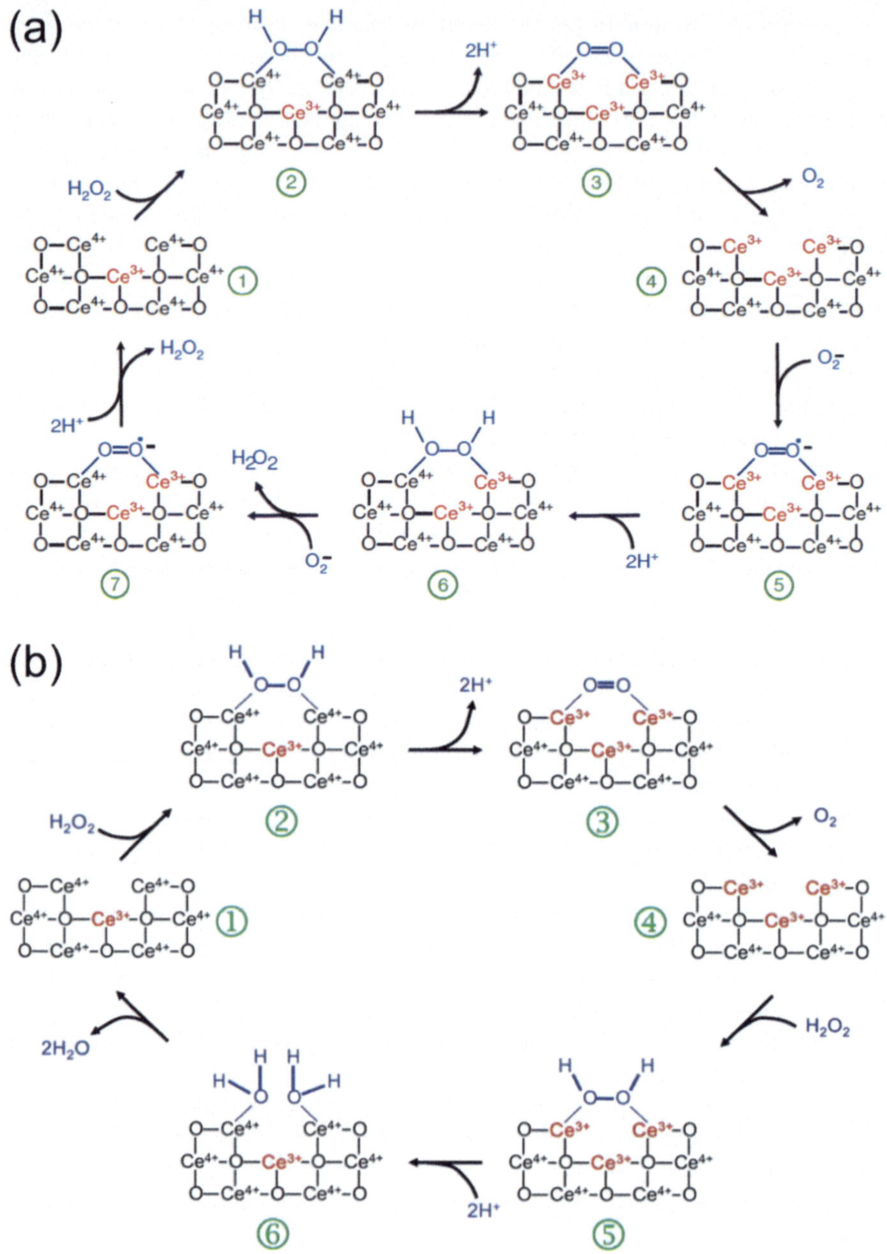

Fig. 6.10 A model of the reaction mechanism for **a** the dismutation of superoxide, and **b** the dismutation of H_2O_2 by ceria NPs. The first four steps in the two cycles are identical. Adapted from Ref. [18] with permission. Copyright © 2011 Royal Society of Chemistry

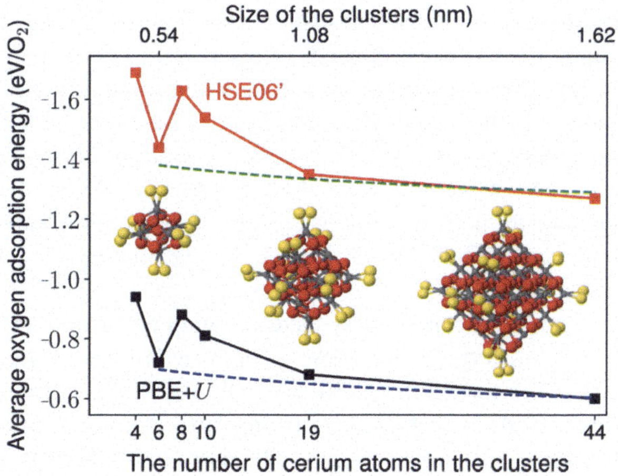

Fig. 6.11 Average ΔE_{ads} for O_2 adsorbed on Ce sites plotted against cluster size using hybrid and PBE + U(5 eV) functionals. Adapted from Ref. [116] with permission. Copyright © 2019 American Chemical Society

on a hydrogenated surface. The fundamental problem with their model is that the pKa of $HO_2^{\cdot}$ is 4.69, whereas SOD assays are performed at neutral pH, where the predominant form of the superoxide is $O_2^{\cdot-}$, not $HO_2^{\cdot}$. Furthermore, by the authors' own calculations H_2O_2 should proceed to dissociate into water, but experimentally significant levels of H_2O_2 were detected for SOD mimetic ceria NPs. The H_2O_2 dismutation mechanism that the same authors proposed [111] involved exchanging H atoms with surface O_{latt} sites and was calculated to have a maximum E_a of 0.82 eV, which is somewhat higher than surmountable at ambient temperature. It has the same feature as the one proposed by Celardo et al. [18] (Fig. 6.10b), i.e., the surface was driven by H_2O_2 to locally oscillate between an oxidized state (stoichiometric, Ce^{4+}) and a reduced state (with adsorbed H atoms, Ce^{3+}), which means that the mechanism should be operative on either an oxidized or a reduced ceria surface. Experimentally, however, Ce^{3+}-rich ceria NPs were found to exhibit no catalase mimetic activity [110].

Clearly, additional work is needed to elucidate the mechanisms of superoxide and H_2O_2 dismutation on ceria NPs. So far, we can state with reasonable certainty that the SOD and catalase mimetic functions occur on different active sites. The active sites are in some ways associated with Ce^{3+} and Ce^{4+} respectively, although what they are is not clear.

6.4 Challenges and Outlook

Ceria has been a mainstay in heterogeneous catalysis from the early days of the discipline, prized for its redox activity and oxygen buffering/storage capabilities. Now there is mounting evidence that it also possesses fascinating catalytic properties that mimic important biological functions performed by natural enzymes. However, the mechanism of ceria's action remains unclear. A central tenant in explaining the biological processes that sustain life is energy cascade through the electron transport chain. Each process consists of a series of redox reactions, in which an enzyme that relies on metal ions to switch between oxidation states plays a key role. It seems natural enough to draw a parallel between redox chemistry in the active center of enzymes and the Ce^{3+}/Ce^{4+} redox couple in ceria. Yet details become sketchy as soon as one goes beyond this simple idea. As we have reviewed in this chapter, enzyme mimetic activity (e.g., dephosphorylation, superoxide or H_2O_2 dismutation) has been attributed to Ce^{3+} or Ce^{4+} by various studies, often in apparent conflict with one another.

The disagreement may have arisen in part from the conflation of different origins of Ce^{3+} in ceria. In our view, as supported by recent theoretical findings, the presence of Ce^{3+} in ceria NPs is unlikely to be associated with open surface oxygen vacancies due to significant energetic driving forces for ubiquitous species such as O_2 and water to occupy oxygen vacancies. The more likely origins include: (1) reductive species, the most common of which being H, (2) under-coordination on the surface of NPs imposed by their geometries; (3) oxygen vacancies or impurities in the bulk.

A hydrogenated surface results in Ce^{3+} in the surface region with O_{latt} sites occupied by H atoms [124], forming embedded hydroxide groups in the surface. For dephosphorylation, if phosphate monoesters are activated via P–O_{latt} bonding on $CeO_2(111)$ as suggested by Zhao et al. [78] then a hydrogenated surface (manifested as high surface Ce^{3+} concentrations, and also significant νO_{latt}–H signals if such samples are investigated using vibrational spectroscopy) should exhibit diminished catalytic activity. A hydrogenated surface can result from the interaction of a defective ceria surface with water, which heal oxygen vacancies and deposit H atoms on the surface at the same time. It may also occur when H_2O_2 is used as the precipitant in synthesis where it leaves residual H atoms, resulting in the formation of ceria NPs with high Ce^{3+}/Ce^{4+} ratios [73]. Procedures that remove surface H atoms (i.e., calcination in oxygen) should restore catalytic activity that depends on O_{latt} sites.

Low-coordination sites such as edges, corners, and apices always exist on the surface of ceria NPs. Constrained by geometry, Ce atoms may not be able to achieve sufficient coordination to O atoms at such sites and preferentially exist in the +3 instead of +4 oxidation state [113, 116]. Such sites may be able to adsorb molecular O_2 by reducing it to a superoxide state [116], or coordinate to water when exposed to an aqueous phase [125]. A pair of adjacent such Ce^{3+} cations may form an active site akin to the model used by Kuchma et al. mentioned above and catalyze dephosphorylation via a separate pathway from that suggested by Zhao et al. [78]. The prevalence of this type of active site should decrease with increasing particle size and increasing crystallinity of NPs.

Our survey of the literature reveals some intertwining issues in current experimental methods. For instance, by following the same synthesis procedures, Manto et al. made nanocubes that contained no Ce^{3+} [46], whereas Tan et al. detected significant amounts of Ce^{3+} in the nanocubes they synthesized [48]. A commonly used technique to spectroscopically detect Ce in different oxidation states, XPS, may also pose problems. Ce^{4+}–O interaction produces multiple peaks that typically dominate the spectra and significantly overlap with the peaks due to Ce^{3+}–O interaction. While some authors took small changes in the relative intensities of the peaks to indicate changes in the relative amounts of Ce^{3+} to Ce^{4+} [74], others ascribed no significance to them [45, 73]. As its detection depth could be several nanometers, XPS may not be a true surface technique when applied to the ceria NPs [48]. Bulk vacancies or impurities such as H [126, 127] or F [128] are expected to be insensitive to surface reactions at ambient conditions, but due to the size of NPs the Ce^{3+} associated with them may be detected by XPS nonetheless.

Furthermore, a widely accepted empirical view regarding the oxidation state of a suspension of ceria NPs warrants a second thought, i.e., that treatment with H_2O_2 oxidizes Ce^{3+} to Ce^{4+} because the color of the solution changes from pale to dark yellow [17, 19, 21, 72, 89]. The conclusion is based on analogy to the colors of aqueous solvated Ce^{3+} and Ce^{4+} ions, but there is a lack of firm evidence that the same colors (clear, yellow) should be displayed by ceria NPs that are rich in Ce^{3+} or Ce^{4+}, respectively. Even in those studies where Ce^{3+} is associated with the active site, the ceria NPs contain significant portion of Ce in the +4 oxidation state, whose role in the reaction cannot be readily ruled out. A more rigorous way to oxidize ceria NPs, i.e., calcination at elevated temperature in oxygen [114], should be more widely adopted by biomedical researchers to eliminate ambiguity. Accurate, surface-specific detection methods for Ce^{3+} versus Ce^{4+} are a key area that needs more progress in order to help the research community converge onto an accurate understanding of the mechanistic action of ceria-based enzyme mimetic systems.

Ceria has not only been widely known to catalyze reactions relevant to chemical manufacture and environmental protection at elevated temperature, but its ambient temperature activity for catalyzing biologically important reactions is now being recognized. On a fundamental level, there is a need to consider mechanisms [78, 112] for enzyme-mimetic functions of ceria that are not solely predicated on direct electron transfer between the substrates and Ce cations. A more general view based on reaction energetics, as is commonly taken in heterogeneous catalysis research, should be adopted. We confidently expect that many more such enzyme-mimetic functions will be discovered and investigated for ceria in coming years. Close interaction between heterogeneous catalysis and biochemistry will prove fruitful to researchers in both communities who are seeking to unlock the technological potentials of ceria.

Acknowledgements This work was supported by the U.S. National Science Foundation under Grant #CHE-1664984 and used high performance computational resources provided by Louisiana State University (hpc.lsu.edu), by the Center for Nanophase Materials Sciences, which is a DOE Office of Science User Facility, and by the National Energy Research Scientific Computing Center, which is supported by the Office of Science of US-DOE under Contract No. DE-AC02-05CH11231.

References

1. Wei H, Wang E (2013) Nanomaterials with enzyme-like characteristics (nanozymes): next-generation artificial enzymes. Chem Soc Rev 42:6060
2. Hooley RJ (2016) Taking on the turnover challenge. Nat Chem 8:202–204
3. Liu B, Liu J (2017) Surface modification of nanozymes. Nano Res 10:1125–1148
4. Huang Y, Ren J, Qu X (2019) Nanozymes: classification, catalytic mechanisms, activity regulation, and applications. Chem Rev 119:4357–4412
5. Wu J et al (2019) Nanomaterials with enzyme-like characteristics (nanozymes): next-generation artificial enzymes (II). Chem Soc Rev 48:1004–1076
6. Zhang R, Fan K, Yan X (2020) Nanozymes: created by learning from nature. Sci China Life Sci. https://doi.org/10.1007/s11427-019-1570-7
7. Jiang D et al (2019) Nanozyme: new horizons for responsive biomedical applications. Chem Soc Rev 48:3683–3704
8. Chen W, Li S, Wang J, Sun K, Si Y (2019) Metal and metal-oxide nanozymes: bioenzymatic characteristics, catalytic mechanism, and eco-environmental applications. Nanoscale 11:15783–15793
9. Cox CR, Lee JZ, Nocera DG, Buonassisi T (2014) Ten-percent solar-to-fuel conversion with nonprecious materials. Proc Natl Acad Sci 111:14057–14061
10. Varley JB, Wang Y, Chan K, Studt F, Nørskov JK (2015) Mechanistic insights into nitrogen fixation by nitrogenase enzymes. Phys Chem Chem Phys 17:29541–29547
11. Komiyama M, Takeda N, Shigekawa H (1999) Hydrolysis of DNA and RNA by lanthanide ions: mechanistic studies leading to new applications. Chem Commun, 1443–1451. https://doi.org/10.1039/a901621j
12. Branum ME, Tipton AK, Zhu S, Que L (2001) Double-strand hydrolysis of plasmid DNA by dicerium complexes at 37 °C. J Am Chem Soc 123:1898–1904
13. Katada H, Seino H, Mizobe Y, Sumaoka J, Komiyama M (2008) Crystal structure of Ce(IV)/dipicolinate complex as catalyst for DNA hydrolysis. JBIC J Biol Inorg Chem 13:249–255
14. Tarnuzzer RW, Colon J, Patil S, Seal S (2005) Vacancy engineered ceria nanostructures for protection from radiation-induced cellular damage. Nano Lett 5:2573–2577
15. Schubert D, Dargusch R, Raitano J, Chan S-W (2006) Cerium and yttrium oxide nanoparticles are neuroprotective. Biochem Biophys Res Commun 342:86–91
16. Silva GA (2006) Seeing the benefits of ceria. Nat Nanotechnol 1:92–94
17. Das M et al (2007) Auto-catalytic ceria nanoparticles offer neuroprotection to adult rat spinal cord neurons. Biomaterials 28:1918–1925
18. Celardo I, Pedersen JZ, Traversa E, Ghibelli L (2011) Pharmacological potential of cerium oxide nanoparticles. Nanoscale 3:1411
19. Xu C, Qu X (2014) Cerium oxide nanoparticle: a remarkably versatile rare earth nanomaterial for biological applications. NPG Asia Mater. 6:e90–e90
20. Walkey C et al (2015) Catalytic properties and biomedical applications of cerium oxide nanoparticles. Environ Sci Nano 2:33–53
21. Singh S (2016) Cerium oxide based nanozymes: redox phenomenon at biointerfaces. Biointerphases 11:04B202
22. Nelson B, Johnson M, Walker M, Riley K, Sims C (2016) Antioxidant cerium oxide nanoparticles in biology and medicine. Antioxidants 5:15
23. Ragg R, Tahir MN, Tremel W (2016) Solids go bio: inorganic nanoparticles as enzyme mimics. Eur J Inorg Chem 2016:1906–1915
24. Duchoň T et al (2017) Covalent versus localized nature of 4f electrons in ceria: resonant angle-resolved photoemission spectroscopy and density functional theory. Phys Rev B 95:165124
25. Trovarelli A (2002) Catalysis by ceria and related materials, vol 2. Published by Imperial College Press and distributed by World Scientific Publishing Co.
26. Kašpar J, Fornasiero P, Graziani M (1999) Use of CeO_2-based oxides in the three-way catalysis. Catal Today 50:285–298

27. Bunluesin T, Gorte RJ, Graham GW (1998) Studies of the water-gas-shift reaction on ceria-supported Pt, Pd, and Rh: Implications for oxygen-storage properties. Appl Catal B Environ 15:107–114
28. Rodriguez JA et al (2007) Activity of CeOx and TiOx nanoparticles grown on Au(111) in the water-gas shift reaction. Science 318:1757–1760
29. Wang S, (Max) Lu G (1998) Role of CeO_2 in Ni/CeO_2–Al_2O_3 catalysts for carbon dioxide reforming of methane. Appl Catal B Environ 19:267–277
30. Deluga GA (2004) Renewable hydrogen from ethanol by autothermal reforming. Science 303:993–997
31. Gutiérrez-Ortiz JI, de Rivas B, López-Fonseca R, González-Velasco JR (2006) Catalytic purification of waste gases containing VOC mixtures with Ce/Zr solid solutions. Appl Catal B Environ 65:191–200
32. Aneggi E, de Leitenburg C, Dolcetti G, Trovarelli A (2006) Promotional effect of rare earths and transition metals in the combustion of diesel soot over CeO_2 and CeO_2–ZrO_2. Catal Today 114:40–47
33. Hendren TS, Dooley KM (2003) Kinetics of catalyzed acid/acid and acid/aldehyde condensation reactions to non-symmetric ketones. Catal Today 85:333–351
34. Gaertner CA, Serrano-Ruiz JC, Braden DJ, Dumesic JA (2009) Catalytic coupling of carboxylic acids by ketonization as a processing step in biomass conversion. J Catal 266:71–78
35. Ma Y et al (2018) Regulating the surface of nanoceria and its applications in heterogeneous catalysis. Surf Sci Rep 73:1–36
36. Feng X (2006) Converting ceria polyhedral nanoparticles into single-crystal nanospheres. Science 312:1504–1508
37. Zhou K, Wang X, Sun X, Peng Q, Li Y (2005) Enhanced catalytic activity of ceria nanorods from well-defined reactive crystal planes. J Catal 229:206–212
38. Mai H-X et al (2005) Shape-selective synthesis and oxygen storage behavior of ceria nanopolyhedra, nanorods, and nanocubes. J Phys Chem B 109:24380–24385
39. Aneggi E, Llorca J, Boaro M, Trovarelli A (2005) Surface-structure sensitivity of CO oxidation over polycrystalline ceria powders. J Catal 234:88–95
40. Yang Z et al (2007) Single-crystalline ceria nanocubes: size-controlled synthesis, characterization and redox property. Nanotechnology 18:185606
41. Si R, Flytzani-Stephanopoulos M (2008) Shape and crystal-plane effects of nanoscale ceria on the activity of Au-CeO_2 catalysts for the water–gas shift reaction. Angew. Chemie 120:2926–2929
42. Yuan Q et al (2009) Controlled synthesis and assembly of ceria-based nanomaterials. J Colloid Interface Sci 335:151–167
43. Yan L, Yu R, Chen J, Xing X (2008) Template-free hydrothermal synthesis of CeO_2 nano-octahedrons and nanorods: investigation of the morphology evolution. Cryst Growth Des 8:1474–1477
44. Wang Z-G et al (2015) Facet-dependent effect of well-defined CeO_2 nanocrystals on the adsorption and dephosphorylation of phosphorylated molecules. Part Part Syst Charact 32:652–660
45. Yang Y et al (2016) Redox enzyme-mimicking activities of CeO_2 nanostructures: intrinsic influence of exposed facets. Sci Rep 6:35344
46. Manto MJ, Xie P, Wang C (2017) Catalytic dephosphorylation using ceria nanocrystals. ACS Catal 7:1931–1938
47. Mann AKP, Wu Z, Calaza FC, Overbury SH (2014) Adsorption and reaction of acetaldehyde on shape-controlled CeO_2 nanocrystals: elucidation of structure-function relationships. ACS Catal 4:2437–2448
48. Tan Z, Wu T-S, Soo Y-L, Peng Y-K (2020) Unravelling the true active site for CeO_2-catalyzed dephosphorylation. Appl Catal B Environ 264:118508
49. Hafner J (2000) Atomic-scale computational materials science. Acta Mater 48:71–92
50. Greeley J, Nørskov JK, Mavrikakis M (2002) Electronic structure and catalysis on metal surfaces. Annu Rev Phys Chem 53:319–348

51. Nørskov JK, Bligaard T, Rossmeisl J, Christensen CH (2009) Towards the computational design of solid catalysts. Nat Chem 1:37–46
52. Knowles JR (1980) Enzyme-catalyzed phosphoryl transfer reactions. Annu Rev Biochem 49:877–919
53. Kaplan MM (1972) Alkaline phosphatase. Gastroenterology 62:452–468
54. Herries D (1981) Alkaline phosphatase. Biochem Educ 9:76
55. Coleman JE (1992) Structure and mechanism of alkaline phosphatase. Annu Rev Biophys Biomol Struct 21:441–483
56. Kim EE, Wyckoff HW (1991) Reaction mechanism of alkaline phosphatase based on crystal structures. J Mol Biol 218:449–464
57. Holtz KM, Stec B, Kantrowitz ER (1999) A model of the transition state in the alkaline phosphatase reaction. J Biol Chem 274:8351–8354
58. Lassila JK, Zalatan JG, Herschlag D (2011) Biological phosphoryl-transfer reactions: understanding mechanism and catalysis. Annu Rev Biochem 80:669–702
59. Gettins P, Metzler M, Coleman JE (1985) Alkaline phosphatase. 31P NMR probes of the mechanism. J Biol Chem 260:2875–2883
60. Kirby AJ, Varvoglis AG (1967) The reactivity of phosphate esters. Monoester hydrolysis. J Am Chem Soc 89:415–423
61. Lad C, Williams NH, Wolfenden R (2003) The rate of hydrolysis of phosphomonoester dianions and the exceptional catalytic proficiencies of protein and inositol phosphatases. Proc Natl Acad Sci 100:5607–5610
62. Cleland WW, Hengge AC (2006) Enzymatic mechanisms of phosphate and sulfate transfer. Chem Rev 106:3252–3278
63. Duarte F, Barrozo A, Åqvist J, Williams NH, Kamerlin SCL (2016) The competing mechanisms of phosphate monoester dianion hydrolysis. J Am Chem Soc 138:10664–10673
64. Florián J, Warshel A (1997) A fundamental assumption about OH–attack in phosphate ester hydrolysis is not fully justified. J Am Chem Soc 119:5473–5474
65. Florián J, Warshel A (1998) Phosphate ester hydrolysis in aqueous solution: associative versus dissociative mechanisms. J Phys Chem B 102:719–734
66. Klähn M, Rosta E, Warshel A (2006) On the mechanism of hydrolysis of phosphate monoesters dianions in solutions and proteins. J Am Chem Soc 128:15310–15323
67. Kamerlin SCL, Williams NH, Warshel A (2008) Dineopentyl phosphate hydrolysis: evidence for stepwise water attack. J Org Chem 73:6960–6969
68. Kamerlin SCL (2011) Theoretical comparison of p-nitrophenyl phosphate and sulfate hydrolysis in aqueous solution: implications for enzyme-catalyzed sulfuryl transfer. J Org Chem 76:9228–9238
69. Duarte F, Åqvist J, Williams NH, Kamerlin SCL (2015) Resolving apparent conflicts between theoretical and experimental models of phosphate monoester hydrolysis. J Am Chem Soc 137:1081–1093
70. Pabis A, Kamerlin SCL (2016) Promiscuity and electrostatic flexibility in the alkaline phosphatase superfamily. Curr Opin Struct Biol 37:14–21
71. Tan F et al (2008) An efficient method for dephosphorylation of phosphopeptides by cerium oxide. J Mass Spectrom 43:628–632
72. Kuchma MH et al (2010) Phosphate ester hydrolysis of biologically relevant molecules by cerium oxide nanoparticles. Nanomed Nanotechnol Biol Med 6:738–744
73. Dhall A et al (2017) Characterizing the phosphatase mimetic activity of cerium oxide nanoparticles and distinguishing its active site from that for catalase mimetic activity using anionic inhibitors. Environ Sci Nano 4:1742–1749
74. Yao T, Tian Z, Zhang Y, Qu Y (2019) Phosphatase-like activity of porous nanorods of CeO_2 for the highly stabilized dephosphorylation under interferences. ACS Appl Mater Interfaces 11:195–201
75. Huang M, Fabris S (2007) Role of surface peroxo and superoxo species in the low-temperature oxygen buffering of ceria: density functional theory calculations. Phys Rev B 75:081404

76. Hansen HA, Wolverton C (2014) Kinetics and thermodynamics of H_2O dissociation on reduced CeO_2(111). J Phys Chem C 118:27402–27414
77. Trotochaud L et al (2017) Spectroscopic and computational investigation of room-temperature decomposition of a chemical warfare agent simulant on polycrystalline cupric oxide. Chem Mater 29:7483–7496
78. Zhao C, Xu Y (2018) Theoretical investigation of dephosphorylation of phosphate monoesters on CeO_2(111). Catal Today 312:141–148
79. Perdew J et al (1992) Atoms, molecules, solids, and surfaces: applications of the generalized gradient approximation for exchange and correlation. Phys Rev B 46:6671–6687
80. Kresse G, Furthmüller J (1996) Efficient iterative schemes for ab initio total-energy calculations using a plane-wave basis set. Phys Rev B 54:11169–11186
81. Dudarev SL, Botton GA, Savrasov SY, Humphreys CJ, Sutton AP (1998) Electron-energy-loss spectra and the structural stability of nickel oxide: an LSDA+U study. Phys Rev B 57:1505–1509
82. Gerward L et al (2005) Bulk modulus of CeO_2 and PrO_2—an experimental and theoretical study. J Alloys Compd 400:56–61
83. Gerward L, Olsen JS (1993) Powder diffraction analysis of cerium dioxide at high pressure. Powder Diffr 8:127–129
84. Henkelman G, Uberuaga BP, Jónsson H (2000) A climbing image nudged elastic band method for finding saddle points and minimum energy paths. J Chem Phys 113:9901–9904
85. Smidstrup S, Pedersen A, Stokbro K, Jónsson H (2014) Improved initial guess for minimum energy path calculations. J Chem Phys 140:214106
86. Henkelman G, Jónsson H (1999) A dimer method for finding saddle points on high dimensional potential surfaces using only first derivatives. J Chem Phys 111:7010–7022
87. Heyden A, Bell AT, Keil FJ (2005) Efficient methods for finding transition states in chemical reactions: comparison of improved dimer method and partitioned rational function optimization method. J Chem Phys 123:224101
88. Śmiechowski M (2009) Theoretical calculation of pKas of phosphoric (V) acid in the polarisable continuum and cluster-continuum models. J Mol Struct 924–926:170–174
89. Singh S et al (2011) A phosphate-dependent shift in redox state of cerium oxide nanoparticles and its effects on catalytic properties. Biomaterials 32:6745–6753
90. Calaza FC, Xu Y, Mullins DR, Overbury SH (2012) Oxygen vacancy-assisted coupling and enolization of acetaldehyde on CeO_2(111). J Am Chem Soc 134:18034–18045
91. Stroppolo ME, Falconi M, Caccuri AM, Desideri A (2001) Superefficient enzymes. Cell Mol Life Sci 58:1451–1460
92. Korsvik C, Patil S, Seal S, Self WT (2007) Superoxide dismutase mimetic properties exhibited by vacancy engineered ceria nanoparticles. Chem Commun, 1056. https://doi.org/10.1039/b615134e
93. Tainer JA, Getzoff ED, Beem KM, Richardson JS, Richardson DC (1982) Determination and analysis of the 2 Å structure of copper, zinc superoxide dismutase. J Mol Biol 160:181–217
94. Getzoff ED et al (1983) Electrostatic recognition between superoxide and copper, zinc superoxide dismutase. Nature 306:287–290
95. Tainer JA, Getzoff ED, Richardson JS, Richardson DC (1983) Structure and mechanism of copper, zinc superoxide dismutase. Nature 306:284–287
96. Pelmenschikov V, Siegbahn PEM (2005) Copper−zinc superoxide dismutase: theoretical insights into the catalytic mechanism. Inorg Chem 44:3311–3320
97. Bordo D et al (2011) Copper-zinc superoxide dismutase in prokaryotes and eukaryotes. In: Encyclopedia of inorganic and bioinorganic chemistry. Wiley, pp 1–17. https://doi.org/10.1002/9781119951438.eibc0621
98. Abreu IA, Cabelli DE (2010) Superoxide dismutases—a review of the metal-associated mechanistic variations. Biochim Biophys Acta—Proteins Proteomics 1804:263–274
99. Perry JJP, Shin DS, Getzoff ED, Tainer JA (2010) The structural biochemistry of the superoxide dismutases. Biochim Biophys Acta—Proteins Proteomics 1804:245–262

100. Sheng Y et al (2014) Superoxide dismutases and superoxide reductases. Chem Rev 114:3854–3918
101. Azadmanesh J, Borgstahl G (2018) A review of the catalytic mechanism of human manganese superoxide dismutase. Antioxidants 7:25
102. Nicholls P, Fita I, Loewen PC (2000) Enzymology and structure of catalases. Adv Inorg Chem 51:51–106
103. Putnam CD, Arvai AS, Bourne Y, Tainer JA (2000) Active and inhibited human catalase structures: ligand and NADPH binding and catalytic mechanism. J Mol Biol 296:295–309
104. Alfonso-Prieto M, Vidossich P, Rovira C (2012) The reaction mechanisms of heme catalases: an atomistic view by ab initio molecular dynamics. Arch Biochem Biophys 525:121–130
105. Nicholls P (2012) Classical catalase: ancient and modern. Arch Biochem Biophys 525:95–101
106. Alfonso-Prieto M, Biarnés X, Vidossich P, Rovira C (2009) The molecular mechanism of the catalase reaction. J Am Chem Soc 131:11751–11761
107. Kalko SG, Gelpí JL, Fita I, Orozco M (2001) Theoretical study of the mechanisms of substrate recognition by catalase. J Am Chem Soc 123:9665–9672
108. Domínguez L, Sosa-Peinado A, Hansberg W (2010) Catalase evolved to concentrate H_2O_2 at its active site. Arch Biochem Biophys 500:82–91
109. Heckert EG, Karakoti AS, Seal S, Self WT (2008) The role of cerium redox state in the SOD mimetic activity of nanoceria. Biomaterials 29:2705–2709
110. Pirmohamed T et al (2010) Nanoceria exhibit redox state-dependent catalase mimetic activity. Chem Commun 46:2736
111. Singh R, Singh S (2015) Role of phosphate on stability and catalase mimetic activity of cerium oxide nanoparticles. Colloids Surfaces B Biointerfaces 132:78–84
112. Wang Z, Shen X, Gao X, Zhao Y (2019) Simultaneous enzyme mimicking and chemical reduction mechanisms for nanoceria as a bio-antioxidant: a catalytic model bridging computations and experiments for nanozymes. Nanoscale 11:13289–13299
113. Xu J et al (2010) Size dependent oxygen buffering capacity of ceria nanocrystals. Chem Commun 46:1887
114. Wu Z, Li M, Howe J, Meyer HM, Overbury SH (2010) Probing defect sites on CeO_2 nanocrystals with well-defined surface planes by Raman spectroscopy and O_2 adsorption. Langmuir 26:16595–16606
115. Preda G et al (2011) Formation of superoxide anions on ceria nanoparticles by interaction of molecular oxygen with Ce^{3+} sites. J Phys Chem C 115:5817–5822
116. Du D, Kullgren J, Hermansson K, Broqvist P (2019) From ceria clusters to nanoparticles: superoxides and supercharging. J Phys Chem C 123:1742–1750
117. Chen H-L, Chen H-T (2010) Role of hydroxyl groups for the O_2 adsorption on CeO_2 surface: a DFT+U study. Chem Phys Lett 493:269–272
118. Li H-Y et al (2009) Multiple configurations of the two excess 4f electrons on defective CeO_2(111): origin and implications. Phys Rev B 79:193401
119. Kullgren J, Hermansson K, Broqvist P (2013) Supercharged low-temperature oxygen storage capacity of ceria at the nanoscale. J Phys Chem Lett 4:604–608
120. Conesa JC (2009) Surface anion vacancies on ceria: Quantum modelling of mutual interactions and oxygen adsorption. Catal Today 143:315–325
121. Gionco C, Giamello E, Mino L, Paganini MC (2014) The interaction of oxygen with the surface of CeO_2–TiO_2 mixed systems: an example of fully reversible surface-to-molecule electron transfer. Phys Chem Chem Phys 16:21438–21445
122. Livraghi S et al (2019) Formation of reversible adducts by adsorption of oxygen on Ce–ZrO_2: an unusual η^2 ionic superoxide. J Phys Chem C 123:27088–27096
123. Polliotto V, Livraghi S, Agnoli S, Granozzi G, Giamello E (2019) Reversible adsorption of oxygen as superoxide ion on cerium doped zirconium titanate. Appl Catal A Gen 580:140–148
124. Vicario G, Balducci G, Fabris S, de Gironcoli S, Baroni S (2006) Interaction of hydrogen with cerium oxide surfaces: a quantum mechanical computational study. J Phys Chem B 110:19380–19385

125. Sayle TXT et al (2013) Environment-mediated structure, surface redox activity and reactivity of ceria nanoparticles. Nanoscale 5:6063
126. Wu Z et al (2017) Direct neutron spectroscopy observation of cerium hydride species on a cerium oxide catalyst. J Am Chem Soc 139:9721–9727
127. Werner K et al (2017) Toward an understanding of selective alkyne hydrogenation on ceria: on the impact of O vacancies on H_2 interaction with CeO_2(111). J Am Chem Soc 139:17608–17616
128. Kullgren J et al (2014) Oxygen vacancies versus fluorine at CeO_2(111): a case of mistaken identity? Phys Rev Lett 112:156102

Chapter 7
Emerging 2D Materials-Based Nanoarchitecture for Water Purification

Shamily Patibandla, Avijit Pramanik, Ye Gao, Kaelin Gates, Manoj K. Shukla, and Paresh Chandra Ray

Abstract Even today, getting safe drinking water is one of the big challenges for society. As per World Health Organization (WHO), several millions of people are lacking drinking water that is free from viruses, toxic chemicals, and bacteria. Here, we discuss the new development of a water filtration systems using a two-dimensional (2D) nanomaterials. Due to atomically thin surface and good mechanical strength, 2D graphene, graphene oxide, as well as transition metal dichalcogenides are considered to be advanced membrane materials. This chapter highlights the recent reports on how emerging material-based membranes have been used to tackle water desalination. Notably, we discuss the synthetic method development for the design of novel membranes which have the capability for the separation of toxic chemicals and pathogens. Finally, we have discussed the future challenges and prospects of current development.

7.1 Introduction

According to the United Nations (UN), 3 out of 10 people in our world do not have safely managed drink water services that are free from toxic chemicals and waterborne pathogens [1, 2]. It is now well documented that common water sources are contaminated by per- and polyfluoroalkyl substances (PFAS), pesticides, and

S. Patibandla · A. Pramanik · Y. Gao · K. Gates · P. C. Ray (✉)
Department of Chemistry and Biochemistry, Jackson State University, Jackson, MS, USA
e-mail: paresh.c.ray@jsums.edu

S. Patibandla
e-mail: shamily.patibandla@students.jsums.edu

A. Pramanik
e-mail: avijit.pramanik@jsums.edu

M. K. Shukla
US Army Engineer Research and Development Center, 3909 Halls Ferry Road, Vicksburg, MS, USA
e-mail: Manoj.K.Shukla@usace.army.mil

M. Shukla et al. (eds.), *Emerging Materials and Environment*,
Challenges and Advances in Computational Chemistry and Physics 37,
https://doi.org/10.1007/978-3-031-39470-6_7

fertilizers [1–4]. According to the World Health Organization (WHO), more than 400 million pounds of antibiotics and other medicines are discarded in the environment each year [3]. According to the *Centers for Disease Control and Prevention* (*CDC*), multidrug-resistant superbugs in water kill one hundred thousand people annually [5–7]. Due to the above facts, scientists have taken initiatives for the development of 2D materials-based architectures which can be used for the separation of PFAS, heavy metals, viruses, and bacteria from drinking water [6–10]. Recently, we and other groups have reported two-dimensional (2D) nanomaterial-based approaches as shown in Fig. 7.1, for the design of membranes with superior properties [11–30].

From 2004, two-dimensional (2D) graphene oxide, transition metal chalcogenides (TMDs), and others have been designed for possible use in physics, chemistry, material science, and biology as membrane material [11–25]. 2D materials have been used as novel membranes for the separation of toxic metals and biological residues [26–40]. Since 2D materials can assemble a structure at a well-defined interlayer distance that acts as separation channels, in the last ten years, 2D materials have received huge attention for the membranes development [20–39]. As several reports indicate that 2D materials membranes exhibit enormous advantages in terms of size-selective separation, it has been used for water purification [11–25]. We and other groups have shown 2D membranes can demonstrate excellent permeation properties, as well as antibacterial, behavior [11–25]. Due to the above advantages, tremendous efforts have been devoted to the design of 2D and heterostructure material-based membranes for the purification of water via the separation of toxic metals and pathogens [26, 30–40].

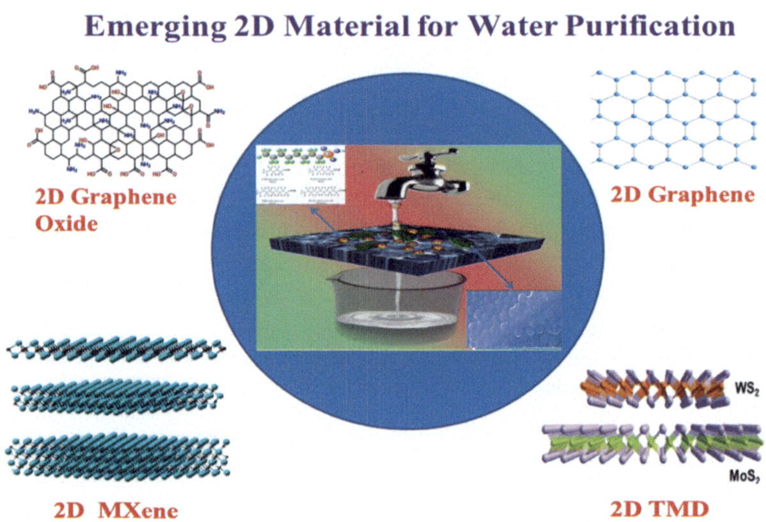

Fig. 7.1 Scheme shows different emerging 2D material-based architectures used for water purification

In this chapter, we have highlighted the recent progress on the development of high-performance 2D membranes and heterostructure material-based membrane design. We discussed the major design criteria used for the development.

7.2 Using 2D Graphene, 2D-GO- and GO-Based Heterostructures for Water Purification

Now it is well documented that graphene is a planar 2D material with a pore size of 0.64 Å as shown in Fig. 7.2a [30]. Since graphene exhibits high mechanical strength (130 GPa) and Young's modulus (1 TPa), scientists have used graphene heavily for the fabrication of novel membranes [26, 30–40]. As we and others have reported, the synthesis and processing of graphene oxide (GO), as well as the reduced graphene oxide (r-GO) are very easy; as a result, GO and its derivatives have been used to develop novel membranes [11–25]. As shown in Fig. 7.2b, the TEM image of GO shows disorder due to the oxidation and it forms a network with small holes [31]. As it is now well reported that GO exhibits a Young's modulus of $\approx$207.6 GPa, which is comparable to that of stainless steel, GO is highly suitable for novel membrane development [25–34].

It is well documented that the graphene oxide surface has negatively charged functional groups which are responsible for the repulsion between sheets as reported in Fig. 7.2c. Due to the above fact, the interlayer distance between GO or r-GO sheets is higher. The high separation between two sheets helps for designing better water passing channel [32]. On the other hand, as shown in Fig. 7.2b, pores on the basal plane have been used for water passing channel. As shown in Fig. 7.2d, molecular dynamics (MD) simulations studies have predicted that graphene can be used for selective separation membranes [29]. Due to all the above exciting properties, Lockheed Martin Company is developing the graphene-based membrane for society.

Although we and others have designed a single-layer GO-based membrane, the instability in water makes it difficult to use for society. To overcome this, we and others have developed different cross-linking methods to stabilize GOs in water. Mi et al. [38] have reported GO-based membrane, where GO nanosheets are cross-linked by 1,3,5-benzenetricarbonyl trichloride, as shown in Fig. 7.3a. Figure 7.3b shows the SEM image of the membrane which indicates that the pores are 10–30 nm in diameter.

Their experimental data as reported in Fig. 7.3 indicates that cross-linking provided good stability for GO nanochannels. Their reported data as shown in Fig. 7.3c indicate that the designed membrane exhibits relatively low rejection for monovalent and divalent salts (6–46%). On the other hand, reported water fluxes are 4–10 times higher than the current nanofiltration membranes available in the market [38].

We have developed GO-based membrane using polydopamine nanoparticle as the cross-linking substance [10]. For this purpose, we have designed graphene oxide

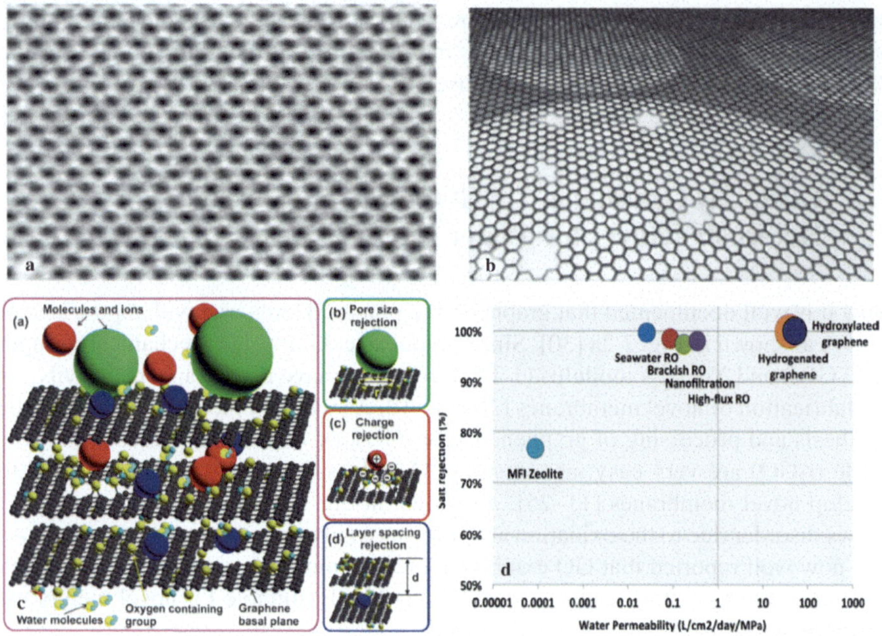

Fig. 7.2 **a** TEM image of a graphene membrane (reproduced and adapted with permission from Ref. [30], American Chemical Society, 2008). **b** Image shows pores are created in the graphene membrane using ion bombardment (reproduced and adapted with permission from Ref. [31], American Chemical Society, 2014). **c** Possible transport mechanism for 2D GO membrane (reproduced and adapted with permission from Ref. [37], American Chemical Society, 2019). **d** Chart shows the comparable performance between graphene and current technologies (reproduced and adapted with permission from Ref. [29], American Chemical Society, 2012)

and polydopamine nanoparticle separately [10]. After that, as shown in Fig. 7.4a, we have designed dopamine nanoparticle conjugated graphene. As shown in Fig. 7.4b, the SEM image shows the pore size of membrane is 2–130 nm. As shown in Fig. 7.4c, membrane has capability to capture superbugs. As shown in Fig. 7.4d, same membrane has capability for the separation of toxic substances from river, lake, and tap water [10]. Experimental data indicates 100% separation and killing of drug-resistant bacteria [10]. We have also developed carbon nanotube (CNT) as cross-linking agent for graphene oxide membrane which can be used for the separation of bacteria and other contaminants from water [16]. For this purpose, as shown in Fig. 7.5a, we have conjugated CNT between GO sheets via chemical coupling between acid and amide groups. As shown in Fig. 7.5b the SEM data shows the pore size for the membrane is 300–500 nm.

As shown in Fig. 7.5c membrane can be used to capture *E. coli* O157:H7 bacteria from water [16]. Similarly, as shown in Fig. 7.5d membrane can be used for the removal of toxic metals. Experimental data reported by us shows that membrane can be used for simultaneous removal bacteria and toxic metals from river water [16].

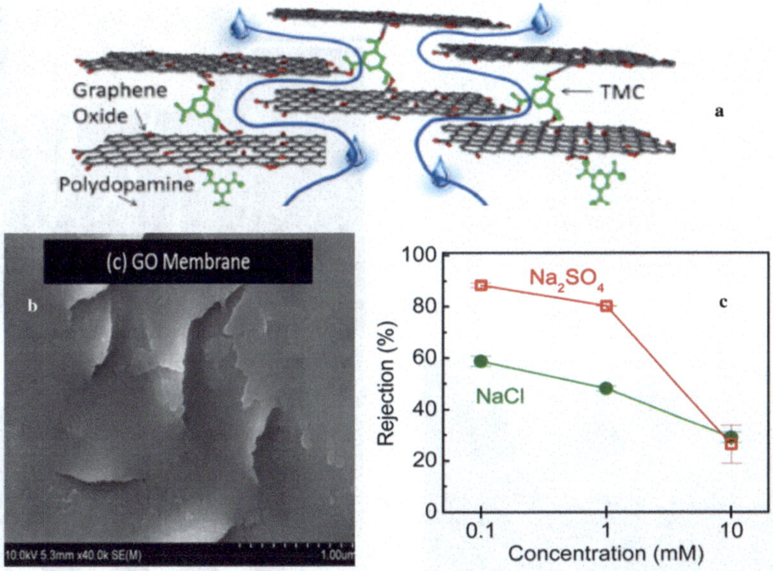

Fig. 7.3 **a** Scheme shows the design of novel GO membrane via 1,3,5-benzenetricarbonyl trichloride cross-linking. **b** SEM image of the membrane. **c** Plot shows how the rejection by the membrane varies with salt concentration (reproduced and adapted with permission from Ref. [38], American Chemical Society, 2013)

As per EPA's recent report [7–9], several surface water sites in USA contain 200–300 parts per million concentrations of PFAS, which is 3–4 orders of magnitude higher than the US EPA limit [1]. For the possible removal of PFAS graphene, GO and r-GO-based new technology has been developed [39]. Ali et al. [39] have developed SiO_2–GO capsules for the removal of PFOA, as shown in Fig. 7.6a. In their design, amine group was added on SiO_2–GO capsules to design capsules as positively charged, as shown in Fig. 7.6b. Reported experimental data, as shown in Fig. 7.6c, shows efficient PFOA removal by the capsules.

In this section, we have discussed 2D-graphene, 2D-GO and heterostructure-based design of novel membrane for water purification. Here we have discussed how 2D material-based membrane can be used for removal of toxic substances from water. To understand the possible mechanism, computational simulation studies on nano-bio surface are necessary.

7.3 Using 2D-TMD for Water Purification

In the last decade, transition metal dichalcogenides (TMDs)-based nano-systems have been designed for various applications [41–50].

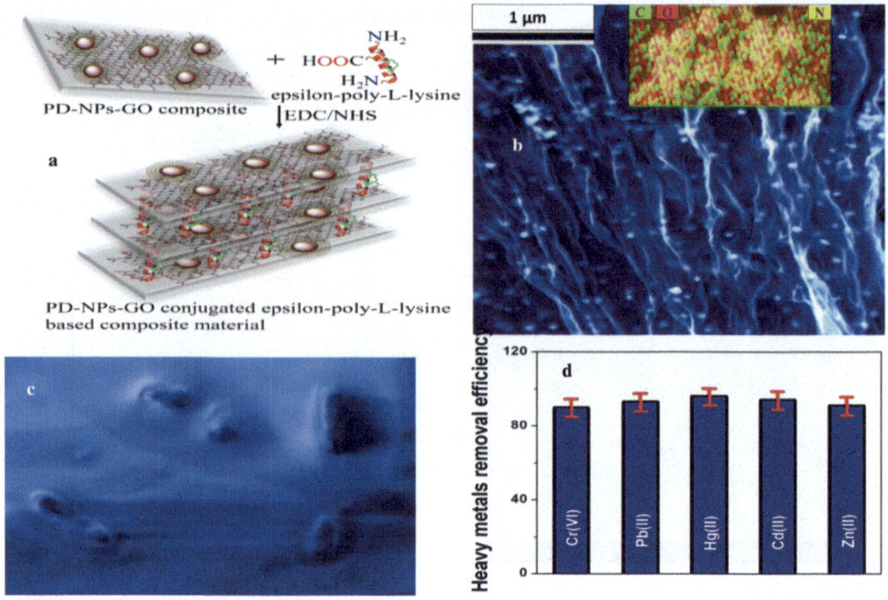

Fig. 7.4 **a** Scheme for the design of based membrane. **b** SEM image of the nanoarchitecture. **c** SEM image shows KPN captured by nanoarchitecture. **d** Plot shows the toxic metals removal capability by nanoarchitecture (reproduced and adapted with permission from Ref. [10], American Chemical Society, 2019)

Recent reports suggest that TMD-based membranes can exhibit better performance than GO-based membranes under similar conditions [41–50]. Reported data indicates 2D TMDs displayed excellent rejection greater than 80% [46–50]. Li et al. has reported an MD study to determine seawater desalination for MoS_2-based filter, as shown in Fig. 7.7 [43]. Reported theoretical data, as shown in Fig. 7.7c, indicate that MoS_2 filter has high water transparency. It also shows vigorous salt filtering capability [43].

Hirunpinyopas et al. have designed ~5 μm thickness laminar membranes using MoS_2, as shown in Fig. 7.8a [48]. For this purpose, MoS_2 flakes were filtered using polyvinylidene difluoride supporting membranes [48]. Using above method, they have produced supported MoS_2 membranes. As shown in Fig. 7.8b, ion-selective filtration has been demonstrated using laminar films of MoS_2. As shown in Fig. 7.8b, they have shown ~99% rejection efficiency and long-term stability for MoS_2 laminar films. Li et al. [47] have reported TMD-based membranes. As shown in Fig. 7.9, designed membrane has only ~7 nm thickness.

As shown in Fig. 7.9, using chemical vapor deposition (CVD), MoS_2 layers were deposited on the porous polymeric supports. Their reported data shows that 2D MoS_2 membranes exhibit high water permeability as well as ionic sieving capability. From their report, that they have concluded that observed excellent properties can be attributed to the intrinsic atomic vacancies in MoS_2 layers.

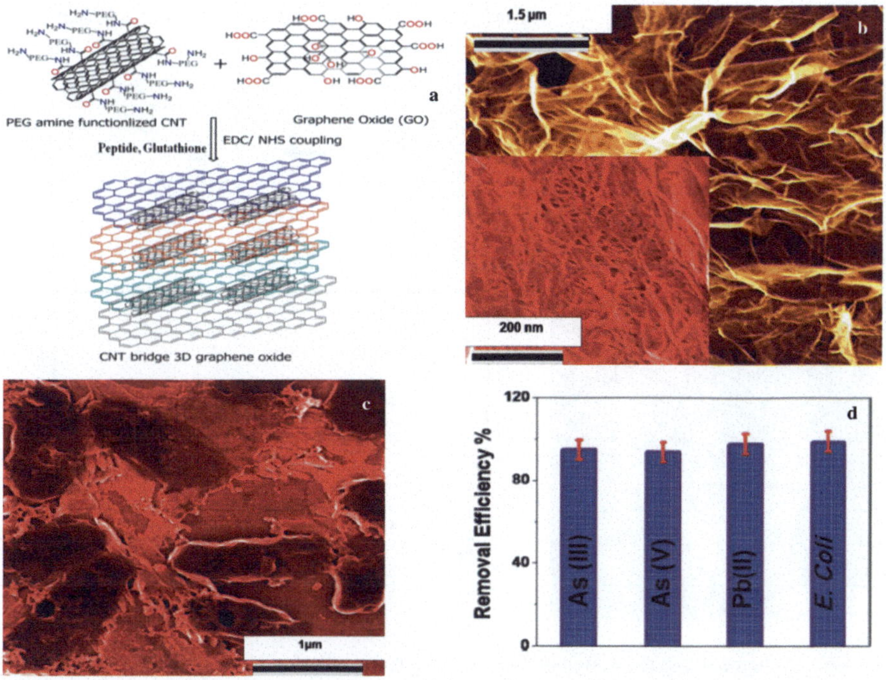

Fig. 7.5 **a** Scheme shows using CNT as cross linker for GO membrane. **b** SEM image of membrane shows the pore sizes is 300–500 nm. **c** SEM image indicates *E. coli* O157:H7 bacteria have been captured by membrane from river water. **d** Toxic metal removal efficiency using membrane from river water (reproduced and adapted with permission from Ref. [16], American Chemical Society, 2015)

Recently we have developed antimicrobial peptide attached MoS_2-based nanoplatform [51]. In our design, as shown in Fig. 7.10a, we have synthesized MoS_2 nanosheets from MoS_2 powder [51]. After that MoS_2 nanosheets were modified with lipoic acid-terminated polyethylene glycol and then melittin peptide was attached [51]. Reported experimental data, as shown in Fig. 7.10b–d, shows superbugs can be killed totally using PEG-MoS_2-AMP nanoplatform [51]. Reported 100% killing can be due to the presence of melittin antimicrobial peptide. We believe that those peptides make pores on the surface of bacteria which help to enhance photodynamic therapy & photothermal therapy (PDT & PTT) efficiency [51].

In this section, we have discussed 2D-TMD-based design of novel membrane for water purification. Here we have discussed how TMD-based architecture can be used for removal of toxic substances from water. To understand the possible mechanism, computational simulation studies on TMD-bio surface are necessary.

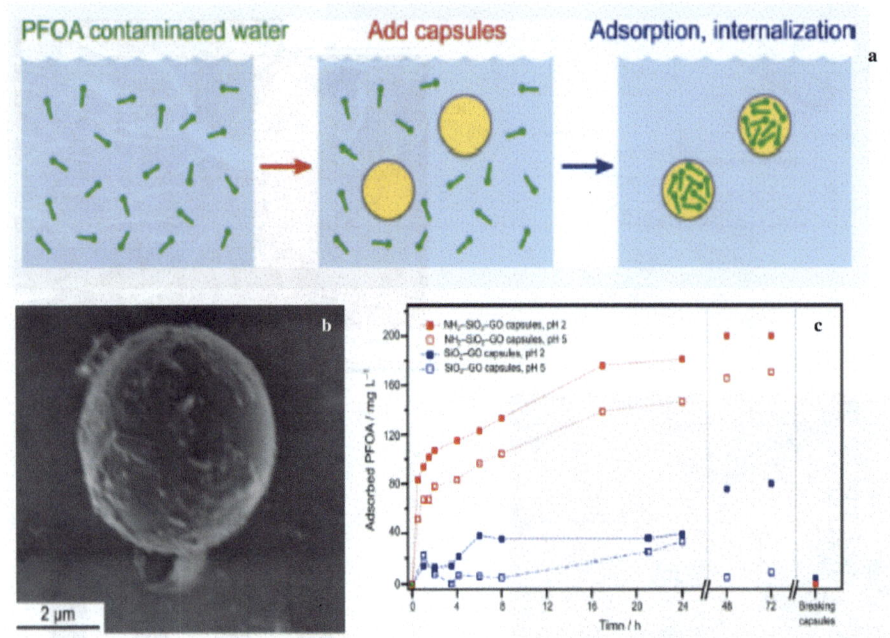

Fig. 7.6 **a** Scheme shows the design of SiO_2–GO microcapsules for PFOA removal. **b** SEM image microcapsules. **c** How pH effects the adsorption efficiency (reproduced and adapted with permission from Ref. [39], American Chemical Society, 2020)

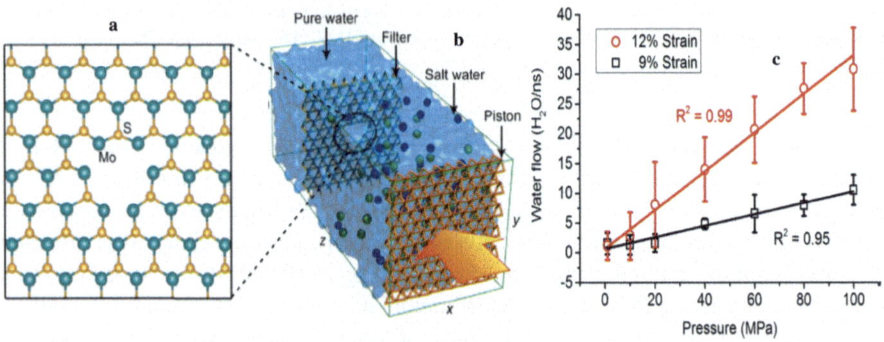

Fig. 7.7 **a** Scheme shows MoS_2 nanopore. **b** Simulation model for MoS_2 nanopore. **c** How water flow varies with pressure (reproduced and adapted with permission from Ref. [43], American Chemical Society, 2016)

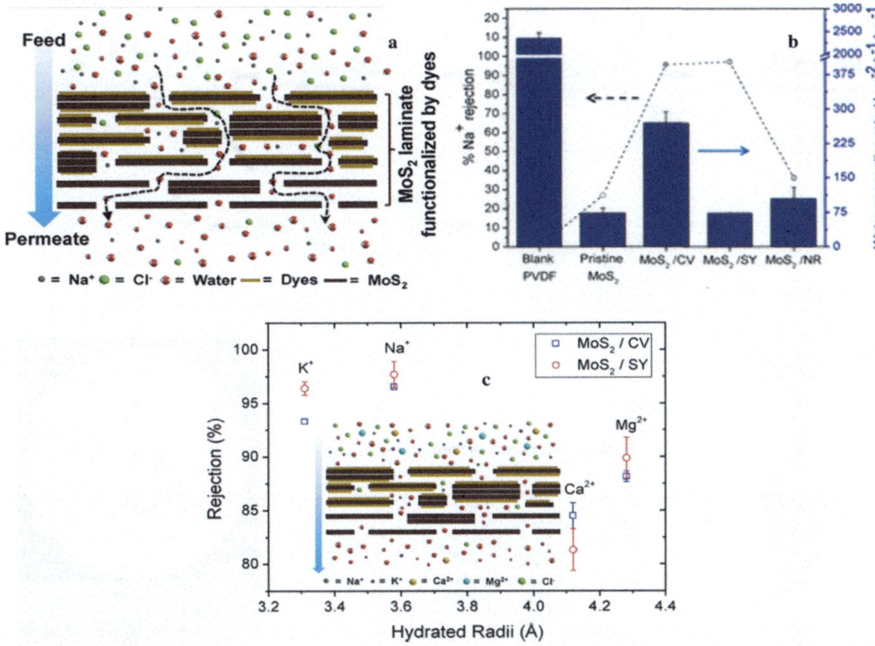

Fig. 7.8 **a** Scheme shows the MoS_2 laminate membrane. **b** Plot shows how sodium ion (Na^+) percentage rejection varies with pressure. **c** Plot shows the rejection efficiency for different ions from seawater (reproduced and adapted with permission from Ref. [48], American Chemical Society, 2017)

7.4 Using 2D-MXene for Water Purification

In last few years, transition metal carbides and nitrides or MXenes is the new family of exciting 2D materials for research and technology [52–63]. In last few years in MXene family, $Ti_3C_2T_x$ has been designed by several groups [52–63]. Since 100% internal light-to heat conversion efficiency is reported for MXenes, several groups have designed $Ti_3C_2T_x$ based membranes as photothermal conversion material for solar desalination [52–63].

Ren et al. has developed 2D $Ti_3C_2T_x$ membrane which shows charge and size dependence rejection capability [57]. For this purpose, vacuum-assisted filtration was used, as shown in Fig. 7.11a [57]. Figures 7.11b and c indicates membrane thickness is ~1 nm [57]. As shown in Fig. 7.11d, their developed membranes exhibited ultrafast water flux [57]. As shown in Fig. 7.11e, their designed membrane exhibited high selectivity toward single-, double- and triple-charged metal cations [57]. Reported results show that larger charge cations exhibit an order of magnitude slower permeation [57].

Lu et al. have reported the design of the anti-swelling membranes via self-cross-linked MXene membrane [56]. As shown in Fig. 7.12a, MXene membranes were

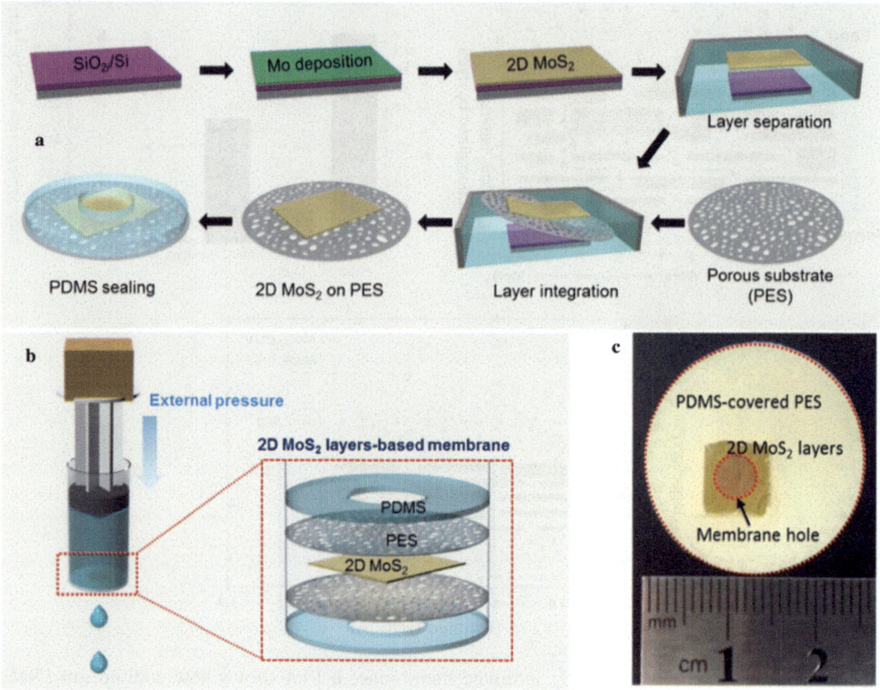

Fig. 7.9 **a** Schematic procedures for few-layer 2D MoS_2 membranes preparation, transfer, and integration. **b** Schematic setup for water permeation measurement. **c** Representative image of a 2D MoS_2 layer-integrated PES support with PDMS sealing. 2D MoS_2 layer-membrane hole is denoted by the inner dotted circle (reproduced and adapted with permission from Ref. [47], American Chemical Society, 2019)

developed using self-cross-linking reaction between MXene nanosheets [56]. SEM and AFM images shown in Fig. 7.12b indicates the morphology which shows the lateral size is around 1 μm [56]. As shown in Fig. 7.12c and d, the permeation rates for metal ions are low [56]. Reported experimental data shows good anti-swelling properties for the membrane [56].

Zha et al. reported the design of photothermal membrane using MXene-coated cellulose [58]. They have demonstrated membrane can be used for solar-driven evaporation [58]. As shown in Fig. 7.13a, MXene/cellulose membranes were designed using dip-coating method. Figure 7.13b shows the membrane exhibited huge steam generation [58]. Figure 7.13c shows the infrared thermal image which shows the surface temperature reaches 44.7 °C in 10 min [58]. As shown in Fig. 7.13d, SEM data show bacteria can be captured using membrane. Experimentally reported data indicate that the membrane exhibits very high antibacterial efficiency due to the photothermal effect.

In this section, we have discussed 2D-MXene-based design of membrane for water purification. Here we have discussed how MXene-based architectures can be

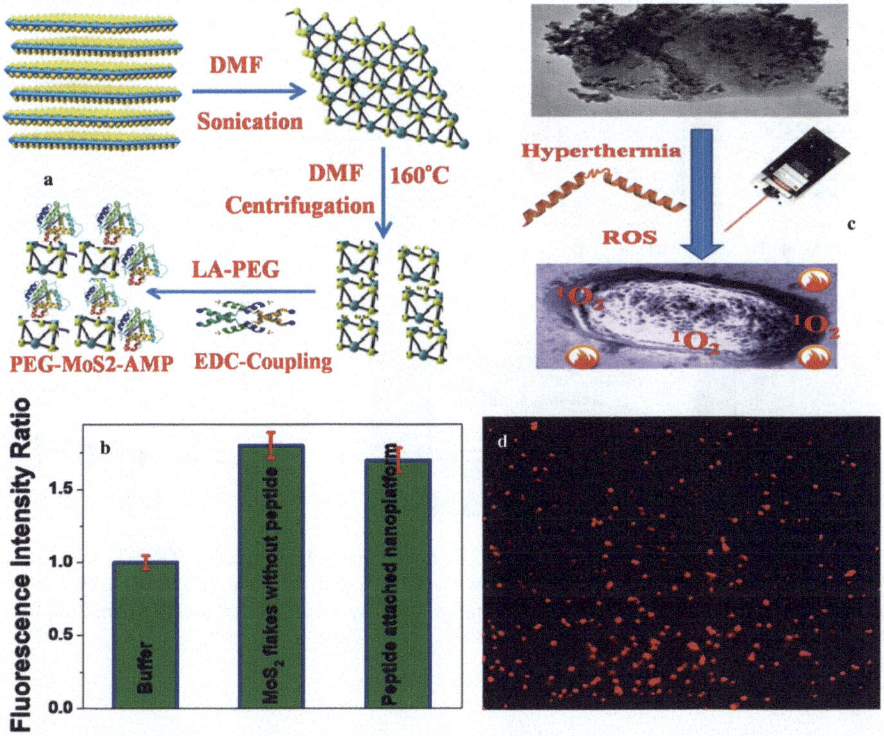

Fig. 7.10 **a** Scheme shows the design of MoS_2 nanoplatform. **b** Plot indicates ROS formation by PEG-MoS_2-AMP nanoplatform in the presence of NIR light. **c** Scheme shows antibacterial activity for PEG-MoS_2-AMP nanoplatform via photo thermal, as well as photodynamic and antimicrobial peptide process. **d** Fluorescence image shows killing of KPN superbugs (reproduced and adapted with permission from Ref. [51], American Chemical Society, 2019)

used for applications for water desalination. Theoretical understanding is necessary to understand the possible mechanism for bacteria killing using photothermal membrane.

7.5 Summary and Outlooks

In conclusion, in this chapter, we have discussed recent advancements on emerging 2D-graphene, 2D-GO, 2D-TMD and 2D-MXene and heterostructure-based design of water desalination nanoarchitectures. Our and others reported data show 2D-materials and heterostructures materials-based membranes can be used for toxic substance separation from water. Several reported data show a huge potential for capturing and killing of bacteria using 2D material-based membranes. We hope that

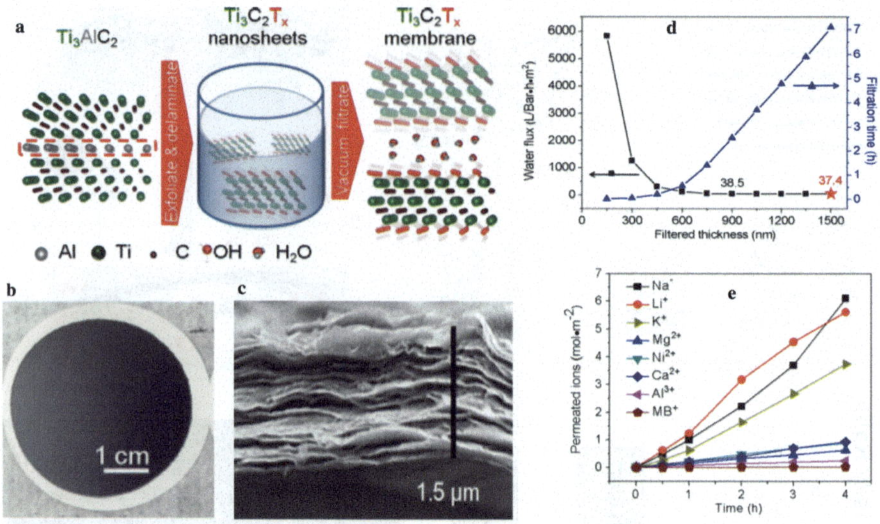

Fig. 7.11 **a** Scheme shows the design of $Ti_3C_2T_x$ membrane. **b** Photograph of $Ti_3C_2T_x$ membrane. **c** SEM image of same membrane. **d** Plot shows water flux data for same membrane. **e** Plot shows cation permeation capability for same membranes (reproduced and adapted with permission from Ref. [57], American Chemical Society, 2015)

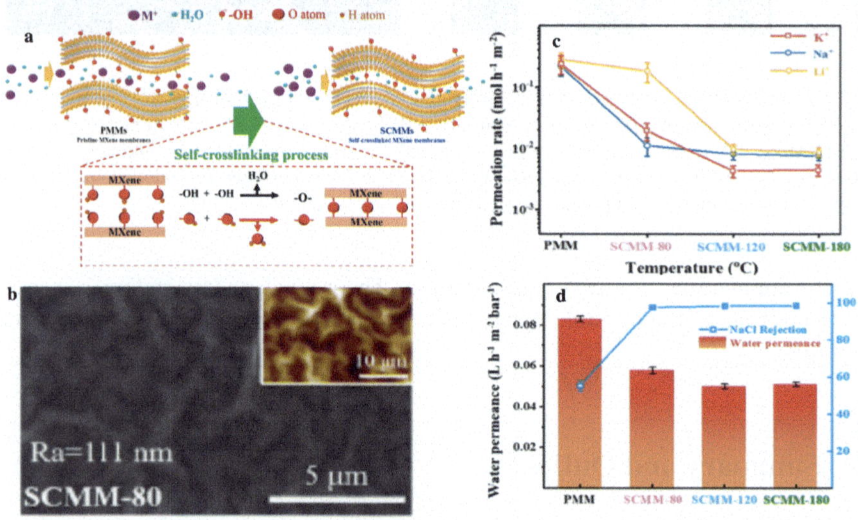

Fig. 7.12 **a** Scheme shows self-cross-linking process have been used for the design of the MXene membranes. **b** SEM and AFM shows the morphology of membrane surface. **c** Plot shows the permeation rates for different cation. **d** Plot shows the membrane capability for water permeance and NaCl rejection (reproduced and adapted with permission from Ref. [56], American Chemical Society, 2019)

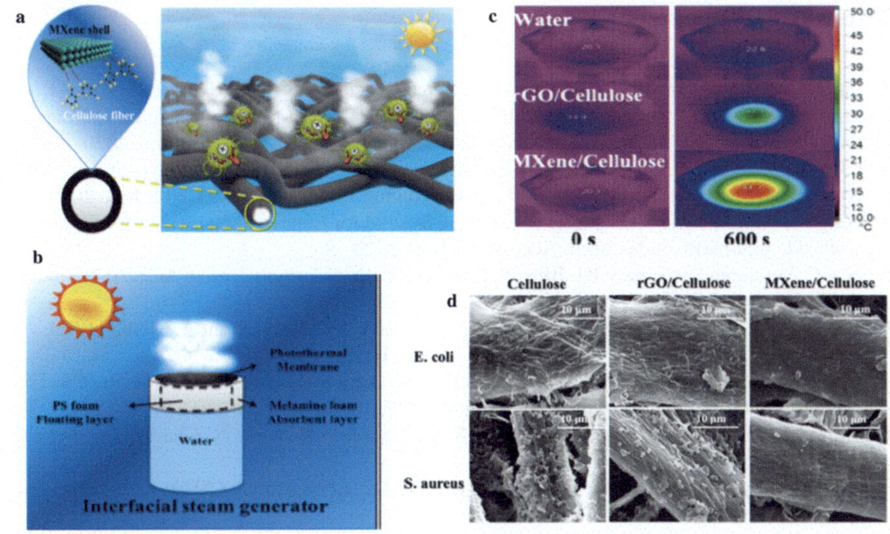

Fig. 7.13 **a** Schematic shows the development of the MXene/cellulose membrane. **b** Schematic shows the steam generator capability of photothermal membrane. **c** IR thermal image of membrane in the presence of light. **d** SEM images for the membrane surfaces in the presence of bacteria (reproduced and adapted with permission from Ref. [59], American Chemical Society, 2019)

the current chapter will enable researchers to realize the potential of 2D material-based membranes for water purification.

Although from the recent reported data we can see exciting results for 2D membranes, it is still too early for real life applications. For this purpose, scientists need to develop low-cost and mass production of different 2D material-based membranes with high reproducibility. Interdisciplinary research collaboration is needed for development and deployment of advanced 2D-material-based membranes. Detailed toxicological and pharmacological profiles are also missing for 2D membranes which is very important before they can be used for society.

Acknowledgements The use of trade, product, or firm names in this report is for descriptive purposes only and does not imply endorsement by the U.S. Government. The tests described and the resulting data presented herein, unless otherwise noted, were obtained from research conducted under the Environmental Quality Technology Program of the United States Army Corps of Engineers and the Environmental Security Technology Certification Program of the Department of Defense by the USAERDC. Permission was granted by the Chief of Engineers to publish this information. The findings of this report are not to be construed as an official Department of the Army position unless so designated by other authorized documents. This document has been approved for public release (Distribution Statement A). Dr. Ray thanks NSF-PREM grant # DMR-1826886, NIH-NIMHD grant # 1U54MD015929-01 and ERDC grant # W912HZ-20-2-0069 for their generous funding.

References

1. http://www.un.org/en/sections/issues-depth/water/. Accessed 8 June 2020
2. https://www.who.int/news-room/detail/12-07-2017-2-1-billion-people-lack-safe-drinking-water-at-home-more-than-twice-as-many-lack-safe-sanitation. Accessed 8 June 2020
3. CDC 2018: Drug-resistant 'nightmare bacteria' pose growing threat. https://nypost.com/2018/04/03/drug-resistant-nightmare-bacteria-pose-growing-threat-cdc. Accessed 8 June 2020
4. National Antimicrobial Resistance Monitoring System for Enteric Bacteria (NARMS). https://www.cdc.gov/narms/index.html. Accessed 8 June 2020
5. Nys C, Versieren L, Cordery KI, Blust R, Smolders E, De Schamphelaere KAC (2017) Environ Sci Technol 51:4615
6. Wujits S, van den Berg HHJL, Miller J, Abebe L, Sobsey M, Andremont A, Medicott KO, van Passel MWJ, de Roda Husman AM (2017) J Water Health 16(2):175
7. EPA's Per- and Polyfluoroalkyl Substances (PFAS) Action Plan. Report 823-R-18-004; U.S. Environmental Protection Agency: Washington, DC, 2019
8. Amin MA, Sobhani Z, Liu Y, Dharmaraja R, Chadalavada S, Naidu R, Chalker JM, Fang C, (2020) Environ Technol Innov 1:100879
9. Kumarasamy E, Manning IM, Collins LB, Coronell O, Leibfarth FA (2020) ACS Central Sci 6:487
10. Pramanik A, Gates K, Gao Y, Zhang Q, Han FX, Rightshell C, Sardar S, Ray PC (2019) ACS Appl Nano Mater 2:3339
11. Vangara A, Pramanik A, Gao Y, Gates K, Begum S, Ray PC (2018) ACS Appl Bio Mater 1:298
12. Befum S, Pramanik A, Gates K, Gao Y, Ray PC (2019) ACS Appl. Bio Mater 2:3339
13. Pramanik A, Jones S, Pedraza F, Vangara A, Sweet C, Williams MS, Ruppa-Kasani V, Risher SE, Sardar D, Ray PC (2017) ACS Omega 2:554
14. Begum S, Pramanik A, Davis D, Patibandla S, Gates K, Gao Y, Ray PC (2020) ACS Omega 5:3116
15. Jones S, Pramanik A, Kanchanapally R, Nellore BPV, Begum S, Sweet C, Ray PC (2017) ACS Sustain Chem Eng 5:7175
16. Viraka Nellore BP, Kanchanapally R, Pedraza F, Sinha SS, Pramanik A, Hamme AT, Arslan Z, Sardar D, Ray PC (2015) ACS Appl Mater Interfaces 7:19210
17. Pramanik A, Hamme JS, Gao Y, Sweet C, Begum S, Shukla MK, Buchanan JP, Moser RD, Ray PC (2017) J Mater Chem B 5:9522
18. Begum S, Pramanik A, Gates K, Gao Y, Ray PC (2019) ACS Appl Bio Mater 2:769
19. Gao Y, Pramanik A, Begum S, Sweet C, Jones S, Alamgir A, Ray PC (2017) ACS Omega 2:7730
20. Sinha SS, Jones S, Pramanik A, Ray PC (2016) Acc Chem Res 49:2725
21. Geim AK, Novoselov KS (2007) Nat Mater 2007(6):183
22. Geim AK (2009) Science 324:1530
23. Nair RR, Wu HA, Jayaram PN, Grigorieva IV, Geim AK (2012) Science 335:442
24. Dreyer DR, Park S, Bielawski CW, Ruoff RS (2010) Chem Soc Rev 39:228
25. Garaj S, Hubbard W, Reina A, Kong J, Branton D, Golovchenko JA (2010) Nature 467:190
26. Surwade SP, Smirnov SN, Vlassiouk IV, Unocic RR, Veith GM, Dai S, Mahurin SM (2015) Nat Nanotechnol 10:459
27. Joshi RK, Carbone P, Wang FC, Kravets VG, Su Y, Grigorieva IV, Wu HA, Geim AK, Nair RR (2014) Science 343:752
28. Rezania B, Severin N, Talyzin AV, Rabe JP (2014) Nano Lett 14:3993
29. Tanugi DC, Grossman JC (2012) Nano Lett 12:3602
30. Meyer C, Kisielowski C, Erni R, Rossell MD, Crommie MF, Zettl A (2008) Nano Lett 8:3582
31. O'Hern SC, Boutilier MSH, Idrobo JC, Song Y, Kong J, Laoui T, Atieh M, Karnik R (2014) Nano Lett 14:1234
32. Wang EN, Karnik R (2012) Nat Nanotechnol 7:552
33. Liu G, Jin W, Xu N (2016) Angew Chem Int Ed 55:13384

34. Zheng Z, Grunker R, Feng X (2016) Adv Mater 28:6529
35. Wang L, Boutilier MSH, Kidambi PR, Jang D, Hadjiconstantinou NG, Karnik R (2017) Nat Nanotechnol 12:509
36. Howorka S (2017) Nat Nanotechnol 12:619
37. Yang T, Lin H, Loh KP, Jia B (2019) Chem Mater 31:1829
38. Hu M, Mi B (2013) Environ Sci Technol 47:3715
39. Ali M, Meaney SP, Giles LW, Holt P, Majumder M, Tabor RF (2020) Environ Sci Technol 54:3549
40. Huang L, Zhang M, Li C, Shi G (2015) J Phys Chem Lett 6:2806
41. Radisavljevic B, Radenovic A, Brivio J, Giacometti V, Kis A (2011) Nat Nanotechnol 6:147
42. Tan C, Zhang H (2015) Chem Soc Rev 44:2713
43. Li W, Yang Y, Weber JK, Zhang G, Zhou R (2016) ACS Nano 10:1829
44. Li H, Wu J, Yin Z, Zhang H (2014) Acc Chem Res 47:1067
45. Heiranian M, Farimani AB, Aluru NR (2015) Nat Commun 6:8616
46. Deng M, Kwac K, Li M, Jung Y, Park HG (2017) Nano Lett 17:2342–2348
47. Li H, Ko TJ, Lee M, Chung HS, Han SS, Oh KH, Sadmani A, Kang H, Jung H (2019) Nano Lett 19:5194
48. Hirunpinyopas W, Prestat E, Worrall SD, Haigh SJ, Dryfe RAW, Bissett MA (2017) ACS Nano 11:11082
49. Ang EH, Chew JW (2019) Chem Mater 31:100002
50. Guo BY, Jiang SD, Tang MJ, Li K, Sun S, Chen PY, Zheng S (2019) J Phys Chem Lett 10:4609
51. Begum S, Pramanik A, Gates K, Gao Y, Ray PC (2019) ACS Appl Bio Mater 2:769
52. Naguib M, Kurtoglu M, Presser V, Lu J, Niu J, Heon M, Hultman L, Gogotsi Y, Barsoum MW (2011) Adv Mater 23:4248
53. Ying Y, Liu Y, Wang X, Mao Y, Cao W, Hu P, Peng X (2015) ACS Appl Mater Interfaces 7:1795
54. Ding L, Wei Y, Wang Y, Chen H, Caro J, Wang HA (2017) Angew Chem Int Ed 56:1825
55. Ding L, Wei Y, Li L, Zhang T, Wang H, Xue J, Ding L-X, Wang S, Caro J, Gogotsi Y (2018) Nat Commun 9:155
56. Lu Z, Wei Y, Deng J, Ding L, Li ZK, Wang H (2019) ACS Nano 13:10535
57. Ren CE, Hatzell KB, Alhabeb M, Ling Z, Mahmoud KA, Gogotsi Y (2015) J Phys Chem Lett 6:4026
58. Zha XJ, Zhao X, Pu JH, Tang LS, Ke K, Bao RY, Bai L, Liu YZ, Yang MB, Yamg W (2019) ACS Appl Mater Interface 11:36589
59. Mojtabavi M, Mohammadi AV, Liang W, Beidaghi M, Wanunu M (2019) ACS Nano 13:3042
60. Liu G, Liu S, Ma K, Wang H, Wang X, Liu G, Jin W (2020) Ind Eng Chem Res 59:4732
61. Gao L, Li C, Huang W, Mei S, Lin H, Ou Q, Zhang Y, Guo J, Zhang F, Xu S, Zhang H (2020) Chem Mater 32:1703
62. Zhang Q, Yi G, Fu Z, Yu H, Chen S, Quan X (2019) ACS Nano 13:13196
63. Soundiraraju B, George BK (2017) ACS Nano 11:8892–8900

Chapter 8
Emergent Materials and Processes for Efficient Environmental Per- and Polyfluoroalkyl Substances Containment

Manoj Kolel-Veetil and Swathi Iyer Ganjigunteramaswamy

Abstract The past several decades have seen persistent and cumulative contamination of the environment by per- and polyfluoroalkyl substances due to the increase in utilization of such compounds in a myriad of applications owing to their inherent stability. While this stability resulting from their exceptional thermodynamically-stable C–F bonds has been a boon in applications, in contrast, with regard to their post-application lives, it has become a bane due to the remarkable recalcitrance of the C–F bonds upon permeation of the environment. Deleterious effects of such compounds on human and bio-organisms include a host of diseases and other environmental problems. Extensive studies on containment strategies for PFASs during the past few decades have included steps such as capture and thermal incineration, various catalytic degradation strategies, and biodegradation efforts using microorganisms leading to partial destruction of these compounds. This review specifically focuses on the most recent emergent techniques in this area.

8.1 Introduction

Per- and polyfluoroalkyl substances or PFASs have been termed collectively as "forever chemicals" due to their extreme recalcitrance to chemical breakdown resulting from thermodynamic stability of constituent carbon–fluorine (C–F) bond which is the strongest known carbon-derived single bond or covalent bond in any organic compound [1–3]. Such exceptional stability of the C–F bond is what propelled the

M. Kolel-Veetil (✉) · S. I. Ganjigunteramaswamy
Chemistry Division, Naval Research Laboratory, Washington, DC 20375, USA
e-mail: manoj.kolel-veetil@nrl.navy.mil

S. I. Ganjigunteramaswamy
Nova Research, Inc., Alexandria, VA 22308, USA

M. Shukla et al. (eds.), *Emerging Materials and Environment*, Challenges and Advances in Computational Chemistry and Physics 37,
https://doi.org/10.1007/978-3-031-39470-6_8

use of the PFAS compounds across various applications in almost every aspects of our life since 1940s. As a consequence, their omnipresence and resistance to degradation have caused major environmental problems [4–6]. PFASs are now known to have contaminated many regions around the globe. As of June 2022, in the US alone, it has been estimated that 2858 locations in 50 states and 2 territories have been affected with PFAS contamination (https://www.ewg.org/interactive-maps/pfas_contamination/). This has translated to more than half the population finding their drinking water supplies being contaminated with PFASs. A US National Health and Nutrition Examination 2011–2012 Survey found that detectable serum PFAS concentrations were observed in 97% of the tested individuals [7]. Human PFAS exposure has been linked to several health-related issues such as cancer, elevated cholesterol, obesity, immunosuppression, and endocrine disruption (https://www.atsdr.cdc.gov/pfas/health-effects/index.html). Shorter chain PFASs that began to be used as a replacement for their longer chain cousins are now being discovered to have even greater impact on human health. They have also been found to be extremely mobile and consequently to be able to contaminate the environment to a greater degree [8]. Until recently, the US DoD has used PFASs heavily in aqueous film forming foams (AFFF) for fire training and emergency response purposes (https://www.defense.gov/Explore/News/Article/Article/2349028/dod-officials-discuss-fire-fighting-foam-replacement-remediation-efforts/). In order to address this legacy environmental problem of the DOD, the US Congress has mandated that all DoD applications involving PFAS chemicals be replaced with fluorine-free substitutes by mid-2024 (https://www.congress.gov/bill/116th-congress/house-bill/535/text). Understandably, the recalcitrance to degradation of these chemicals has elicited an enormous amount of investigations into their destruction from researchers in academia, industry, and the Government, especially in the past three decades. Such approaches have been diverse ranging from capture of PFASs in adsorbents and their subsequent thermal destruction to biodegradation methods to catalytic degradation involving both reduction and oxidation of these compounds. Each such method has its own pros and cons, and several reviews have appeared involving a comparison of these methods [1, 2].

We will discuss briefly some remediation efforts as examples. Starting with capture of PFAS by adsorbents, which is a relatively simple process; the efficiency of such capture depends on the adsorbents. This is because PFASs are generally hydrophobic in nature and gets easily expelled from more hydrophilic surfaces found in common adsorbents such as granulated activated carbon (GAC) [9, 10]. GAC is a versatile sorbent and an effective contaminant removal media due to its complex porous structure and high surface area. Obviously, once captured, PFASs have to be degraded as the option to simply dispense the PFAS-containing adsorbent in a landfill or some designated area does not solve the problem on a long-term basis, and stricter regulations are being promulgated routinely against such disposals. This is also the issue with treatment strategies such as nanofiltration (NF) as ultimately such filter-separated PFASs have to be treated and degraded to less harmful versions [11]. Furthermore, the incineration of these adsorbent-captured PFASs has been seen to produce harmful compounds including active greenhouse gases such as hexafluoroethane (CF_3–CF_3) and tetrafluoromethane (CF_4) with global-warming potentials of 5700 and 11,900 and lifetimes of 50,000 and 10,000 years, respectively [12–15] Incineration is also an energy intensive process, and one has to consider the cost and the chemical footprint of the utilized energy.

Since PFASs are typically found as mixtures in water or entrapped in soil, the obvious choice is to induce and drive their degradation in/by water-based catalytic reactions. In this regard, advanced oxidation and reduction methods [16, 17] such as photo [18]- or electro [16]-Fenton [19], oxidation and persulfate (PSf) [20, 21] oxidation, and aqueous electron-based reductions have been explored and determined to be useful. Also, other chemical catalytic options are available for such degradations including in electrochemical systems or in the presence of a photocatalyst capable of producing aqueous reactive oxygen species (ROS) such as peroxo and hydroxyl radicals [22]. Alternatively, concentration of PFASs into water-free pure phases by filtration and extraction can enable the degradation of these chemicals in non-aqueous phases. An example of such degradation is seen in the silylium-carborane-based catalytic degradation of PFASs [23].

Biodegradation is another option wherein microorganisms in their native states could possibly degrade PFASs as their natural response while coexisting with such chemicals [24]. This could be both for energy and evolutionary purposes/reasons of the microorganisms. In this regard, a few iron (Fe) containing microorganisms that contain the metal in their metabolic machinery or as an enzyme component or co-factor are seen to effect such degradation. They include *Pseudomonas plecoglossicida* (oxidation of Fe(II) to Fe(III)) [25], *Acidimicrobium* sp. Strain A6 (reduction of Fe(III) to Fe(II)) [26], and the microalgae diatoms (with 0.5–2% Fe oxides) [27]. In fact, the option of biodegradation is an important one, and it has started to gain increased attention as of late.

Other unique and creative degradation methods such as sonochemical [28, 29] and plasma-based [2, 30–33] degradation systems are coming to the fore and appear to be quite promising for the degradation of PFASs. The efficacies of most of these processes are being evaluated with regard to practicability including the associated cost of such processes. Interestingly, even simple techniques such as ball milling of PFASs with media containing Si and other metals have recently been observed to be effective for the degradation of PFASs [34], and the development of such simple and relatively inexpensive processes will certainly add to the excitement of discovering efficient containment strategies of PFASs in the near future.

Finally, artificial intelligence/machine learning methods are coming to the fore in aiding the classification of various PFASs compounds with an eye toward expeditiously predicting bond dissociation energies of various PFASs bonds [35] and their proclivity to destruction by a specific destruction methodology. In this review, we attempt to highlight emergent materials and processes that have appeared in the past few years in this fast-moving field for the efficient containment of per- and polyfluoroalkyl substances.

8.1.1 Capture and/or Concentration of PFASs via Filtration Facilitated by Adsorption and/or Ion Exchange

The obvious processes for the containment of any toxic substance from its dissolved version in water would be to either catalytically degrade it in that state or to concentrate it by filtration, extraction, and/or separation by adsorption. While degradation of PFASs can be effected in aqueous solutions using appropriate chemical reagents, extraction into a non-aqueous phase provides the option to degrade the extracted PFASs using catalysts that are typically sensitive to water. If PFASs are adsorbed/captured, they can be further degraded in the captured matrix by inputting thermal energy or the adsorbed PFASs can be eluted from the adsorbents using appropriate organic solvents for further degradative treatment as discussed above. In capturing and concentrating PFASs using adsorbents, the interactions that enable such processes can be based on both chemical and physical interactions. Specifically, hydrophobic and fluorophilic interactions generated by functional groups on both PFAS and the adsorbent in question can lead to either attractive or repulsive forces. Such interactions result, in addition to others, largely due to dispersion (van der Waals), polar, and hydrogen bonding scenarios. Molecular sizes and shapes of PFASs can also have an effect in the adsorption and filtration science being directly related to the hydration/solvation sphere of the PFAS molecule. While the PFAS molecules are hydrophobic and fluorophilic in nature, the non-fluorophilic sites of the adsorbents such as granulated activated carbon (GAC), on the other hand, are still capable of capturing PFASs because the PFASs' carboxylic or sulfonic acid head groups and their anionic versions can interact with the hydroxyl, carboxyl, epoxide, or amine functionalities on the GAC surfaces. Furthermore, depending on the adsorbent pore sizes and size of the PFAS molecules, there can be size exclusion effects. The option of filtering out PFASs compounds from waste streams using specific membranes can involve both chemical processes such as with ion-exchange membranes involving the swapping of chemical bonds and physical interactions such as those involving van der Waals forces and other dispersive forces such as described above. While a majority of the adsorption-driven examples are geared toward subsequent incineration of the captured PFAS, the below recent examples relate only to the capture of PFASs.

GAC and GAC-derived systems: GAC, being the most commonly used adsorbent for PFASs, has been established to be a more effective adsorbent-based treatment technology for long-chained PFASs than for short-chained ones. The salient features that aid GAC in PFAS and other adsorption processes are the high surface area of its porous structure and the innate dispersive interactions of its surface functionalities. In this regard, the adsorption characteristics of super-fine powder activated carbon (SPAC; particle diameter <1 μm; typical GAC diameters are 45 μm–4.75 mm) were investigated in a continuous flow ceramic microfiltration system for PFAS removal. The results showed that SPAC had nearly double the adsorption potential that GAC has with PFASs. The adsorption potential was further observed to depend on the adsorbent size, pore content, and PFAS chain length at 10% breakthrough of PFASs during filtration [9].

In addition to its high surface area, chemical modifications of GAC surfaces can increase the uptake of PFASs. Treatment of GAC with acid, base, or other oxidant can effect such modifications. In producing perfluorooctanoic acid (PFOA) and perfluorohexanesulfonic acid (PFHxS) adsorption, two types of GAC from Calgon Carbon Corporation (F400 and CBC) were treated with acid (HCl), base (NaOH), heat activated PSf, and hydrogen peroxide catalyzed with iron (H2O2/Fe). The adsorption characteristics of the HCl-treated GAC versus untreated GAC showed that after 2-, 5-, and 10-day reaction times the extent of adsorption of both PFHxS and PFOA increased (7–8% in F400 and 6–9% in CBC) likely due to increase in positive charge density. No significant change was observed with NaOH-treated version, and a decrease in adsorption was seen with PSf (22–25% in F400 and 27–35% in CBC) and H_2O_2/Fe (4–8% in F400 and 12–13% in CBC) treated PFAS samples. Clearly, specific surface functionalizations of GAC surfaces and prior treatment of PFAS samples elicit distinct adsorption responses in GAC [36].

In addition to surface chemical characteristics, the impact of pore surface area on PFAS adsorption by GAC was recently investigated. For example, a group of bituminous coal-derived GACs (i.e., F400, Carbsorb 40, HPC, and CMR400) was studied for the adsorption breakthrough behavior of nine PFASs ($C = 4, 5, 6, 7, 8, 10,$ or 12 PFOAs and C_4 and C_8 perfluorooctane sulfonic acids (PFOSs)) in groundwaters. The study showed that the adsorption capacity for F400 specifically depended on the hydrophobicity of PFASs. At a given pH, the diffusivity of PFASs was seen to be independent of adsorbent diameter for three differently sized F400 at a given condition. Among the tested carbons, the positively charged adsorbents (F400, HPC, and CMR400) were observed to exhibit higher apparent adsorption capacities for hydrophilic and marginally hydrophobic PFASs than the negatively charged adsorbent (Carbsorb 40). In addition, the former adsorbents with higher micropore surface areas were found to exhibit higher apparent adsorption capacities of hydrophilic and marginally hydrophobic PFASs. Further, the mesoporous HPC displayed an increasingly larger adsorption capacity for more hydrophobic PFASs compared to the microporous F400 [37].

The role of the air–water interface in enhancing the removal of long-chain PFASs through combined oxygenation and adsorption on GAC is an important phenomenon. Oxygenation enhances the removal of long-chain PFASs through incorporation into the air–water interface and subsequent adsorption of the PFASs-enriched air bubbles to GAC. The removal of long-chain PFAS carboxylic (C_nF_{2n} + 1COOH, $n \geq 7$) and sulfonic (C_nF_{2n} + $1SO_3H$, $n \geq 6$) acids was found to increase with the assistance of oxygenation, depending on the perfluoroalkyl chain length. It was determined that aeration is more effective in enhancing long-chain perfluoroalkyl acids (PFAA) removal as air–water interface absorption increased with PFAA chain length due to higher surface activity (Fig. 8.1). When adsorbed air bubbles were removed by centrifugation, up to 80% of the long-chain PFASs were able to desorb from the sorbent confirming the contribution of the air–water interface to the sorption of PFASs on GAC [38].

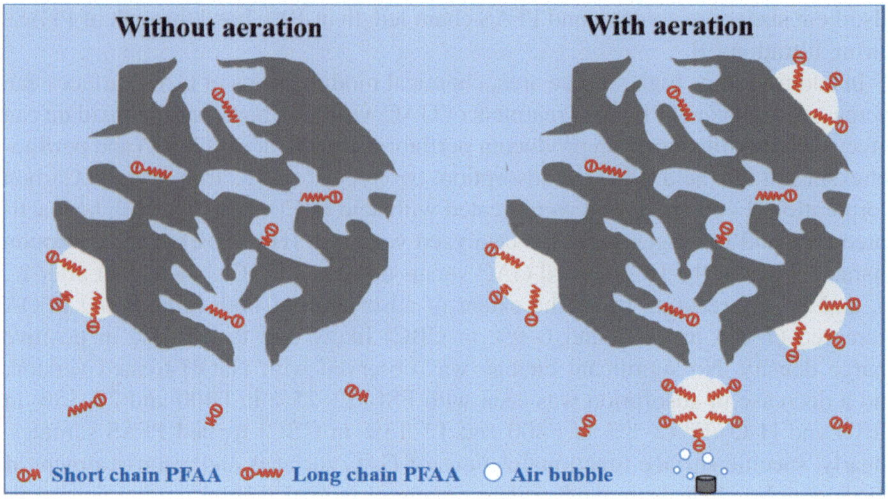

Fig. 8.1 Schematic description of the adsorption mechanism of PFAAs on AC with the contribution of air–water interface adsorption. Reproduced from Ref. [38] and reprinted with permission from Elsevier

The above examples clearly show that changes in pore sizes and chemistries of GAC and associated interfacial interactions have a significant impact on PFASs adsorption processes, and one needs to consider all such factors together when attempting to understand the underlying science.

Ion-exchange (IX) systems: Besides GAC, other innovative filtration and anion-exchange (AIX) systems have the ability to capture PFASs. Such filtration media are also amenable to surface functionalization to alter the filtration dynamics. In fact, the conversion of the interior channels of the filtration media to more *fluorophilic* versions can indeed enhance the dispersive and electronegative interactions between the PFAS entities and the interior channel sites. For example, in studies using linear fluorinated silane-functionalized aluminum oxide hydroxide (γ-AlOOH) filters, the removal of perfluro octane sulfonic acid (PFOS) and PFOA from drinking water [39] occurred at >90% efficiency at very high fluxes. However, due to the hydrophobicity of these linear fluorinated silanes, high pressure drop was noted across the membrane thickness during the filtration process. To reduce the back-pressure drop, linear fluorinated silanes were attached with hydrophilic poly(ethylene glycol) units, and such new *hydrophilic* versions substantially reduced the pressure drop across the filter while still maintaining ~99.9% PFOS and PFOA removal. MD simulations involving PFOS and PFOA revealed that the higher adsorption of PFOS was attributed to its extra fluorine atoms facilitating more fluorine-fluorine interactions with the adsorbent [40, 41].

In a similar fluorine-functionalized filtration study, ionic fluorogel resins were found to effectively remove a chemically diverse mixture of PFASs from water. This study created a material library of such resins based on systematic variations in fluorous and ionic components, leading to the identification of an *optimized resin*, that was able to be regenerated and reused multiple times, for a highly selective rapid removal of PFASs in the presence of nonfluorous contaminants commonly found in groundwater [42].

In addition to surface functionalization, the choice of polymer composition and mobile counter ion can have a great impact on the adsorption capability of an anion-exchange resin (AER), specifically for anionic acids such as PFOA and PFOS. Recently, batch experiments were conducted comparing polystyrene and polyacrylic AER in both chloride- and sulfate-forms using natural groundwater spiked with natural organic matter (NOM) and/or six PFAAs. The hydrophobic polystyrene resin was found to be more effective for PFAA removal, and the hydrophilic polyacrylic resin was found to be more effective for NOM removal. Additionally, polystyrene AER showed greater removal of PFOS than perfluorobutanoic acid (PFBA). Removal of NOM and PFAAs by both resin polymer compositions was found to be greater when sulfate was the mobile counter ion instead of chloride [43]. These interesting findings call for a systematic down selection of such AER characteristics.

Recently, the competitive adsorption and mechanistic details of six PFAS contaminants were studied with a commercial acrylic gel-polyamine anion-exchange resin (IRA67) in aqueous solutions (Fig. 8.2). Long-chain PFASs were observed to adsorb more than short-chain PFASs leading to a higher equilibrium of adsorbed long-chain PFAS. However, this equilibrium was shown to be concentration-dependent. At low concentrations, such as ~0.016 mmol/L, the competition between the removal of short- and long-chained PFASs was less apparent. This is due to the presence of accessible adsorption sites suitable for both PFAS contaminants. At higher concentrations, such as ~0.077 mmol/L, the adsorption favorability toward long-chain PFASs was profound. For instance, using IRA67 the removal decreased for PFBA and perfluorobutane sulfonic acid (PFBS) by 77.78% and 72.09%, respectively. This decrease in short-chain PFAS adsorption involves competitive replacement by long-chain PFASs which is driven by favorable interactions such as hydrophobicity and carbon backbone and polar head group characteristics. Therefore, the study found that long-chain PFASs are likely to replace adsorbed short-chain PFASs on anion-exchange resins, leading to the effective removal of long-chain PFAS from aqueous solutions. Furthermore, short-chain PFASs have the potential to be removed with a higher dosage of anion-exchange resins to increase available adsorption sites or in specific scenarios where pH is decreased [44].

Generally, magnetic ion-exchange (MIEX) resins are employed in filtration applications. MIEX resins typically involve fast agglomeration rates due to the magnetic attraction between resin particles leading to increased sedimentation required for filtration. In a recent study, an MIEX resin was evaluated at environmentally relevant PFAS concentrations and in the presence of dissolved organic matter to access the removal of nine PFASs (including both carboxylic and sulfonic acids). An apparent equilibrium constant was derived for PFAS adsorption to serve as an indicator for MIEX adsorption capacity in dilute solutions. Furthermore, this study identified that MIEX resins could be efficiently washed with a 10% w/w NaCl solution, resulting in regeneration of the capacity for PFAS adsorption. The regenerative wash leads to the

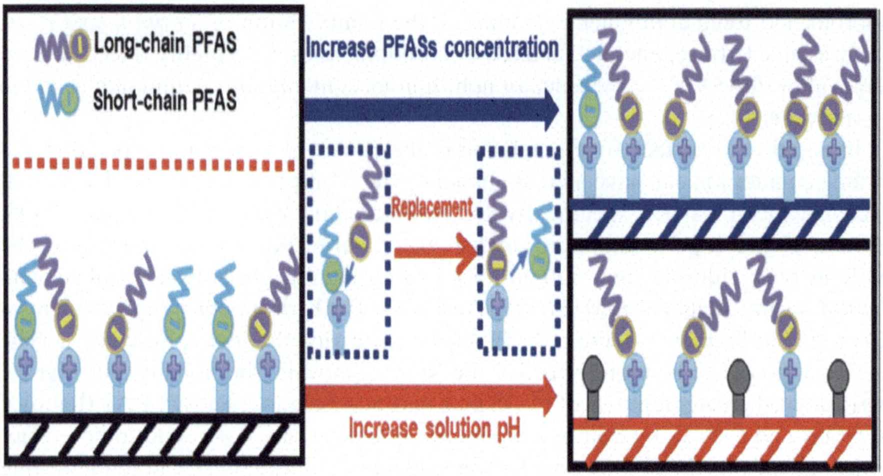

Fig. 8.2 Schematic diagram for the PFASs competitive adsorption at different concentrations and pH on the anion-exchange resin. Reproduced from Ref. [44] and reprinted with permission from Elsevier

desorption of dissolved organic matter, thereby freeing up sorptive sites for PFAS, and is achieved in ~30 min [37].

With regard to investigating optimized regeneration strategies for anionic ion-exchange (IX) resins during the removal of persistent PFAS, including GenX (Note: GenX is a Chemours trademark name for a synthetic, short-chain organofluorine chemical compound, the ammonium salt of hexafluoropropylene oxide dimer acid (HFPO-DA) fluoride), from surface and treated wastewater effluents, a regeneration process with 10% NaCl solution with a contact time of 2 h was found to be optimal for IX operations [45].

Finally, in a recent study, GAC and an AIX system were used in combination with a nanofiltration (NF) system to improve the efficiency of PFAS capture and removal. With such a combination, an enhancement in the removal efficiency of up to 99% of both long-chain and short-chain PFASs was observed. The PFAS removal efficiencies for concentrated NF waste streams were found to increase up to fourfold than those for dilute raw feed water. Thus, this combination of NF and AIX adsorbent systems was found to generate drinkable water by the effective treatment of PFAS-laden waste streams. Furthermore, the GAC and AIX adsorbents were also found to be regenerable making this system suitable for repeated use [11].

Metal organic frameworks: Metal organic frameworks (MOFs) are a group of exceptional materials that are used in many applications including catalysis and gas separation. MOFs are formed from metal ions and clusters by coordination to organic ligands resulting in highly ordered and porous 1D, 2D, or 3D coordination polymers. The extraction of PFAS from aqueous media using such highly porous and tunable MOF materials is appealing for targeted liquid phase sorption. In this

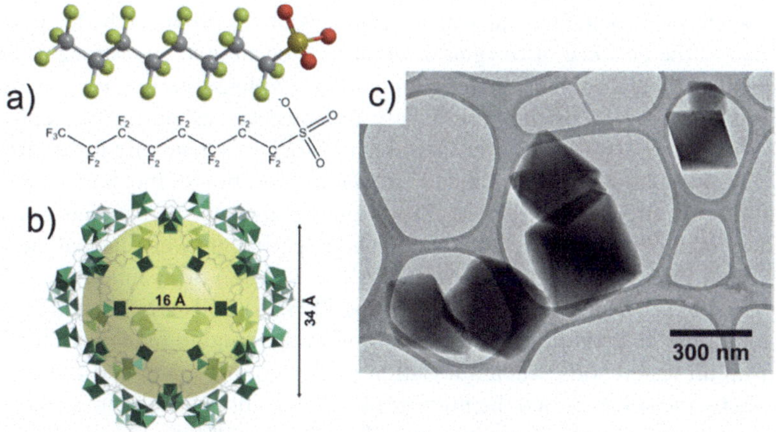

Fig. 8.3 **a** Structure of a perfluorooctanesulfonate anion. **b** Mesoporous cage of the MIL-101 framework. **c** Representative transmission electron microscope image of Cr-MIL-101 particles. Reproduced from Ref. [46] and reprinted with permission from The American Chemical Society

regard, the excellent capture of PFOS using both the chromium (Fig. 8.3) and iron analogs of the cage-like MIL-101 framework revealed unique differences in sorptive properties between these two analogs, providing key implications for future PFOS sorbent design. X-ray photoelectron spectroscopy (XPS) analysis indicated strong interaction between sulfur atoms of the polar head group of PFOS and the metal center of the framework and also with the fluorinated nonpolar tail. In addition, in situ 19F NMR revealed higher PFOS affinity for Cr-MIL-101 than for Fe-MIL-101 based on sorption affinities [46].

In another study, an anionic iron(II) tetrahedral molecular cage (FeMOP) was studied for its ability to interact with various PFASs in aqueous media. LC–MS studies revealed that longer chain length (≥C6) perfluorocarboxylic, -sulfonic, and fluorotelomers were removed from solution. In contrast, bulky CH_3-substituted fluorosulfonamido acetic acid PFASs had reduced association with the cage. Solution binding studies in D_2O using 19F NMR titrations showed a 1:1 binding stoichiometry for perfluorohexanoic acid (PFHxA) and perfluoroheptanoic acid (PFHpA) with an association constant (K_a) of <103 and thus a favorable free energy of association ($\Delta G°$). Perfluorononanoic acid (PFNA), on the other hand, was found to form an insoluble host–guest complex with FeMOP with a 1:1 host–guest ratio. The solid-state host–guest complexes formed from solution complexation of PFHxA, PFHpA, and PFNA with the cage were characterized by Fourier transform infrared (FTIR) spectroscopy which revealed an interaction between the fluorocarbon groups of PFASs to the phenylsulfonate functional groups of the ligand. A computational docking model also indicated that this possible interaction results with the PFASs adsorbing onto the surface of the cage rather than forming a true host–guest complex within the internal cavity [47]. Fittingly, MOFs are beginning to be looked at as effective adsorbents for PFASs and appear to have untapped potential in that capacity.

Biochar systems: Biochar is a carbonaceous matrix that is produced by the pyrolysis of biomass in the absence of oxygen or in an oxygen-limited environment. Due to this, typical biochars contain reduced amount of oxidized groups (i.e., hydroxyl, epoxide, carboxyl, etc.) as they are mostly liberated as carbon dioxide and other gases during pyrolysis. Also, the surface area of biochar is generally lower than that of GAC. A recent study looked at the effect that the addition of biochars to soils has on the immobilization of PFASs. Serial sorption experiments were undertaken using a pine-derived biochar produced at 750 °C. All experiments were carried out either on individual PFASs (solution with one PFAS at 5 mg/L) or as mixtures (solution with 5 mg/L of each: PFOS, PFOA, PFHxS, and PFHxA), and carried out in 2:1 water to soil solutions. The addition of biochar to soils was found to reduce desorption of PFAS from all tested soils. Such alleviation of desorption was found to be largely influenced by the soil type and the number of PFAS components in the sample [48].

In a comparative study, the sorption of PFASs, i.e., PFOA, PFOS, PFBA, and PFBS, on GAC and softwood-derived biochar was studied. While GAC reached sorption equilibrium of PFASs within 3–24 h, the corresponding time for biochar was 12–48 h; both following a pseudo second-order model. In the case of sorption isotherms, the PFAS sorption capacity was chain length dependent with the following order: PFOS > PFOA > PFBS > PFBA. GAC exhibited greater maximum Langmuir sorption capacity for both PFOS and PFOA, i.e., 43% and 39.6%, respectively, than biochar. On GAC, PFBS (48.3 μmol g^{-1}) had a higher maximum sorption capacity than PFBA (31.4 μmol g^{-1}), while the opposite sorption trend was observed for biochar. The sorption mechanisms were found to involve both electrostatic attraction and hydrophobic interaction. The sorption of PFASs was found to increase with a decrease in pH. The results indicated that both GAC and biochar are effective adsorbents for PFASs removal from wastewater [49].

Proteinaceous materials such as hemp protein powder, i.e., *Cannabis Sativa* L, have been shown to be highly effective in removing PFAS from contaminated groundwater. Hemp was found to remove more PFAS compared to soy, lupin, whey, pea, and egg proteins when normalized for protein content. Reaction kinetics showed that rapid PFAS removal with a high degree of removal (>98%) was attained in approximately one hour for PFOS. In the presence of hemp protein powder, increasing salinity appeared to favor PFAS removal. FTIR analysis showed that hydrogen bonding and hydrophobic interactions were influential in the PFAS-protein binding reaction [50].

Finally, *designed* biochars are being developed to optimize the remediation of organic and inorganic contaminant-polluted soils, including PFAS contaminants. It was suggested that the utility for such "designer" biochars involving processes such as iron enrichment or activation should be explored depending on the total organic content (TOC) of the soil, the type of contaminants, and remediation goals [51].

Other adsorbent systems: Soils adsorb cationic and zwitterionic PFASs by a nonlinear process due to their innate chemical characteristics and porous structure. The reversible sorption of cationic PFAS was found to correlate strongly with the soil organic matter (SOM) content, while sorption of zwitterionic PFAS showed a concentration-dependent hysteresis in soils with a low SOM content. Electrostatic interactions with negatively charged soil constituents and the hydrophobic effect were found to be the major sorption driving forces for cationic/zwitterionic PFAS at low and high concentrations, respectively. Density functional theory (DFT)-derived maximum electrostatic potential of PFAS ions was found to be a useful predictor of the sorption propensity of ionic PFAS species [52].

The sol–gel method creates porous organosilica adsorbents, with fluorophilic amide or fluoroalkyl functionalized porous matrix together with quaternary groups for ion exchange. These adsorbents were found to efficiently remove PFASs from water. A distinguishing aspect of the adsorbents was the ability of their flexible porous architectures to volumetrically substantially swell in the presence of organic liquids creating expanded mesopores which were hypothesized to yield greater adsorption capacity for PFASs [53].

Interestingly, a recent study evaluated the adsorption of PFAS analytes onto a variety of laboratory-ware such as filter units and centrifuge tubes (glass and plastics). Substantial adsorption of PFOA was observed on tubes made from polypropylene (PP), polystyrene (PS), polycarbonate (PC), and glass where losses observed were between 32–45%, 27–35%, 16–31% and 14–24%, respectively. Contrary to expectations, the results indicated that the greatest sorption losses for PFOA occurred on PP, whereas losses on glass tubes were found to be much lower [54].

In another recent study, modified clay produced by intercalating quaternary ammonium cations in the exchangeable interlayer sites of smectite clay was found to effectively remove PFAS pollutants in groundwater via strong adsorption. The performance of the modified clay (with removal efficiencies 95–99%) was found to be superior to those of GAC or hard-wood biochar and was found to be comparable to that of an IE resin. Based on MD simulations, the anionic PFASs was found to first occupy the highly polarized bare interlayer edge sites leading to a linear isotherm, and then, the interlayer surface sites yielding a *Langmuir* isotherm. The ionic interactions between the cationic intercalant (N^+) and the terminal oxygen atoms of carboxylate or sulfonate groups of PFASs were found to play a dominant role in adsorption, and the lateral interaction was found to increase fluorophilic attraction to accelerate the adsorption of PFASs [55].

Membrane filtration systems: In addition to the above examples of filtration, a recent study reported the degradation of PFOA in a microwave (MW)-assisted catalytic ceramic membrane filtration system. Water permeation through such pristine and catalyst-coated membranes under the influence of MW irradiation suggested that the coating layer and water temperature produced an increase in flux permeate by the Carman-Kozeny and Hagen-Posieulle (non-slipping and slit-like) modes. The PFOA containing permeate first adsorbed on membrane and catalyst materials and then fully penetrated the membrane filter after reaching adsorption equilibrium. Under microwave irradiation (7.2 W cm^{-2}), more than two-thirds of PFOA in the feed solution was found to be degraded within couple of minutes due to the MW-initiated Fenton-like reactions [56].

The use of an electrodialytic remediation method was recently applied to remove PFASs from contaminated soil. Of the 23 PFASs tested, significant PFAS electromigration toward the anode was observed for predominantly negatively charged C_3–C_7 perfluoroalkyl carboxylates (PFCAs) (PFBA, perfluoropentanoic acid (PFPeA), PFH_xA, PFOA) and C_4, C_6, and C_8 perfluoroalkane sulfonates (PFSAs) (perfluorobutane sulfonic acid (PFBS), PFHxS, PFOS. In contrast to the electromigration of the charged PFASs, N-methyl perfluorooctane sulfonamide (MeFOSA), perfluorooctane sulfonamidoacetic acid (FOSAA), and ethyl FOSAA (EtFOSAA) showed significant transport toward the cathode, which was attributed to electro-osmotic flow of these predominantly neutral PFASs [57].

Recently, direct contact membrane distillation (DCMD) was used for concentrating and removing PFPeA compounds using poly (tetrafluoroethylene) (PTFE) membranes. The process showed efficacy for PFAS removal opening a new route to concentrate and remove PFASs [58].

8.1.2 Processes for the Degradation of PFASs

Once PFASs are captured using adsorbents or are separated and concentrated, they can then be destroyed by direct incineration or degraded via catalytic processes including reductive and oxidative processes, by using plasma radiation, sonochemistry, photochemistry, biodegradation, and other means. The following sections describe recent advances in some of these processes. While there are numerous PFAS degradation techniques such as advanced oxidation [16, 17], bioremediation [24], high temperature (>1100 °C) incineration [12–14], etc., these methods have limitations. For example, the bioremediation process has substantially reduced PFAS remediation efficacy because PFASs are biologically inert. Also, the high-temperature incineration, when not properly employed, can produce toxic intermediates in incineration methods. Alternatively, advanced oxidative processes (AOP) that include Fenton reaction [19], electrochemical treatment [16], photocatalysis [18], PSf oxidation [20, 21], and sonochemical or ultrasound oxidation have been garnering greater attention of late.

8.1.3 Catalytic Systems for Degrading PFASs

Oxidative and reductive processes have been observed to efficiently degrade PFAS compounds. The reductive processes involve systems that can produce hydrated electrons that further react with PFASs to cause their destruction. Typically, UV radiation is used to generate hydrated electrons from photosenstive materials/substances in aqueous solutions to cause such degradations. In oxidative processes, persulfate and other oxidants are commonly used to initiate advanced oxidation conditions.

Reductive processes: Recently, Tenorio et al. applied the UV-sulfite treatment, generating strongly reducing hydrated electrons (e^{aq-}; NHE = −2.9 V), to a diluted AFFF comprising fifteen different PFASs to study e^{aq-} reactions with these PFASs. Results showed that reactivity varied widely among PFASs, but reaction rates observed for individual PFASs in AFFF were similar to rates observed in single-solute experiments. While some structures, including long-chain perfluoroalkyl sulfonic acids (PFSAs) and perfluoroalkyl carboxylic acids (PFCAs) were readily degraded, other structures, most notably short-chain PFSAs and fluorotelomer sulfonic acids (FTSs), were found to be more recalcitrant. This resulted in incomplete fluoride ion release (up to 53% of the F content in AFFF) during reactions producing transient intermediates or unreactive end-products via e^{aq-} reactions with the initial PFAS entities in AFFF [59].

Ultraviolet (UV) radiation has been used to generate electrons in water leading to their solvated versions, i.e., aqueous electrons or hydrated electrons. Recently, in a systematic study, Bentel et al. investigated structure−reactivity relationships within several PFASs undergoing defluorination with such UV-generated hydrated electrons produced with Na_2SO_3 [12]. They investigated both perfluoroalkyl acids and telomeric PFAS compounds and studied the dependence of chain length in these two types of PFASs on the reductive defluorination rates. No chain length dependence was found for perfluorocarboxylates (PFCAs, C_nF_{2n+1}–COO–) as similar deflourination rates were achieved for $n = 2$ to $n = 10$ PFCAs. Conversely, deflourination rates for perfluorotelomeric substrates (i.e., C_nF_{2n+1}–CH_2CH_2–COO–) and perfluorosulfonates (PFSAs, C_nF_{2n+1}–SO_3–) were found to be dependent on the fluoroalkyl chain length (Fig. 8.4). Calculation of the C−F bond dissociation energies (BDEs) using density functional theory (DFT) and high-resolution mass spectrometry product analyses highlighted the strong relationships between experimental defluorination, BDE, and chemical structure (alkyl chain length and polar head group) as spontaneous C–F bond dissociation occurs upon reduction. The weakest C–F bond, as determined from the BDE analysis, resulted in the observed alkyl chain length dependence. In the case of PFCAs, regardless of chain length, reduction leads to the dissociation of C–F bond on the carbon adjacent to the head group (lowest C–F BDE). In the case of perfluorotelomers and PFSAs, the weakest C–F bond changes based on alkyl chain length and polar head group and therefore results in a length-dependent defluorination [12].

Electron transfer (ET) from a photosensitive indole acetic acid (IAA) to PFAS was investigated recently using a ternary self-assembled micelle. The micelle composite consisted of IAA, cetyltrimethylammonium bromide (a cationic surfactant), and PFOA. Upon UV irradiation, efficient ET and degradation of PFOA were observed under ambient conditions. The micelle composite and degradation process showed reasonable pH stability (ranging from pH 4–8) highlighting the suitability of the micelle system across solution conditions. Furthermore, after 2.5 h of UV irradiation, the PFOA concentration decreased from 10 mg L^{-1} to ~60 ng L^{-1}, leading to efficient photo-induced degradation [60].

Heterogeneous reductive defluorination of PFOS has been explored using UV-generated hydrated electrons from surface-modified carbon nanotube electrodes. PFOS adsorbed to the nanotube surface was degraded using direct ET coupled

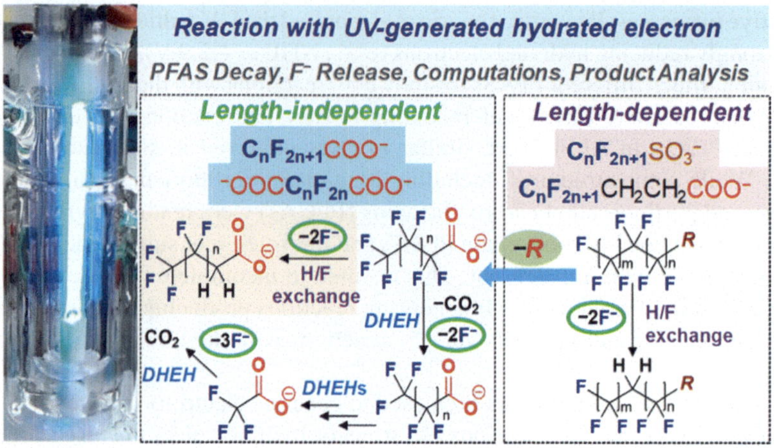

Fig. 8.4 Length-independent and length-dependent reaction of PFAS molecule with UV-generated hydrated electron. Reproduced from Ref. [12] and reprinted with permission from The American Chemical Society

with a low electrode potential. The ET resulted in faster PFOS defluorination and chain shortening reaction rates leading to the formation of shorter chain (C_3–C_7) perfluorinated acids [61].

A recent computational study detailed the early events in PFAS reductive defluorination and presented a methodology for predicting the standard reduction potential of PFASs. Using DFT, the authors showcased how to use calculations to predict the effect of reduction on PFCAs and PFSAs by probing electron attachment sites and defluorination reaction intermediates, establishing free energy relationships, and evaluating the inherent structural diversity of PFAS. Similar to the results of Bentel et al., electron attachment to linear PFCAs occurred at the α-carbon (with respect to the polar head group) and results in defluorination of the α-C–F bond. Resonance stabilization of the resulting carbon-centered radical from the carboxylic head group ensures that the α-carbon will always be energetically the most favorable defluorination position, supporting the independent chain length nature of PFCAs. On the contrary, electron attachment to linear PFSAs generally occurs at C–F bonds along the inner portion of the alkyl chain. Therefore, PFOSs are more likely to exhibit a chain length dependence because as the chain length becomes shorter or longer, the location for electron attachment differs. Thus, reductive defluorination of PFOS and PFOA differs based on the predicted locations of C–F dissociation [62].

Lastly, another theoretical study investigated the reductive-induced degradation of PFAS in the presence of excess electrons using self-interaction-corrected Born–Oppenheimer molecular dynamics (MD) simulations [63]. Results showed that the initial phase of the degradation involved defluorination and transformation of an alkane-type C–C bond into an alkene-type C=C bond in the PFAS molecule. This occurs via sequential elimination of fluorine atoms on adjacent carbon atoms. Similar reactivity has been proposed by Van Hoomissen et al. [62] Suffices to say, additional detailed and nuanced research is warranted to further explore the underpinning science of such reductive degradations of PFASs.

Oxidative processes: Advanced oxidative processes including PSf-initiated oxidation and Fenton reactions of PFAS are well known. The potential of a modified Fenton system, i.e., homogeneous PSf conjugated with silver (Ag), to decompose PFAS under ambient conditions was recently reported. Among many combinations of common oxidants and transition metals tested, only the PSf/Ag pair was seen to produce significant reactivity at 20 °C exclusively toward carboxylic PFAS including PFOA, resulting in fluoride ion release and reaction by-product formation. However, the activity of this reaction to a sulfonic PFAS was seen to be limited [64].

In a recent study, a transition metal catalyst such as iron (Fe(0)) particles was used together with common oxidants such as hydrogen peroxide, PSf, and peroxymonosulfate for the removal of PFOS. The type and concentration of the oxidant and the temperature and pH of the reaction were controlled to dictate the various oxidizing and reducing reactive species that were generated during the catalytic reactions. In a majority of the reactions, a substantial amount of PFOS was observed to be removed. PFOA was also seen to be well decomposed, producing many expected intermediates such as shorter chain PFASs [65].

Sulfate radical ($SO_4^{\cdot -}$)-based advanced oxidation processes have proven to be effective for degrading certain PFASs and recently were applied in a cyclic treatment approach for addressing PFOA removal. Each cycle included IX for purifying considerable PFOA-polluted water and concentrating trace PFOA on a small quantity of the IX resin. The PFOA-sorbed resins were further regenerated with heat activation of peroxodisulfate (PDS). Results showed that cumulative PFOA removal linearly increased with the cycle number for regenerated resins [66].

Recently, a dual-frequency ultrasound (US) was combined with PSf to cooperatively degrade PFAS in water and soil with a 62–71% degradation efficiencies for fourteen PFASs. The pathways and kinetics of PFOA destruction were studied by monitoring the degradation intermediates. The results indicated that the combination of the dual-frequency US/PSf was not simply additive but had a synergistic effect on PFOA degradation [29]. A more detailed section on PFAS degradation science by US can be found below.

8.1.4 Nanomaterials-Based

By careful manipulation of the particle size and structure in the nanometer regime, several unique enhancement of properties can be achieved that is not feasible with bulk materials. The increase in surface area due to decrease in the particle size allows tunable electronic, optical, and surface properties which makes nanomaterials an interesting option for PFAS remediation. Toward this, the use of nanomaterials to remove PFAS/PFOS from water can be broadly classified into two categories (a) non-destructive and (b) destructive. The passive or non-destructive physical approach has been applied over the past decade which primarily depends on removing PFASs through surface adsorption or membrane filtration. However, the destructive (non-passive approach) utilizes the electro-/photochemical properties of the nanomaterials to fragment the perfluoro compound. Also, suitable choice of nanomaterials can allow the simultaneous application of both passive and destructive remediations [67].

Physical Approach (non-destructive): Physical adsorption techniques are considered as the most environmentally friendly method that is used for the removal of organic pollutants from aqueous streams, due to its low cost, simplicity, abundance, etc. The hydrophobic nature of the carbon-based materials and the lyophilic PFAS molecules lead to the dispersive interaction that allows the selective adsorption in PFAS on the surface of carbon materials. Due to this reason, among several materials that are used for physical adsorption, carbon-based materials are most common. Bulk activated carbons have been used over the past few decades due to affordability in cost, availability, and maturity of the technology, and the process has shown great efficiency in removing organic pollutants from aqueous media [68]. On the other hand, GAC provides additional benefits in PFAS removal (more than 90% efficiency due to its increased surface area as compared to activated carbon). However, due to the shorter retention of PFAS on its surface, it needs more media for effective removal of PFASs. Also, these approaches require frequent regeneration of adsorbent, when deployed in large scale, thereby leading to increase in cost [69].

Carbon nanomaterials (carbon nanotubes (CNTs), graphene, graphene oxide etc.,) have been of great interest for PFAS remediation in recent past due to the excellent properties these materials offer. The high specific surface area, excellent porosity (low density), ionic adsorption capability, mechanical and chemical stability, etc., are some of the key properties of carbon nanomaterials that render them attractive. The recent advancement in large scale manufacturing and selective surface engineering of these materials makes it much more convenient than ever for remediation applications possibly at an affordable cost [69].

Owing to their intense adsorption affinity toward organic moieties, CNTs are widely examined for PFAS remediation [70]. Pristine multi-walled and single-walled CNTs have been evaluated for PFAS removal and in general, the specific surface area, accessibility of surface to the PFAS molecules (percolation) dominate their PFAS removal efficiency. Though surface area on multi-walled carbon nanotubes (MWCNTs) can be increased by decreasing the diameter of the individual tubes, the higher packing density (due to decrease in tube diameter) increases the resistance to PFAS in aqueous media to reach the CNTs for effective adsorption and hence decreases the remediation efficiency. Based on the above, it has been shown that single-walled carbon nanotubes (SWCNTs) perform more effectively than MWCNTs. For instance, decrease in diameter of the MWCNTs increases the effective surface area, but that did not translate to PFAS removal efficiency [71], while positive trend (increase in remediation) was noticed with decrease in the SWCNTs diameter [72]. Most of the CNTs growth techniques utilize catalyst-assisted processes, leaving the metal catalyst as a decoration [73]. The effect of the MWCNTs with and without metal catalyst demonstrated that the metal decorated MWCNTs have sixfold higher adsorption of PFOS compared to pristine MWCNTs, and this increase was accredited to the electrostatic interaction due to metal particle decoration. On similar lines, nano-copper-decorated MWCNTs showed 18 times more PFOS adsorption. Also, recent research suggested that altering the surface properties of the CNTs have significant impact on their adsorption efficiency. For instance, pristine (hydrophobic) CNTs showed threefold increase in PFOS adsorption capacity relative to –COOH and –OH surface-functionalized MWCNTs [73].

Providing electrostatic/polarization forces in the form of electric field enhances the adsorption (electro-adsorption) efficiency of CNTs, and toward this CNT composite electrodes have been extensively studied. This approach on average showed a 10–100 fold increase in PFOS/PFAS adsorption with an electric potential of 0.5–5 V relative to the unbiased pristine powder counterpart. In addition to electro-adsorption, it is also possible to induce electro-oxidative degradation using this approach. Recently toward electro-adsorption, CNT sponges were utilized, which showed ~10-fold increase in efficiency over simple MWCNTs. In a similar vein, new nanocomposite electrodes made of graphene platelets and CNTs are also being actively pursued for PFAS remediation [68].

Destructive PFAS removal: The destructive mode of PFAS remediation depends on the electrolysis reaction pathways where solvated electron initiates desulfonation or decarboxylation of $-SO_2OH$ or –COOH groups of PFOS and PFOA molecules, respectively. Tunability of the electro-optical properties of nanomaterials enable electron generation through an external stimulus, either through electric field or via photon-initiated process. By engineering the particle size, the bandgap can be tuned, thus enabling maximum electron generation which in turn degrades PFASs. Most common and well-studied among the above is photochemical reaction degradation route [70, 74].

TiO_2 is one of the well-studied nanomaterial for aqueous pollutant removal because of its ability to initiate oxidization, hydrophobicity, chemical stability, etc. Photoelectrons are generated in size-engineered TiO_2 nanoparticles by UV irradiation, which through decarboxylation process fragments PFASs. The energy of the photoelectron, its loss mechanism and proximity to pollutant, and the density of the pollutant determine the overall process efficiency. The practical application of TiO_2 for PFAS remediation is very limited due to the difficulty in achieving the combination of all of the above [75]. To address these deficiencies, hybrids of TiO_2 nanoparticles with carbon-based nanomaterials were attempted with success. The addition of carbon nanomaterials enables effective PFAS adsorption on its surface and also serves as a transportation media for photoelectrons that are created by the TiO_2 particles, which increase the contact and also the chance of PFAS-electron interaction and hence can lead to efficient degradation of PFASs. Toward this, TiO_2 nanoparticle-MWCNT composites were studied using 365 nm mercury lamp that showed complete degradation of PFAS in 1.6 g/L sample solutions. TiO_2-graphene composites also showed similar kind of increase in the efficiency [76, 77].

Similar to TiO_2, alternate wideband gap oxides such as Ga_2O_3 (4.8 eV), In_2O_3, metal oxide (MO) doped In_2O_3, in simple nanomaterial state or in composite form with carbon nanomaterials, were attempted and shown to have efficiency increase in PFAS remediation [78–80].

Although the nanotechnology-based PFAS remediation shows promise through adsorption, photochemical degradation, and their combinations, this approach still suffers from major issues in terms of effectiveness in real life scenario, where other organic contamination can effectively inhibit the above-mentioned mechanisms. Furthermore, most of the fundamental research works have been focused on utilizing DI water contaminated with PFASs, while the interference from pH-induced redox potential in real world samples can lower the estimated efficiencies. Other aspect that needs to be considered for the implementation of this approach is the environmental part and cost associated with the safe disposal of remediated PFAS and nanomaterials that were used in the process along with the scalability of the process.

8.1.5 Polymer-Based

Polymers, the ubiquitous long-range molecularly ordered systems, have provided innumerable contributions in utilitarian applications. With regard to the science of PFASs, typically, polymers have been used as direct sorbent materials, as substrates to tether and improve existing PFAS adsorbent materials, and possibly also as detectors for the presence of PFASs. As adsorbents, the inherent molecular reactivities of polymers can provide a means to convert them to hydrogels, aerogels, and other crosslinked intricate structures such as via molecular imprinting, etc. In this regard, cyclodextrins have been used as pendant groups on various polymer backbones for adsorbent development for PFAS remediation treatment.

Cyclodextrin-based: Cyclodextrin, a family of cyclic oligosaccharides, is a group of molecular vessels consisting of a macrocyclic ring glucose subunit joined by α-1,4 glycosidic bonds that form host–guest complexes. The characteristics of the guest are fundamentally influenced by the free volume and chemical nature of internal cavity of the host cyclodextrin; i.e., hospitality as driven by innate molecular characteristics! Recently, investigations of cyclodextrin-based polymers as adsorbents for PFASs have originated from several labs. A recent study explored the science of the crosslinkers in decafluorobiphenyl-crosslinked cyclodextrin (DFB-CDP) (Fig. 8.5) in promoting PFAS affinity. The study investigated three DFB-CDP derivatives and two β-CD polymers crosslinked by either epichlorohydrin or 2-isocyanatoethyl methacrylate. The β-CD polymers crosslinked by perfluorinated aromatics with low degrees of phenolation was found to have high propensity for PFAS adsorption [81].

Recently, the functionalization of β-CD with positively charged amine groups was observed to influence the strength of the β-CD:PFAS complex formed with legacy and short-chain PFASs while a negatively charged functionality decreased the complexation effect. The enhanced complexation was assigned to electrostatic attraction

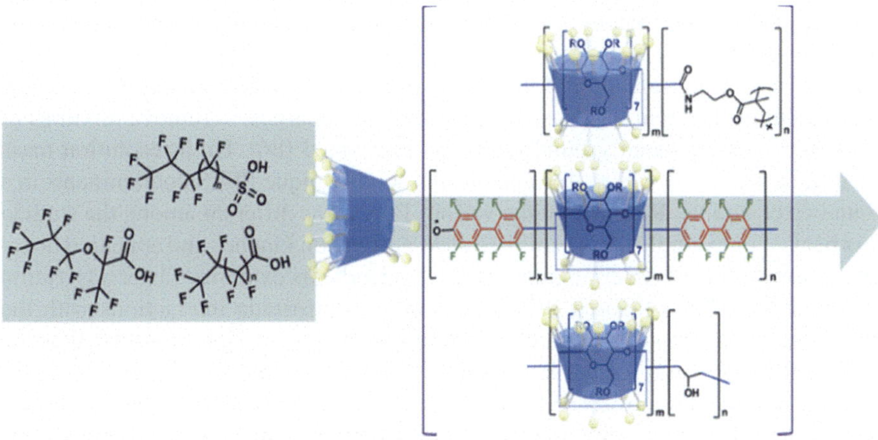

Fig. 8.5 Schematic of adsorption of PFAS molecules using crosslinked *β*-cyclodextrin polymers. Reproduced from Ref. [74] and reprinted with permission from The American Chemical Society

between the negatively charged PFAS head group and the positively charged β-CD derivative making them promising candidates for remediation of short-chain PFASs [82].

Nitrile or primary amine groups derived from the nitrile groups in the *β*-cyclodextrin polymer tetrafluoroterephthalonitrile (TFN-CDP) can have distinct interactions with PFASs. While TFN-CDP exhibited adsorption distribution coefficients (log K_D values) of 2–3 for cationic PFASs and −0.5 to 1.5 for anionic PFASs, the amine-containing TFN-CDP exhibited log K_D values of −0.5 to 1.5 and 2 to 4, respectively, with high affinity toward anionic PFASs [83]. In this study, introducing amino groups into the crosslinkers of a β-CD polymer network was found to improve the binding of many anionic PFASs. In another study involving β-CD-based adsorbents crosslinked with amino or amido tripodal crosslinkers, the polymers containing amine functionality exhibited superior anionic PFAS removal ability [84]. Compared to activated carbon (AC), both β-CD polymers have superior PFOA removal, consistent with the β-CD:PFOA inclusion complexes being prominent for efficient removal. Therefore, the design of β-CD polymers and resulting guest–host complexes requires synergistic electrostatic interactions and host–guest interactions to effectively remove PFOA from aqueous solutions [84].

Recently, the removal of PFAS using activated carbon (AC) materials and *β*-cyclodextrin polymers (CDPs) was compared using a principal component analysis [85]. The major differences between the two materials involve the nature of the surface interactions with PFAS and the differing chemistry of the bonding sites. PFAS adsorption to AC materials was found to be dominated by hydrophobic interactions between the perfluoroalkyl chain and the AC surface. In the case of CDPs, anionic PFAS adsorption occurs via favorable electrostatic interactions with functional groups along the surface, however, this can be limited due to the higher pH found in groundwater resulting in deprotonation and changing of the electrostatic interactions [85].

Furthermore, an evaluation of the ability of five adsorbents, namely AC, an AIX resin, and three different CDPs with varying surface charges, to remove mixtures of PFAS from contaminated groundwater was performed [86]. High-resolution mass spectrometry analysis was performed on sixty-eight unique PFAS contaminants in a groundwater sample. Removal of the various PFAS was different among the studied adsorbents. For example, ACs exhibited slow adsorption kinetics and appeared nonselective toward the PFAS contaminants. CDP adsorbents exhibited faster adsorption kinetics than AC and favored PFAS-containing electrostatic interactions with the CDP surface functional group charge. Analysis of the results identified that the adsorption mode/nature of the adsorbent is crucial for removal of specific classes of PFAS. Improved removal can be associated with synergistic hydrophobic surface interactions, where increasing the length of the perfluorinated tail improves adsorption, or with electrostatic interactions, where the charge and properties of the polar head group of PFAS are important for surface interaction [86].

Other polymer-based examples: Anionic PFAS contaminants showed high affinity toward the copolymer $p(\text{TMA}_x\text{-}co\text{-TMPMA}_{1-x})$, containing amine functional groups. However, changing the amine functionality to nitro-groups changes the mode of active adsorption. The redox-active nitroxyl radical groups were found to undergo electrochemically-controlled capture and release of PFASs. As such, controlling the ratio of amines to nitroxyl radicals is a possible avenue for tuning the redox-activity, hydrophobicity, and binding affinity of the copolymer, to enhance PFAS adsorption and polymer regeneration [87].

Poly (N-[3-(dimethylamino) propylacrylamide, methyl chloride quaternary, (DMAPAA-Q)) hydrogel matrix has been used recently as an effective sorbent for sequestering PFAS from various water matrices. The selective removal of sixteen PFASs from distinct classes using DMAPAA-Q polymer was demonstrated in surface waters with environmentally relevant concentration (i.e., <1000 ng/L). The results showed fast removal kinetics with equilibrium time of 60–120 min. Further, regardless of their chain lengths, a greater extent of removal of sulfonated than carboxylic PFASs was also observed. The impact of background anions on PFAS removal by poly DMAPAA-Q hydrogel was tested and found to follow the order of $SO_4^{2-} > Cl^- > NO_3^{-\cdot}$ with the activity being maintained in six consecutive adsorption/regeneration cycles to remove PFASs [88].

A recent investigation of polydiallyldimethylammonium chloride (polyDADMAC)-functionalized powdered activated carbon (PAC) demonstrated that substantial adsorption of PFOA and PFOS occurred at polyDADMAC-PAC, yielding Freundlich adsorption coefficients of 156 and 629 L/g $^{-n}$, respectively [89]. Such polymer functionalization of porous adsorbents can be a useful method to enhance PFAS adsorption.

Polymer systems have also been used for the detection of PFASs. A molecularly imprinted polymer (MIP) electrode, formed by the anodic deposition of o-phenylenediamine in the presence of PFOS template molecules on a glassy carbon macroelectrode, was used for the detection of PFOS [90]. The performance of the resulting MIP electrode was evaluated by the current obtained from the oxidation of ferrocene carboxylic acid as the electrochemical probe. The MIP electrode was found to be able to detect PFOS with a detection limit of 0.05 nM, which is lower than the health advisory limit of 0.14 nM prescribed by the EPA. Thus, MIP systems provides an exciting route for detection of PFASs [90].

In a recent study, another simple polymeric detection scheme for PFAS was investigated based on crosslinked, thermoresponsive poly(N-isopropylacrylamide) (PNIPAM) hydrogels. Combining swelling caused by PFASs with the fluorimetric response of a solvachromatic dye, nile red, it was revealed that the fluorosurfactant initiated observable aggregation at 0.05 mM and reached saturation at 0.5 mM. The fluorosurfactant was found to homogeneously distribute throughout the polymer matrix from energy dispersive X-ray spectroscopy evaluations, rendering the swelling response as a marker of fluorinated interfacial positioning and delocalized electrostatic repulsion. Thus, the physiochemical response of PNIPAM can be used to assess TPFOS's concentration [91].

8.1.6 Sonochemical/Microwave

In general, sonochemistry utilizes high-frequency ultrasonic (US) irradiation or waves to initiate and generate a series of chemical reactions in the aqueous media. The US waves disintegrate the bubbles or particles generated in the liquid by cavitation. The cavitation theory is based on the process, wherein the US waves irradiate the liquid through hydraulic vibrations to create microbubbles also referred to as cavities. The repeated expansion and contraction of these cavities by the high-energy US waves result in high vapor temperature leading to the collapse of water/bubble interface known as cavitation. During the cavitation process, intense collapse of these microbubbles creates high vapor temperature (4000–10,000 K) [92], bubble-water interface temperature in the range of 1000–1500 K, and adiabatic pressure up to several hundred atmospheres [93]. This high pressure and temperature in the ultrasonication are sufficient to cause pyrolytic decomposition of the PFASs at such microbubble-water interface leading to their removal. Ultra-sonochemical process is considered as an effective advanced oxidative process (AOP) [28], among oxidative processes as it produces hydroxyl radicals (OH·) at the bubble-water interface leading to oxidation of PFASs during the US cavitation process. Hence, by US irradiation, the PFASs compounds can either be directly removed by high-temperature pyrolysis or indirectly by OH· oxidation [29].

The ultrasonication treatment has been demonstrated to reduce the chain length of the long-chained PFASs, $C_nF_{2n+1}SO_3H$ ($n > 8$) into smaller carbon chained ones, thereby facilitating their degradation. Typically, US frequencies ranging from 20 to 1000 kHz or megasonic (>0.5 MHz) are applied during the sonochemical treatment of PFASs. To enhance the efficiency of PFAS degradation, the US is mostly coupled with additives such as photocatalytic products, carbonate, PSf, periodate, and microwave. Among the additives, PSf has been reported to be most beneficial due to its high stability, aqueous solubility, low cost, and longer shelf life. More importantly, PSf on ultrasonication can generate sulfate radicals ($SO_4^{-\cdot}$) which are strong oxidants that facilitate oxidative degradation of PFASs. There have been several efforts targeted to the degradation and removal of different types of PFASs via sonication, where it becomes necessary to optimize sonication conditions and parameters that not only dictate the efficiency of PFASs degradation but are imperative for translating the research efforts to practical applications. Hence, the following sections will focus on the ongoing sonochemical efforts, specifically effects of different US operation variables and additives for PFASs remediation.

The US degradation depends on several operational parameters that directly affects the efficacy of PFASs removal via sonochemical treatment, which include US power density, frequency and temperature, and treatment atmosphere. Among these parameters, US frequency plays the most significant role in the bubble-water dynamics.

Ultrasound Frequency: The frequency of the sound wave that is generated to degrade the PFASs is critically important as the periodic oscillations determine the critical size of the cavitation bubbles [94] when they grow in size and collapse as they can no longer absorb the US energy from the sound wave. There have been several studies of PFAS degradation using different US frequencies. Most of the initial studies were targeted at different PFASs keeping the US power constant while varying the sonolytic rates. For example, Campbell et al. used US frequencies (202, 358, 610, and 1060 kHz) for the degradation of PFBA, PFBS, PFH_xS, PFOA, and PFOS [95]. The US frequency plays a critical role in the efficacy of the sonication process as frequency rates determines the resonant radius of the acoustic bubble and therefore their collapse time. Accordingly, higher US frequencies lead to smaller bubbles and reduced collapse time leading to increase in cavitation rate per unit time and therefore an enhancement in the sonolytic process. Furthermore, the higher frequency also increased the active sites of the cavitation bubble interface, whereby the mass transfer from the bulk solute to the cavitation interface resulted in extensive mass degradation/decomposition through interfacial pyrolysis. A detailed summary for the degradation of PFASs at different US frequencies is described in Table 1-1 of Cao et al. [28] The US frequency was optimized to break the carbon chain length of the PFOS/PFAS materials and while 358 kHz was found to be effective for carbon chain length greater than 5 (PFHxA, PFHxS, PFOA and PFOS), 610 kHz was more suitable for smaller chain lengths (perfluorobutanoic acid (PFBA) and perfluorobutane sulfonic acid (PFBS)).

While the sonochemical rates of decomposition of the longer chain carbon PFASs were dependent on the rate of bubble cavitation created by the higher US frequency, the lower chain carbon PFASs were found to be mostly degraded by extensive mass transfer of the bulk solute using lower frequencies. Overall, for the efficient removal of the PFOS and PFOA, a synergistic effect of a combination of frequencies as compared to a single frequency was found to be more effective and led to the use of dual-frequency ultrasonication [92, 95]. Thus, using a combination of lower and higher frequencies caused interference effect, resulting in a larger effect generating a higher cavity growth and eventual intense bubble cavitation collapse in unit time per bubble population, leading to enhanced sonochemical reactions and therefore degradation as compared to using a single frequency [29, 95]. The dual-frequency was shown to improve the degradation efficiency of various contaminants in aqueous solution [92] including PFASs. Further, the dual-frequency sonication was combined by adding PSf for additional improvement in the degradation.

Power: The intensity of ultrasonication is determined by the sonication power referred to as the power that is delivered to per square cm of the radiating area. The power intensity is directly related to the cavitation bubble size, the rate of collapse, the transient temperature and pressure within the bubble during the collapse. Tuning the US power allowed controlling the rate of collapse of the bubble cavities and therefore the sonochemical activity [95]. Hence, for a constant volume, the power intensity increased linearly with power density and also linear correlation between the applied power and the sonochemical reaction can be established [28]. Furthermore, at fixed frequency, increasing the power density resulted in an increase in the size, number, and rate of the active bubbles generated and also facilitated enhanced mixing strength and rate of mass transfer in the aqueous media due to the agitation caused by the cavitation bubble collapse [96, 97]. Finally, the power density plays a critical role in the rate of degradation, as higher power density can increase the bubble density in the container, which can; (i) scatter the sound waves off the container walls; (ii) create bubble cloud on the liquid surface that can scatter and decay the sound waves; (iii) create higher input energy; (iv) affect energy transfer; and (v) cause bubble clustering leading to bubble merger into larger bubbles, impacting the sonochemical reaction, and hence the degradation.

Additives to ultrasound: While the power and frequency play a critical role in the cavitation process, sonication alone is known to consume energy and adding additive salts to the ultrasonication process can generate large amounts of free radicals during the sonochemical reactions that can reduce the need for a high amount of input energy accelerating the rate of PFAS degradation [29]. Addition of different types of these salts into the aqueous PFAS solution changed the ionic strength and surface tension of the aqueous phase, in turn changing the PFAS concentration in the interfacial region of the cavitation bubbles [98]. Furthermore, to enhance the efficiency of PFAS degradation, US is also coupled with additives such as photocatalytic products, carbonate, PSf, periodate, and microwave among which adding PSf has been reported to be most beneficial due to its high stability, aqueous solubility, low cost, and longer shelf life. More importantly, the PSf on ultrasonication can generate sulfate radicals ($SO_4^{-\cdot}$) which are strong oxidants that facilitate oxidative degradation.

To further improve PFAS degradation, sonochemical methods were combined with other conventional techniques such as ultraviolet (UV), photocatalysts, and temperature. The UV light increased the defluorination and degradation of the PFASs substantially. Also, increasing or decreasing the temperature of the solvent during ultrasonication enhanced or retarded the decomposition and defluorination [99]. At elevated temperatures, for example in PFOA, the surface tension of PFOA solution decreased therefore decreasing the adsorption of PFOA to the bubble interface. This lowered their movement at the air–water interface resulting in lower mass transfer rate of PFOA to the cavitation bubble resulting in reduced rates of PFAS defluorination and decomposition and hence, ultimate removal. Therefore, while the additional steps incorporated to the US are critical, it is highly important to suitably tune them for efficient degradation, decomposition, and removal of the PFASs from their aqueous solution.

Finally, the sonochemical approaches have demonstrated significant potential for PFASs degradation and remediation, however, with associated limitations. Firstly, the scale of the reactors is small and limits the scalability of the sonochemical reactions for PFAS removal. The scalability requires consideration of the US parameters such as the frequency, power density, energy consumption, and geometry of the reactors. Secondly, the degradation cannot be selective in such physicochemical reactions as PFASs can be originating from different environments. Finally, the cavitation process can produce toxic intermediates, which can corrode the power transducers or walls of the sonicators, when upscaling and this can turn out to be very expensive. However, some of these limitations can be improved by combining the ultrasonication process with conventional techniques such as including additives, microwave, UV and temperature, and ozone-induced reactions to enhance the PFAS degradation due to synergistic effects. Thus, there is a constant effort to scale up the sonochemical process while overcoming the limitations to effectively decompose/degrade the PFASs and enhance the remediation process.

8.1.7 Sub-/Supercritical Fluids (Water)-Based Degradation

Degradation of PFASs in subcritical water system was initially explored around 2006 for decomposing perfluoroalkylsulfonates bearing shorter chain (C_2–C_6) perfluoroalkyl groups such as PFH_xS contained in an antireflective coating agent used in semiconductor manufacturing. Although PFH_xS demonstrated little reactivity in pure subcritical water, addition of zerovalent metals to the reaction system was found to enhance PFH_xS decomposition to form F^- ions, with an increasing order of activity of metal Al < Cu < Zn, Fe [100].

Recently, fluorinated surfactants have been found to be degradable under supercritical and subcritical water conditions. An original fluorinated surfactant, 3-hydroxy-2-(trifluoromethyl) propanoic acid, that was part of a latex of poly vinylidene fluoride was shown to be easily decomposed in subcritical water, releasing fluoride anions [101].

Also, recently, complete mineralization of fluorinated ionic liquids, although they are not typical PFASs, was achieved in subcritical water in the presence of potassium permanganate. Reactions of $[Me_3PrN][(CF_3SO_2)_{2N}]$ and $[C_3mpip][(CF_3SO_2)_{2N}]$ (C3mpip = 1-methyl-1-propylpiperidinium) in subcritical water were investigated to develop a technique for recycling fluorine element. By adding $KMnO_4$ to the system, quasi-complete mineralization of the ionic liquids was achieved at 300 °C suppressing the formation of environmentally undesirable CHF_3 [102].

A US patent application was recently awarded to Battelle Corporation (PCT/US2020/040196) for a technology involving PFAS destruction by oxidation in supercritical conditions. By this technology, preconcentrated PFAS in water is annihilated in supercritical conditions to levels below 5 parts per trillion (ppt), and the water effluent was used to recover heat, returned to subcritical conditions, and then released back into the environment.

8.1.8 Plasma Treatment of PFAS Degradation and Removal

Plasma, generally termed as the fourth state of matter is an excited state of gas that is made up of free electrons, active ions, excited atoms, and photons (emitted when the excited atoms are relaxing to ground state). This excited state of matter is broadly classified as (a) cold (non-equilibrium) plasma where the electrons have higher energy (temperature) than the positively charged ions and (b) hot (equilibrium) plasma, where both electrons and ions have similar energy. At suitable condition(s), cold plasma can be excited with less energy in a confined space with minimum footprint, while high-temperature plasma typically require expensive setup, larger footprint, and extensive maintenance, which makes cold non-equilibrium plasma an attractive option for several applications. Since the applications of interest mainly use cold plasma, our focus will be on cold plasma henceforth.

By applying an electric field across two parallel electrodes that are located in pressure/gas controlled ambient environment, breakdown of the gas can be initiated, i.e., the electrons from the negative electrode on elastic collision with a molecule initiate ionization, which then avalanches into a cascade of ionization events establishing electric flow between the electrodes. When the gas flow or pressure and the electron density and energy are balanced in this process of ionization in a controlled way, this leads to a steady state called a glow discharge or plasma.

$$V_{\mathrm{B}} = \frac{Apd}{\ln(pd) + B}$$

Paschen's law (equation above) provides the relationship between the breakdown voltage (V_{B}), distance between the electrodes (d), pressure (p), and nature of the gas (A and B constants) [103], which guides the selection of the suitable plasma parameters for the required application. For example, low pressure with inert gas allows larger electrode separation, and hence, plasma can be sustained with non-inert gases and fraction of reactive precursors for plasma-based surface engineering and thin films deposition. At atmospheric pressure, flowing inert gas at high rate between the electrodes placed close to each other can create and sustain plasma with an extended afterglow [104].

This extended afterglow primarily consists of electrons, excited gas molecules and ions. This unusual chemistry has enabled successful application of remote plasma treatment process for plasma enhanced chemical vapor deposition, plasma polymerization, surface sterilization/activation, etc., and the recent focus in treating water to remove microbes and contaminations [105].

While the early research and development were strongly focused on understanding and utilizing the plasma chemistry at lower pressures for etching, thin film deposition, etc., the interest over the last decade has shifted toward atmospheric pressure (AP) plasma and addressing the challenges in creating AP-plasma and its application. The relentless quest to explore the fundamental interactions of the plasma species with various matter (solids, liquids, and gases; all of different chemistry) has helped in establishing the effectiveness of individual plasma species. Hence, recent research on atmospheric pressure plasma (APP) has focused on creating critical plasma species using inexpensive approaches, such as by utilizing filamentary arc discharges, corona, and dielectric barrier discharges. Operating plasma at atmospheric pressure or slightly higher pressure also enables plasma application on the surface or within the liquid allowing environmental remediation, such as water treatment (breakdown and removal of organic and inorganic contaminants) and disinfection (removal of bacteria, viruses, spores, and protozoa) [106].

It has been established that, in general, the free electrons (e^-) and the active radicals in the plasma interact with water at the interface to provide oxidative (OH^*, O, H_2O_2, H_2O^*, O_2^-, O_3) and reductive species (solvated electrons (e_{aq}^-), H^*, Ar^+, H^+, Ar^*). These highly reductive free electrons and solvated electrons have been shown to influence the defluorination process of PFASs. The active free electrons, the solvated electrons, and Ar^* radicals attack the $–SO_2OH$/–COOH functional group of PFASs, initiating HF elimination reactions to subsequently reduce the PFASs [12, 32]. PFASs are predicted to undergo chain shortening processes via decarboxylation/desulfonation-hydroxylation-elimination-hydrolysis, by losing CO_2 or SO_2, etc., leading to defluorination. Furthermore, during such degradation/defluorination, elimination of $–CF_2–$ groups from the evolving fluorocarbons can also be facilitated by such electrons. Length dependence has also been observed in the distinctive natures of the available degradation pathways when telomeric versions of PFASs are compared with their natural versions [5, 6]. Thus, free electrons, the resultant solvated electrons, $H^\cdot$, and $OH^\cdot$ radicals play a vital role in the PFAS/PFOS degradation process efficiency.

Recently, it was demonstrated that using reactive gases such as O_2 in direct current (DC) plasma created both oxidative and reactive species, thus, substantially improving PFAS degradation efficiency [30]. In a similar pioneering effort, Lewis et al. utilized a reverse vortex flow gliding arc plasma system that is submerged in PFAS contaminated water (Fig. 8.6), where interactions between circulating water and plasma species including oxygen radicals resulted in 75% PFAS removal. The plasma power used, 255 W (918 kJ/L or 918 KWh/m^3), in this study was relatively higher than in other plasma approaches [31]. Further, Singh et al. used a pilot-scale spark discharge plasma reactor setup (66×48 cm), where the short-lived corona streamers generated above the water interacted with Ar that extended further and

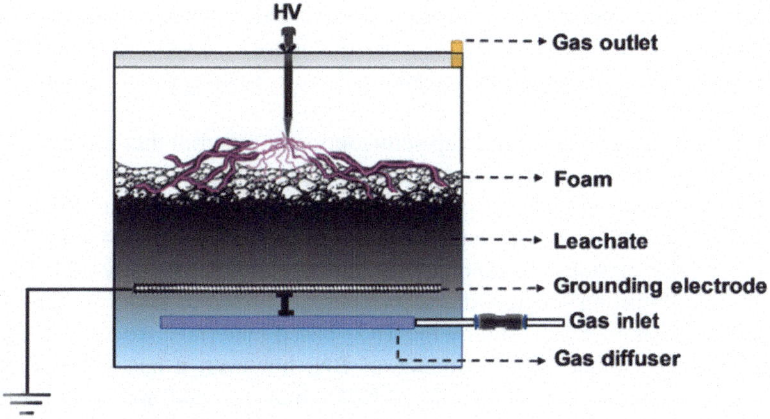

Fig. 8.6 Schematic of the plasma reactor. Reproduced from Ref. [31] and reprinted with permission from The Royal Society of Chemistry

created active radicals, which together with the streamers on contact with bubbling setup initiated the PFOS/PFOA defragmentation [33]. Unlike other plasma techniques, the corona streamers produced with this method increased the water temperature; hence, it was imperative to constantly recirculate water to avoid vaporization. This approach showed 36–99% overall PFAS removal efficiency and energy (2–60 KWh/m^3) required was one order lower than the glide arc plasma discharge approach. In comparison, other competing PFAS remediation approaches such as PSf, photochemical oxidation, and sonolytic processes require ~5000 KWh/m^3 energy, which is 1–2 orders higher than that required by plasma-based approaches [2].

Among all the pursued PFAS remediation approaches, plasma-based approach is in its infancy and has already shown significant promise and improvement in terms of cost and efficiency. While scale-up and cost still remains a challenge for this route, incorporating other remediation approaches in combination with plasma is expected to bring out balance in the cost and efficiency and scale-up.

8.1.9 Photocatalysis for PFAS Degradation

Photocatalysis or photolysis is an advanced reduction/oxidation process where incident photons of light generate holes and electrons which then react with water molecules to form reactive oxygen species ($OH^{\cdot}$ and $O_2{}^{-}$) and other hydrated species such as hydrated electrons that react with PFASs to reduce them into shorter chain segments. Highly effective photosensitive catalytic material is a critical requirement for photocatalysis of PFASs. Toward this, the most used TiO_2-based photocatalytic materials were extensively studied for PFOA/PFOS remediation with little success due to limited electron/hole separation. Alternatively, wide band gap semiconductors that have better electron/hole separation efficiencies have been attempted in the recent past including boron nitride (BN), iron-oxides, Ga_2O_3, In_2O_3, $Bi_3O(OH)(PO_4)_2$ (BiOHP), etc., either as a single-phase material or in combination (in composites) to enhance the efficiency.

BN is a well-known wide band gap semiconductor material that is recently being used for water treatment such as photocatalytic water splitting, where BN is believed to have higher adsorption affinity toward organic moieties such as PFOA/PFOS [107]. Recently, it was shown that under UV illumination (at 254 nm, for 2 h), the commercially procured hexagonal BN (bulk) with intentionally added surface defects (via ball milling) degraded a carboxylic acid form of PFAS by 20% which was significantly higher than that by the TiO_2-based materials. Added defects were attributed for the enhanced performance of BN; however, it has to be noted that careful surface engineering of exfoliated h-BN may provide higher PFOA/PFOS degradation efficiencies [108].

Bismuth-based photocatalytic materials such as $BiPO_4$, BiOCl, and $BiFeO_3$ offer higher efficiency compared with TiO_2 in PFOS/PFOA degradation [109]. In the presence of suitable UV irradiance, BiOHP was able to degrade PFOA 15 times faster than TiO_2, however, an increase in pH value decreased adsorption efficiency and can have significant impact on the performance of BiOHP. In order to utilize the benefits of BiOHP, while mitigating the adsorption limitation, nanocomposites of BiOHP with carbonaceous materials were explored. Toward this, BiOHP-carbon nanosphere composites were fabricated that showed more than 90% adsorption of PFOA and its subsequent degradation (32.5% fluorine converted to fluoride) via electron exchange [110].

Introduction of Fe(III) sensitive to 254 nm (UV) has shown great promise (47% efficiency) in PFOA/PFOS degradation, where the Fe(III) was found to form complexes with the PFOA to enable an efficient electron transfer from PFOA to Fe(III) and the subsequent breakdown of PFOA [22, 111]. Relative to Fe(III), iron hydroxides have the ability to absorb photons from UV-600 nm and have greater affinity (absorption) toward PFOA molecules, thus facilitating better electron transfer from PFOA than for Fe(III) and photodegradation of PFOA [66]. The attractive tendency of PFOS/PFOA molecules toward carbonaceous materials has become a natural choice for researchers to blend the photocatalytic material with carbon to improve the electron transfer efficiency and readily decompose PFOS/PFOA molecules. Xu et al. used FeO with carbon sphere nanocomposite to demonstrate ~57% conversion of fluorine to fluoride under lab-created solar light [34]. The above approach demonstrated that the photocatalytic approach could effectively decompose PFOS/PFOA at normal solar irradiance; however, this technology is still limited due to challenges in scale-up and the ability of sun light reaching the active site in a turbulent solution, requirement of reactivation of carbonaceous materials, etc. Recent research explored the possibility of using planar carbon structures such as reduced graphene oxide (r-GO) and graphene to effect such electron transfers. In one such study, a composite catalyst based on TiO_2 and r-GO (95% TiO_2/5% r-GO) was used for photolytic decomposition of PFOA (Fig. 8.7). The study found that r-GO provided the appropriate band structure and band gap and possibly acted as an

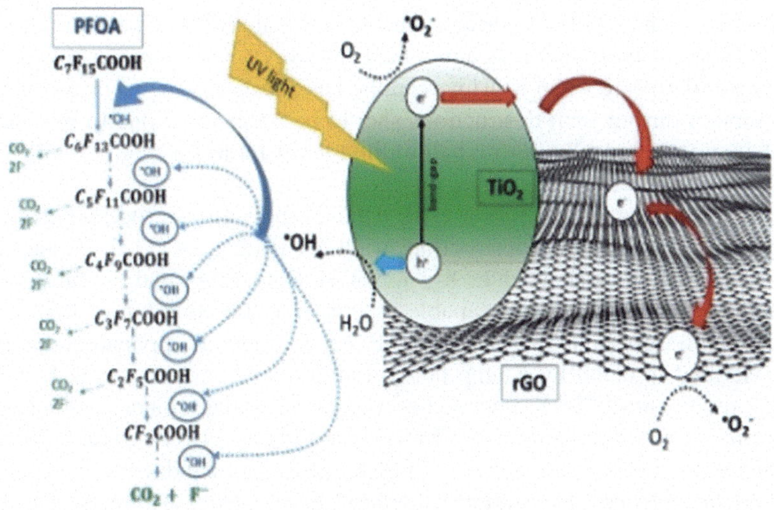

Fig. 8.7 Photocatalytic pathways of PFOA decomposition using the TiO_2–r-GO catalyst. Reproduced from Ref. [18] and reprinted with permission from Elsevier

electron acceptor, thus avoiding a high recombination of electron/hole pairs, thereby enhancing the photolytic decomposition of PFOA.

8.1.10 Ball Milling

Ball milling or the mechanically-enabled chemical transformation of materials to their end fates is one of the simplest and time-tested method in use for centuries. The application of this technique to PFAS destruction is slowly gaining steam and is a significant development, since a fundamental understanding of the chemical transformations that occur in a PFAS molecule as a function of the milling speed, temperature, and media (i.e., the material comprising the balls) is essential. Additionally, the material of the container and the atmosphere in which the milling occurs are both important as these factors also control the science of milling. In this method, a cylindrical container housing milling media such as steel balls and a solid sample of the material under study is rotated, and the impact of balls with solid particles reduces the particle sizes, and a reaction takes place at the surface of the ball mill [112]. Zhang et al. applied this method to soil samples containing PFOS and PFOA as a mechanochemical means to achieve the destruction of these compounds [112]. They observed that 100% destruction of PFOA occurred at 3 h and 99.88% destruction of PFOS occurred at 6 h. In fact, the use of potassium hydroxide (KOH) as a co-reagent during ball milling was found to cause the complete destruction of both PFOA and PFOS and result in higher fluoride yields. X-ray diffraction (XRD) analysis showed that the final products of PFOS destruction contained KF and K_2SO_4.

Other PFASs, such as PFH_xS, perfluorobutane sulfonate (PFBS), perfluorodecanoic acid (PFDA), and perfluorododecanoic acid (PFDoA), were also destroyed effectively with ball milling when KOH was applied as a reagent. The mechanochemical transformation during such destructions should be accelerated due to the input of thermal energy as typically during ball milling temperatures as high as 450 °C can result in the container.

In fact, a recent study by Turner et al. detailed the various processes that occur during ball milling-enabled destruction of PFAS (Fig. 8.8) [113]. Turner et al. suggested that co-reagents, such as KOH, NaOH, CaO, SiO_2, and Fe–SiO_2 as used by Zhang et al. and others, are capable of initiating the emission of high-energy particles required for bond dissociation of PFAS compounds, a phenomenon previously described in detail by Kaupp [114]. Furthermore, in the account by Turner et al., the treatment of PFAS-amended porous media using a planetary ball mill in the presence of water and absence of KOH co-milling agent is discussed. The exact mechanism(s) and reaction pathway(s) of PFAS degradation induced via ball milling are unknown; however, the authors suggested numerous possibilities. The physical impact of grinding balls is believed to activate inorganic co-milling reagents and facilitate electron transfer processes that decompose PFAS and other organic molecules. The authors hypothesized that fracture of PFAS-amended porous media resulted in free electrons, high-energy neutrals, ions, photons, radical species, and/or dangling bonds which generate a reductive environment to attack and degrade PFASs. Furthermore, the authors suggest that the breaking of oxygen vacancies in SiO_2 and the formation of dangling bonds lead to the effective generation of free electrons and radicals required for destroying PFOA via ball milling. The generation of dangling bonds is suggested to lead to highly reactive radicals or non-bridging oxygen-hole centers, which in addition to oxygen vacancies paired with Lewis acid sites in the co-milling agent alumina, are effective in generating free electrons or radicals and are reported to be why alumina is a successful co-milling agent for PFOA destruction [115].

Interestingly, the use of ball milling to modify GAC has been performed to introduce magnetic components such as Fe_3O_4 into GAC adsorbents to facilitate the separation of PFAS adsorbed GAC. These magnetically-modified adsorbents (MAC) can then be easily separated from mixtures in contact with an external magnetic source. Furthermore, the generation of MAC using a 1:3 ratio of Fe_3O_4:GAC was found to have optimal adsorption capacity for PFOS, PFOA, PFHxS and PFBS [116]. Thus, opportunities exist to incorporate ball milling into PFAS remediation efforts, however, more research at all levels is required for practical and effective application toward PFAS.

8.1.11 Biodegradation

The following provocative question was posed in a recent article by Lim in ACS Central Science! It asked: “Can Microbes Save Us from PFAS?” [117]. Setting aside

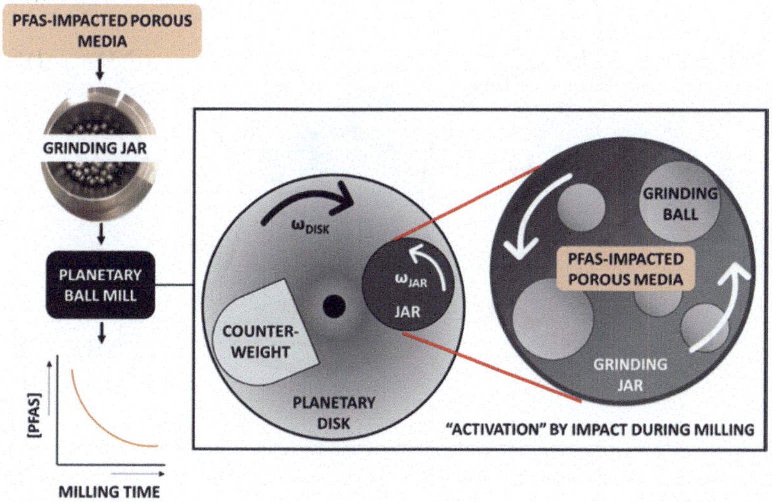

Fig. 8.8 Schematic of the interaction of PFAS-impacted porous media and its activation during planetary ball milling. Reproduced from Ref. [113] and reprinted with permission from Elsevier

skepticism of whether bioremediation of PFAS is indeed viable, this article cited two of the recent works on bioremediation of PFASs. Firstly, a study by Alvarez and coworkers from Rice University investigated the generation of superoxide from xanthine oxidase found in various superoxide-generating Pseudomonas bacteria. The results indicated that both superoxide and hydroxyl radicals generated from decomposing hydrogen peroxide did not induce the degradation of PFOA [118]. On the contrary, a second study performed by Peter Jaffe's team from Princeton involved the study of A_6, a strain of the microbe Acidimicrobium, and application toward PFAS degradation. Their study observed the microbe/autotroph undergo the Feammox reaction for PFAS defluorination (Fig. 8.9) [119]. As shown, the microbe facilitates the transfer of electrons from ammonium ions to iron (III) ions in acidic soil, oxidizing ammonium to nitrite while reducing ferric iron. Incubations with pure and PFAS enrichment cultures of A_6 conducted in the presence of PFOA or PFOS at 0.1 mg/L and 100 mg/L, resulted in defluorination of the PFAS substrate (either PFOA or PFOS). Therefore, Jaffe's team showed that A6 can indeed defluorinate PFOA/PFOS, leading to the possibility that microbes could be employed to save us from PFAS contamination [26].

Recently, Men and coworkers reported on C–F bond cleavage and fluoride production in two hexyl per- and polyfluorinated compounds via reductive defluorination by microbes using lactate as the electron donor, and the PFAS as the electron acceptor [120]. The minor phylogenetic groups in the community other than Dehalococcoides was found to be responsible for the reductive defluorination.

Studies on the PFAS-impacted legacy of the fire-fighting training area in Canada have been expanded to develop an in-depth assessment of the relationship between PFAS and in situ microbial communities. These studies showed that lineages within

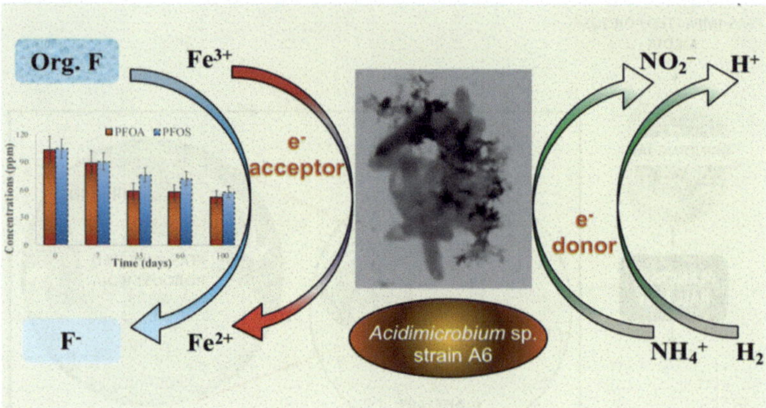

Fig. 8.9 Schematic representation of cometabolic Feammox and PFAS defluorination by *Acidimicrobium* sp. Strain A6. Reproduced from Ref. [112] and reprinted with permission from The American Chemical Society

the Oxalobacteraceae family had strong negative correlations with PFAS, while the Desulfococcus genus has strong positive correlations. Results also suggested that a range of genera might have been stimulated at low to mid-range concentrations (e.g., Gordonia and Acidimicrobium), with some genera being potentially inhibited at high PFAS concentrations. Such distinctive correlations may provide unique insights such as positive correlations for the development of biodegradation technologies for PFAS-impacted sites and negative correlations for the potential negative effects of PFAS on ecosystem health [121].

8.1.11.1 AI/ML Approaches

In the past decade, the advancement of artificial intelligence/machine learning (AI/ML) algorithms has been impressive virtually in every imaginable field including finance, commerce, medicine, manufacturing, and many scientific disciplines. However, the explicit use of a greater degree of ML algorithms has only recently gained steam in synthetic and reactive organic chemistry applications. Other than for use in classification of organic compounds, the thrust has also been to utilize ML algorithms to predict stabilities and reactivities of chemical compounds quicker than by DFT calculations and molecular modeling. With regard to classification of PFAS compounds, a recent ML study addressed the inconsistency in the terminology recommendations for naming PFAS based on broad categorization rather than by simple molecular structures. Recent advancements in chemical analysis were leveraged and used in combination with expert knowledge and cheminformatics approaches to develop a workflow to classify PFAS. Firstly, PFAS compounds are subdivided based on the following chemical structure C_nF_{2n+1}–X–R, where X is CO, SO_2, CH_2, and CH_2CH_2, using the "splitPFAS" approach. Secondly, the "splitPFAS" output was used with the "Classy Fire" approach to categorize and name PFAS using five scenarios of original and simplified structures. The described workflow was applied to 770 PFAS compounds from the OECD PFAS list. While some mixed results exist, the results of the new PFAS classification workflow showed that chem-informatics approaches can have the potential for facilitating consistent categorizing of PFAS [41].

In another ML study, quantitative structure–activity relationship (QSAR) models were developed to explore the potential hazards of PFAS, such as bioactivity, bioaccumulation, and toxicity. A database of PFAS-specific bioactivity was developed based on twenty-six bioassays and contains information on ~1000 PFAS. Using the OECD list, biologically active PFAS compounds were categorized into two classes, fluorotelomer-based PFAS, and PFCA and PFCA precursors. Furthermore, five different ML models were prepared using a variety of conventional methods such as random forest, multitask neural network (MNN), and advanced graph-based models such as graph convolutional network. The models were trained using the developed PFAS database and tested using a validation dataset. The study found that the best performance was achieved using the MNN and graph-based ML models [122].

A recent study used supervised ML classifiers for allocating the source of PFAS contamination based on patterns identified in component concentrations. A dataset containing PFAS component concentrations in 1197 environmental water samples was assembled based on data from sites from around the world and tested with three conventional ML classifiers, namely extra trees, support vector machines, K-neighbors, and one multilayer perceptron feedforward deep neural network. Of the methods tested, the deep neural network and extra trees were found to exhibit particularly high performance at classification of samples from a range of sources. Thus, methods that function on completely different principles providing similar predictions supported the hypothesis that patterns exist in PFAS water sample data that can allow forensic source allocation. Furthermore, the results of the work supported the notion that supervised machine learning may have substantial promise as a tool for forensic source allocation [40].

An important first application of ML on PFAS for predicting and rationalizing C–F bond dissociation energies to aid in their efficient treatment and removal was recently reported (Fig. 8.10) [35]. Using a variety of ML algorithms including random forest, least absolute shrinkage and selection operator regression, and feedforward neural networks, extremely accurate predictions for C–F bond dissociation energies (with deviations less than 0.70 kcal/mol) that are within chemical accuracy of the PFAS reference data were obtained. This ML approach was found to be extremely efficient, requiring less than 10 min to train the data and less than a second to predict the C–F bond dissociation energy of a new compound using only simple chemical connectivity in a PFAS structure without recourse to a computationally expensive quantum mechanical calculation or a three-dimensional structure. The authors also presented an unsupervised ML algorithm that could automatically classify and rationalize chemical trends in PFAS structures that would otherwise have been difficult to humanly visualize or process manually [35].

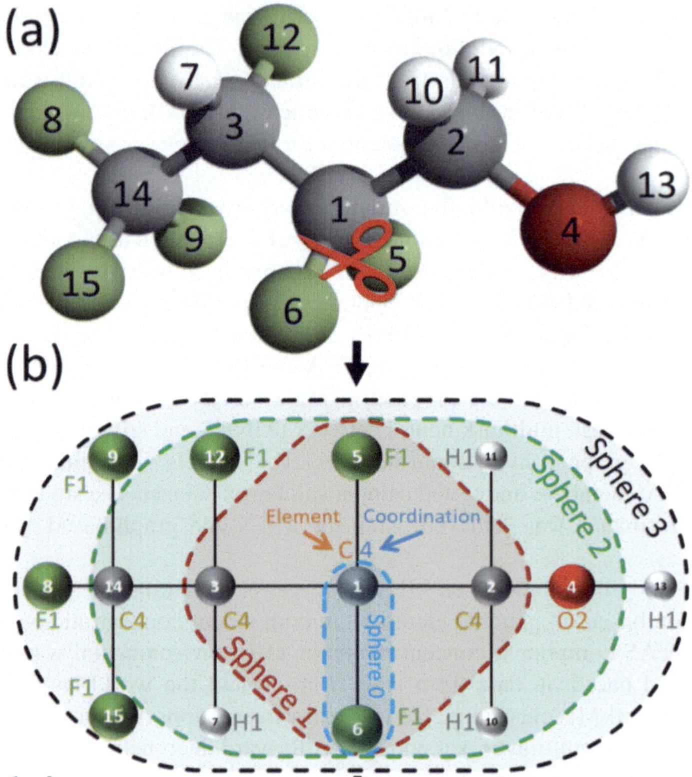

(c)

	C2	C3	C4	H1	N1	N2	N3	N4	O1	O2	S1	S2	S3	S4	F1
Sphere 0			1												1
Sphere 1			2												1
Sphere 2			1	3						1					1
Sphere 3				1											3

Fig. 8.10 Schematic of the chemical descriptors and data used by machine learning algorithms to autonomously and rapidly process C–F bond dissociation energies in general PFAS structures. Panel **a** depicts a specific target C–F bond to be dissociated, and panel **b** shows the atom labeling scheme and spheres used to construct the point descriptor table in panel **c**. Reproduced from Ref. [35] and reprinted with permission from The American Chemical Society

8.2 Conclusions

Recent advances in various degradation methods for remediating the collective recalcitrant group of PFAS contaminants give hope that more efficacious, scalable, and therefore, practically viable technologies will be developed in the near future. This will certainly be a positive step toward containing these pernicious environmental pollutants which became introduced in applications in the first place due to their impressive chemical stability. Understandably, the methods that can overcome the chemical stability of PFASs will have to exploit any kinetic and thermodynamic aspects of reactions that will enable the destruction of these materials. Also, such science should lead to new insights into the fundamentals of the stability of organic compounds in general and could open avenues to produce new stable compounds with more nuanced properties. This could also lead to designing of new PFAS replacements with sufficient enough chemical stability for intended chemical applications.

References

1. Yang L, He L, Xue J, Ma Y, Xie Z, Wu L, Huang M, Zhang Z (2020) J Hazard Mater 393:122405
2. Nzeribe BN, Crimi M, Mededovic Thagard S, Holsen TM (2019) Crit Rev Environ Sci Technol 49:866
3. Berger R, Resnati G, Metrangolo P, Weber E, Hulliger J (2011) Chem Soc Rev 40:3496
4. Bhavsar SP, Fowler C, Day S, Petro S, Gandhi N, Gewurtz SB, Hao C, Zhao X, Drouillard KG, Morse D (2016) Environ Int 94:415
5. Lindstrom AB, Strynar MJ, Libelo EL (2011) Environ Sci Technol 45:7954
6. Miralles-Marco A, Harrad S (2015) Environ Int 77:148
7. Lewis RC, Johns LE, Meeker JD (2015) Int J Env Res Public Health 12:6098
8. Kabadi SV, Fisher JW, Doerge DR, Mehta D, Aungst J, Rice P (2020) Toxicol Appl Pharmacol 388:114878
9. Murray CC, Vatankhah H, McDonough CA, Nickerson A, Hedtke TT, Cath TY, Higgins CP, Bellona CL (2019) J Hazard Mater 366:160
10. Liu CJ, Werner D, Bellona C (2019) Environ Sci: Water Res Technol 5:1844
11. Franke V, McCleaf P, Lindegren K, Ahrens L (2019) Environ Sci: Water Res Technol 5:1836
12. Bentel MJ, Yu Y, Xu L, Li Z, Wong BM, Men Y, Liu J (2019) Environ Sci Technol 53:3718
13. Maga D, Aryan V, Bruzzano S (2021) Environ Toxicol Chem 40:947
14. Solo-Gabriele HM, Jones AS, Lindstrom AB, Lang JR (2020) Waste Manage 107:191
15. Mühle J, Ganesan AL, Miller BR, Salameh PK, Harth CM, Greally BR, Rigby M, Porter LW, Steele LP, Trudinger CM, Krummel PB, O'Doherty S, Fraser PJ, Simmonds PG, Prinn RG, Weiss RF (2010) Atmos Chem Phys 10:5145
16. Fang C, Sobhani Z, Niu J, Naidu R (2019) Chemosphere 219:36
17. Wang S, Yang Q, Chen F, Sun J, Luo K, Yao F, Wang X, Wang D, Li X, Zeng G (2017) Chem Eng J 328:927
18. Gomez-Ruiz B, Ribao P, Diban N, Rivero MJ, Ortiz I, Urtiaga A (2018) J Hazard Mater 344:950
19. Mitchell SM, Ahmad M, Teel AL, Watts RJ (2014) Environ Sci Technol Lett 1:117
20. Bruton TA, Sedlak DL (2017) Environ Sci Technol 51:13878
21. Bruton TA, Sedlak DL (2018) Chemosphere 206:457
22. Ohno M, Ito M, Ohkura R, Mino AER, Kose T, Okuda T, Nakai S, Kawata K, Nishijima W (2014) J Hazard Mater 268:150
23. Douvris C, Ozerov OV (2008) Science 321:1188
24. Luo Q, Yan X, Lu J, Huang Q (2018) Environ Sci Technol 52:10617
25. Presentato A, Lampis S, Vantini A, Manea F, Daprà F, Zuccoli S, Vallini G (2020) Microorganisms 8:92
26. Huang S, Jaffé PR (2019) Environ Sci Technol 53:11410
27. Albert K, Hsieh P-Y, Chen T-H, Hou C-H, Hsu H-Y (2020) J Hazard Mater 384:121491

28. Cao H, Zhang W, Wang C, Liang Y (2020) Ultrason Sonochem 69:105245
29. Lei Y-J, Tian Y, Sobhani Z, Naidu R, Fang C (2020) Chem Eng J 388:124215
30. Hayashi R, Obo H, Takeuchi N, Yasuoka K (2015) Electrical Eng Japan 190:9
31. Lewis AJ, Joyce T, Hadaya M, Ebrahimi F, Dragiev I, Giardetti N, Yang J, Fridman G, Rabinovich A, Fridman AA, McKenzie ER, Sales CM (2020) Environ Sci: Water Res Technol 6:1044
32. Singh RK, Fernando S, Baygi SF, Multari N, Thagard SM, Holsen TM (2019) Environ Sci Technol 53:2731
33. Singh RK, Multari N, Nau-Hix C, Anderson RH, Richardson SD, Holsen TM, Mededovic Thagard S (2019) Environ Sci Technol 53:11375
34. Xu T, Ji H, Gu Y, Tong T, Xia Y, Zhang L, Zhao D (2020) Chem Eng J 388:124230
35. Raza A, Bardhan S, Xu L, Yamijala SSRKC, Lian C, Kwon H, Wong BM (2019) Environ Sci Technol Lett 6:624
36. Siriwardene DP, Crimi M, Holsen TM, Bellona C, Divine C, Dickenson E (2019) Environ Eng Sci 36:453
37. Park M, Wu S, Lopez IJ, Chang JY, Karanfil T, Snyder SA (2020) Water Res 170:115364
38. Meng P, Jiang X, Wang B, Huang J, Wang Y, Yu G, Cousins IT, Deng S (2020) J Hazard Mater 386:121981
39. Johnson JK, Hoffman CM, Smith DA, Xia Z (2019) ACS Omega 4:8001
40. Kibbey TCG, Jabrzemski R, O'Carroll DM (2020) Chemosphere 252:126593
41. Sha B, Schymanski EL, Ruttkies C, Cousins IT, Wang Z (2019) Environ Sci: Proc Impacts 21:1835
42. Kumarasamy E, Manning IM, Collins LB, Coronell O, Leibfarth FA (2020) ACS Cent Sci 6:487
43. Laura del Moral L, Choi YJ, Boyer TH (2020) Water Res 178:115846
44. Wang W, Mi X, Zhou Z, Zhou S, Li C, Hu X, Qi D, Deng S (2019) J Colloid Interface Sci 557:655
45. Dixit F, Barbeau B, Mostafavi SG, Mohseni M (2020) Water Res 183:116098
46. Barpaga D, Zheng J, Han KS, Soltis JA, Shutthanandan V, Basuray S, McGrail BP, Chatterjee S, Motkuri RK (2019) Inorg Chem 58:8339
47. Fulong CRP, Guardian MGE, Aga DS, Cook TR (2020) Inorg Chem 59:6697
48. Askeland M, Clarke BO, Cheema SA, Mendez A, Gasco G, Paz-Ferreiro J (2020) Chemosphere 249:126072
49. Zhang D, He Q, Wang M, Zhang W, Liang Y (2021) Environ Technol 42:1798
50. Turner BD, Sloan SW, Currell GR (2019) Chemosphere 229:22
51. Silvani L, Cornelissen G, Botnen Smebye A, Zhang Y, Okkenhaug G, Zimmerman AR, Thune G, Sævarsson H, Hale SE (2019) Sci Total Environ 694:133693
52. Xiao F, Jin B, Golovko SA, Golovko MY, Xing B (2019) Environ Sci Technol 53:11818
53. Stebel EK, Pike KA, Nguyen H, Hartmann HA, Klonowski MJ, Lawrence MG, Collins RM, Hefner CE, Edmiston PL (2019) Environ Sci: Water Res Technol 5:1854
54. Lath S, Knight ER, Navarro DA, Kookana RS, McLaughlin MJ (2019) Chemosphere 222:671
55. Yan B, Munoz G, Sauvé S, Liu J (2020) Water Res 184:116166
56. Liu F, Hua L, Zhang W (2020) Environ Int 143:105969
57. Sörengård M, Niarchos G, Jensen PE, Ahrens L (2019) Chemosphere 232:224
58. Chen X, Vanangamudi A, Wang J, Jegatheesan J, Mishra V, Sharma R, Gray SR, Kujawa J, Kujawski W, Wicaksana F, Dumée LF (2020) Water Res 182:116010
59. Tenorio R, Liu J, Xiao X, Maizel A, Higgins CP, Schaefer CE, Strathmann TJ (2020) Environ Sci Technol 54:6957
60. Chen Z, Li C, Gao J, Dong H, Chen Y, Wu B, Gu C (2020) Environ Sci Technol 54:5178
61. Su Y, Rao U, Khor CM, Jensen MG, Teesch LM, Wong BM, Cwiertny DM, Jassby D (2019) ACS Appl Mater Interfaces 11:33913
62. Van Hoomissen DJ, Vyas S (2019) Environ Sci Technol Lett 6:365
63. Yamijala SSRKC, Shinde R, Wong BM (2020) Phy Chem Chem Phys 22:6804
64. Parenky AC, Souza NGd, Nguyen HH, Jeon J, Choi H (2020) J Environ Eng 146:06020003

65. Parenky AC, de Souza NG, Asgari P, Jeon J, Nadagouda M, Choi H (2020) Environ Eng Sci 37:472
66. Gao X, Chorover J (2012) Environ Chem 9:148
67. Zhang W, Zhang D, Liang Y (2019) Environ Pollut 247:266
68. Liu L, Liu Y, Gao B, Ji R, Li C, Wang S (2020) Crit Rev Environ Sci Technol 50:2379
69. Ateia M, Attia MF, Maroli A, Tharayil N, Alexis F, Whitehead DC, Karanfil T (2018) Environ Sci Technol 5:764
70. Saleh NB, Khalid A, Tian Y, Ayres C, Sabaraya IV, Pietari J, Hanigan D, Chowdhury I, Apul OG (2019) Environ Sci: Water Res Technol 5:198
71. Deng S, Zhang Q, Nie Y, Wei H, Wang B, Huang J, Yu G, Xing B (2012) Environ Pollut 168:138
72. Chen X, Xia X, Wang X, Qiao J, Chen H (2011) Chemosphere 83:1313
73. Liu L, Liu Y, Li C, Ji R, Tian X (2018) Environ Sci and Pollut Res 25:15507
74. Nakata K, Fujishima A (2012) J Photochem Photobiol C: Photochem Rev 13:169
75. Dillert R, Bahnemann D, Hidaka H (2007) Chemosphere 67:785
76. Yao Y, Li G, Ciston S, Lueptow RM, Gray KA (2008) Environ Sci Technol 42:4952
77. Panchangam SC, Yellatur CS, Yang J-S, Loka SS, Lin AYC, Vemula V (2018) J Environ Chem Eng 6:6359
78. Pearton SJ, Yang J, CaryIV PH, Ren F, Kim J, Tadjer MJ, Mastro MA (2018) Appl Phys Rev 5:011301
79. Lopes da Silva F, Laitinen T, Pirilä M, Keiski RL, Ojala S (2017) Top Catal 60:1345
80. Mukherji A, Seger B, Lu GQ, Wang L (2011) ACS Nano 5:3483
81. Xiao L, Ching C, Ling Y, Nasiri M, Klemes MJ, Reineke TM, Helbling DE, Dichtel WR (2019) Macromolecules 52:3747
82. Weiss-Errico MJ, O'Shea KE (2019) J Inclusion Phen Macrocyclic Chem 95:111
83. Klemes MJ, Ling Y, Ching C, Wu C, Xiao L, Helbling DE, Dichtel WR (2019) Angew Chem Int Ed 58:12049
84. Yang A, Ching C, Easler M, Helbling DE, Dichtel WR (2020) ACS Mater Lett 2:1240
85. Wu C, Klemes MJ, Trang B, Dichtel WR, Helbling DE (2020) Water Res 182:115950
86. Wang R, Ching C, Dichtel WR, Helbling DE (2020) Environ Sci Technol Lett 7:954
87. Kim K, Baldaguez Medina P, Elbert J, Kayiwa E, Cusick RD, Men Y, Su X (2020) Adv Funct Mater 30:2004635
88. Ateia M, Erdem CU, Ersan MS, Ceccato M, Karanfil T (2019) Water Res 156:168
89. Liu C, Hatton J, Arnold WA, Simcik MF, Pennell KD (2020) Environ Sci Technol 54:6929
90. Kazemi R, Potts EI, Dick JE (2020) Anal Chem 92:10597
91. Savage DT, Briot NJ, Hilt JZ, Dziubla TD (2021) J Polymer Sci 59:289
92. Gole VL, Sierra-Alvarez R, Peng H, Giesy JP, Deymier P, Keswani M (2018) Ultrason Sonochem 45:213
93. McNamara WB, Didenko YT, Suslick KS (2003) The J Phy Chem B 107:7303
94. Suslick KS (1989) Sci Am 260:80
95. Campbell T, Hoffmann MR (2015) Sep Purif Technol 156:1019
96. Kang J-W, Hung H-M, Lin A, Hoffmann MR (1999) Environ Sci Technol 33:3199
97. Hung H-M, Ling FH, Hoffmann MR (2000) Environ Sci Technol 34:1758
98. Lin J-C, Hu C-Y, Lo S-L (2016) Ultrason Sonochem 28:130
99. Xu B, Ahmed MB, Zhou JL, Altaee A, Wu M, Xu G (2017) Chemosphere 189:717
100. Hori H, Nagaoka Y, Yamamoto A, Sano T, Yamashita N, Taniyasu S, Kutsuna S, Osaka I, Arakawa R (2006) Environ Sci Technol 40:1049
101. Banerjee S, Schmidt J, Talmon Y, Hori H, Asai T, Ameduri B (2018) Chem Commun 54:11399
102. Hori H, Oishi S, Kato H, Kodama R (2020) Indus Eng Chem Res 59:5566
103. Ohring M (2002) Chapter 4—Discharges, plasmas, and ion-surface interactions. In: Ohring M (ed) Materials science of thin films, 2nd edn. Academic Press, San Diego, p 145
104. Hicks RF, Xiawan Y, Mooravej M, Nowling G, Babayan S (2003) Low temperature, atmospheric pressure plasma processing of materials. Digest of Papers Microprocesses and Nanotechnol 2003. 2003 international microprocesses and nanotechnology conference, 29–31 Oct. 2003; 2003; p 40.

105. Patrick V, Anton YN, Christophe L (2016) Progress in physical states and chemical reactions. IntechOpen Limited, London. https://doi.org/10.5772/61830
106. Bruggeman PJ, Kushner MJ, Locke BR, Gardeniers JGE, Graham WG, Graves DB, Hofman-Caris RCHM, Maric D, Reid JP, Ceriani E, Fernandez Rivas D, Foster JE, Garrick SC, Gorbanev Y, Hamaguchi S, Iza F, Jablonowski H, Klimova E, Kolb J, Krcma F, Lukes P, Machala Z, Marinov I, Mariotti D, Mededovic Thagard S, Minakata D, Neyts EC, Pawlat J, Petrovic ZL, Pflieger R, Reuter S, Schram DC, Schröter S, Shiraiwa M, Tarabová B, Tsai PA, Verlet JRR, von Woedtke T, Wilson KR, Yasui K, Zvereva G (2016) Plasma Sources Sci Technol 25:053002
107. Singh B, kaur G, Singh P, Singh K, Sharma J, Kumar M, Bala R, Meena R, Sharma SK, Kumar A (2017) New J Chem 41:11640
108. Duan L, Wang B, Heck K, Guo S, Clark CA, Arredondo J, Wang M, Senftle TP, Westerhoff P, Wen X, Song Y, Wong MS (2020) Environ Sci Technol Lett 7:613
109. Sahu SP, Qanbarzadeh M, Ateia M, Torkzadeh H, Maroli AS, Cates EL (2018) Environ Sci Technol Lett 5:533
110. Xu T, Zhu Y, Duan J, Xia Y, Tong T, Zhang L, Zhao D (2020) Chem Eng J 395:124991
111. Wang Y, Zhang P, Pan G, Chen H (2008) J Hazard Mater 160:181
112. Zhang K, Huang J, Yu G, Zhang Q, Deng S, Wang B (2013) Environ Sci Technol 47:6471
113. Turner LP, Kueper BH, Jaansalu KM, Patch DJ, Battye N, El-Sharnouby O, Mumford KG, Weber KP (2021) Sci Total Environ 765:142722
114. Kaupp G (2009) Cryst Eng Comm 11:388
115. Lv H, Wang N, Zhu L, Zhou Y, Li W, Tang H (2018) Green Chem 20:2526
116. Meng P, Fang X, Maimaiti A, Yu G, Deng S (2019) Chemosphere 224:187
117. Lim X (2021) ACS Central Sci 7:3
118. Javed H, Metz J, Eraslan TC, Mathieu J, Wang B, Wu G, Tsai A-L, Wong MS, Alvarez PJJ (2020) Environ Sci Technol Lett 7:653
119. Clement J-C, Shrestha J, Ehrenfeld JG, Jaffe PR (2005) Soil Biol Biochem 37(12):2323
120. Yu Y, Zhang K, Li Z, Ren C, Chen J, Lin Y-H, Liu J, Men Y (2020) Environ Sci Technol 54:14393
121. O'Carroll DM, Jeffries TC, Lee MJ, Le ST, Yeung A, Wallace S, Battye N, Patch DJ, Manefield MJ, Weber KP (2020) Sci Total Environ 712:135994
122. Cheng W, Ng CA (2019) Environ Sci Technol 53:13970

Chapter 9
Life Cycle Considerations for Per- And Polyfluoroalkyl Substances (PFASs) and the Evolution of Society's Perspective on Their Usage

Asa E. Carre-Burritt and Shubham Vyas

Abstract Per- and polyfluoroalkyl substances (PFASs) are an expansive class of highly-fluorinated anthropogenic organic compounds that were developed for demanding speciality uses. Exceptional chemical properties led them to be incorporated in consumer products, industrial products, and industrial processes. Their extensive production and widespread application ultimately led to environmental release. Certain members of these compounds were found to be ubiquitous throughout the environment and in biota, having been transported to even the most remote locations. Findings of negative health outcomes associated with the biological occurrence of two particular PFASs (perfluorooctanesulfonic acid, or PFOS, and perfluorooctanoic acid) spurred a legislative movement to cope with the potential hazards of PFAS usage. Meanwhile, the scientific community embarked on a diverse research effort to understand PFAS life cycle considerations including production levels and utilization, environmental release and occurrence, environmental transformation, biological exposure and occurrence, and epidemiology. Furthermore, significant efforts are underway to understand how PFAS release can be prevented, and how to remediate contaminated matrices. Many of the properties making PFASs useful cause unique challenges for their ongoing management. An equilibrium has yet to be fully established between the clear utility provided by using PFASs and the associated risk.

9.1 Introduction

The per- and polyfluoroalkyl substance (PFAS) lifecycle begins when fluorite (CaF_2) is mined from the earths crust. Fluorite (Fig. 9.1) is treated with sulfuric acid to generate hydrofluoric acid (HF), the lifeblood of organofluorine chemistry. From there, hydrofluoric acid is used–sometimes with intermediate steps–to transform carbonaceous materials into a great variety of highly-fluorinated per- and polyfluoroalkyl substances, or PFASs. These anthropogenic compounds posses remarkable proper-

A. E. Carre-Burritt · S. Vyas (✉)
Department of Chemistry, Colorado School of Mines, 1500 Illinois St., Golden, CO 80401, USA
e-mail: svyas@mines.edu

M. Shukla et al. (eds.), *Emerging Materials and Environment*,
Challenges and Advances in Computational Chemistry and Physics 37,
https://doi.org/10.1007/978-3-031-39470-6_9

Fig. 9.1 A fluorite (CaF_2) specimen. Although this sample is purple, pure fluorite is colorless and is useful for certain optical applications

ties such as high thermal stability, resistance to oxidation and reduction, low surface energy, and low dielectric constants. Behind the scenes, PFASs have altered the course of society ever since their advent in the early 20th century. Their innumerable applications have ranged from the uranium enrichment process used to create the atomic bomb, to aqueous film forming foams (AFFFs) critical for putting out hydrocarbon fires at civilian airports and military bases, all the way to nonstick food packaging [1, 64].

Meanwhile, on the basis of their production, widespread application, and disposal, great quantities of PFASs made their way into the environment. And, once there, the characteristics that made them so useful caused them to persist. Although many PFASs partially transform in the environment, with the weaker sections of the molecule breaking apart, they leave behind perfluoalkyl acids and perfluoroether carboxylic acids (PFAAs and PFECAs) as dead-end daughter products. Direct PFAA and PFECA production and emission also accounts for a large part of environmental occurrence [8, 32].

Around the turn of the 21 st century, reports of the presence and persistence of one PFAA, perfluorooctanesulfonic acid (PFOS), in human blood and the environment, in conjunction with findings of mammalian toxicity, heralded a rapid shift in perspective on the ecological and epidemiological hazards of PFAS use. Thereafter, a cascade of research poured forth, covering many important PFAS lifecycle considerations including environmental release, transport, and inventory; biological uptake and toxicity; and remediation needs and approaches.

Despite considerable effort, a concrete path forward for PFAS use has not yet been established. This goal is complicated by an incomplete understanding of the behavior and occurrence of the best-studied legacy PFAAs not to mention the changing landscape of other poorly understood PFASs that are being used in their place, or that have also been used historically. PFASs do not have an equivalent for many of their specialty applications and so society must cope with what level of risk and uncertainty is acceptable in exchange for the clear utility they provide [76]. Within this chapter we discuss what PFASs are, how they came to be, and what many of the considerations are that guide their ongoing management. It is our hope to provide a breadth of perspective on the matter while referencing useful reviews and primary research that the reader may consult should they require more depth within a topic.

9.2 PFAS Scope and Naming

Before delving into the genre of PFASs, it is helpful to define the scope of PFASs and name several prominent categories (Fig. 9.2) and individual compounds (Fig. 9.3). In the most general sense, PFASs are alkanes for which at least one carbon center has had its hydrogen atoms replaced with fluorine in the case of polyfluoroalkyl substances, and all the alkyl carbon centers have been fluorinated in the case of perfluoroalkyl substances [8]. PFASs are united by the suite of physicochemical properties imparted by their high degree of fluorination (*vide infra*, Sect. 9.5). However, PFASs can incorporate the vast array of functional groups familiar to organic chemistry, providing a great diversity of structures and individual characteristics [1, 64].

Because the perfluoroalkyl chain is hydro- and lipophobic, addition of a polar group or hydrocarbon moiety provides amphiphilic character to the compound making it surface active, and so there are many PFAS surfactants (as is elbaorated on in Sect. 9.5.1) [18]. Within this category are perfluoroalkanes substituted with sulfonic, carboxylic, or phosphonic acid head groups and these are called perfluoroalkyl acids (PFAAs). Perfluoroalkylsulfonic acids (PFSAs) and perfluoroalkylcarboxylic acids (PFCAs) in particular are incredibly stable and many other PFASs can be transformed into these compounds as degradation products (*vide infra*, Sect. 9.8.1.1). One 8-carbon (C8) PFCA, perfluorooctanoic acid (PFOA), and one 8-carbon PFSA, perfluorooctanesulfonic acid (PFOS) have been of particular industrial and envi-

Classification of Per- and Polyfluoroalkyl Substances (PFAS)

Non-polymeric

Perfluoroalkyl Substances

All alkyl carbons are fluorinated

- Perfluoroalkanes
- Perfluoroalkyl acids (carboxylic, sulfonic, phosphonic and phisphinic acids)
- Perfluoroalkane sulfonamide
- Perfluoroalkane carbonyl fluorides
- Perfluoroalkane sulfonyl fluorides
- Perfluoropolyethers

Polyfluoroalkyl Substances

Some alkyl carbons are fluorinated

- Fluorotelomer alcohols
- Polyfluorinated alkyl acids
- Derivatives of Perfluoroalkane sulfonamides
- Polyfluorinated alkanes

Polymeric

- Perfluoroalkyl polymers (polytetrafluoroethylene, PTFE) (hexafluoropropylene, HFP)
- Polyfluoroalkyl polymers (polyvinylidene fluoride, PVDF)
- Perfluoroalkyl polymers with ether groups
- Mixed fluoropolymers (Viton, Teflon FEP)

Fig. 9.2 PFAS classification hierarchy with some example compound types

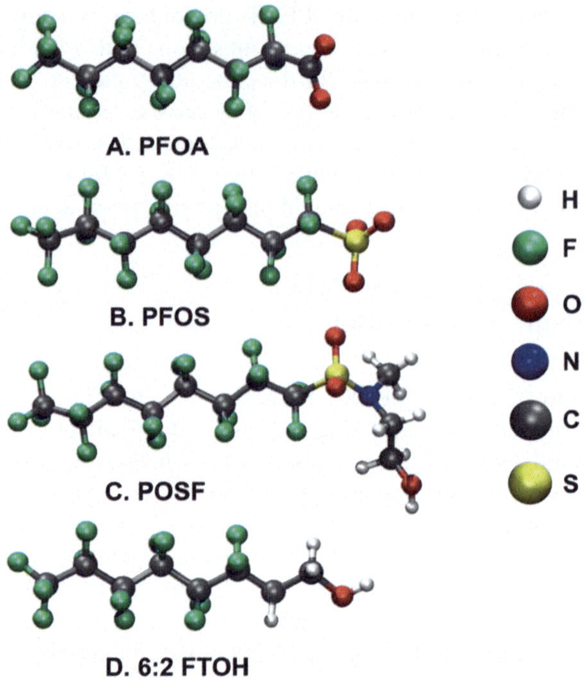

Fig. 9.3 Depictions of several PFASs. **A.** Perfluorooctanoic acid (PFOA), a perfluoroalkyl acid (PFAA). **B.** Perfluorooctanesulfonic acid (PFOS), another example of a PFAA. Both PFOA and PFOS are shown in their deprotonated form in consideration of their low pK_as, which means that they exist as conjugate bases under environmental conditions. **C.** A perfluorooctane sulfonamide derivative, which is an example of a perfluorooctanesulfonyl fluoride-derived compound (POSF). **C.** 6:2 Fluorotelomer alcohol (6:2 FTOH): one of many fluorotelomer compounds. Note the helicity along the fluorinated alkyl chain that is shown for these compounds, which is a characteristic of the perfluoroalkyl moiety [12, 34]

ronmental importance. Although PFAAs are strong acids that are predominantly speciated as conjugate base anions, throughout this chapter they are referred to as acids regardless of their protonation state for the sake of clarity. Perfluorooctanesulfonyl fluoride-derived compounds (POSFs) are some of those which can degrade to PFOS and have been used extensively [8].

Another PFAS that should be given special note is the fluorotelomer alchohol (FTOH), abbreviated as m:n FTOH, where **m** describes the number of fluorinated carbon centers and **n** describes number of intervening methylene units between the perfluoroalkyl moiety and the hydroxyl group. These are christened as such because they are ultimately derived from the telemorization of tetrafluoroethylene with iodopentafluoroethane (*vide infra*, Sect. 9.4) [8].

Beyond small molecules, polymers also make up a large group of PFASs. There are several main categories: first those with a fluorinated carbon backbone, such as polytetrafluoroethylene (PTFE), which are fluoropolymers. Polymers that have a backbone consisting of alternating perfluoroalkyl segments and oxygen atoms are perfluoropolyethers and these are generally oligomeric. Then there are side-chain fluorinated polymers which are organic polymers that have been decorated with PFAS side-chains. An excellent framework of PFAS classification is provided by Buck et al. [8].

9.3 Discoveries Leading to PFAS Synthesis

Developing the synthetic tools for commercial PFAS production took well over a century. During this time, progress was beset by the voracious reactivity of F_2 and the hazards presented by handling HF. The following are a few important landmarks during this journey [1]. As early as 1764, attempts were made to commercially produce anhydrous HF by treating CaF_2 with H_2SO_4 to yield 2 HF and $CaSO_4$. The first organofluoride, CFH_3, was produced in 1835 by Dumas and Péligot. In 1860, the first reported halogen exchange reaction–perhaps the most important C-F bond-forming approach, even contemporarily–occurred when Borodin treated benzoyl chloride with KHF_2 in a platinum retort. A momentous accomplishment for science came in 1886 when Moissan and Frémy generated F_2 by electrolysis of anhydrous HF in the presence of KF. Around 1890, F. Swarts began impactful work on halogen exchange of alkyl chlorides and bromides by treating them with SbF_3Br_2. This work was advanced by Midgley, Henne, and McNary who selectively synthesized CF_2Cl_2 as a refrigerant, later known as Freon®, which was an excellent refrigerant that eventually became notorious for depleting the ozone layer. F. Swartz also established an olefin synthesis route by haloethane dehalogenation using zinc, which preferentially eliminates halogens other than fluorine. Later on, this process was important for the discovery of polytetrafluoroethylene (PTFE) (*vide infra*).

Only 3 perfluoroalkanes and several perfluoroalkyl derivatives had been isolated and appropriately characterized by 1937 when Joe Simmons developed a process for HgF_2-mediated reaction of carbon with F_2 to produce mixtures of saturated C1-C7 linear or branched perfluoroalkanes [1]. Simmons discovered these compounds had high thermal and oxidative stability, and in 1940 Simmons sent his ca. 2 mL stock of perfluoroalkanes to Manhattan project scientists who verified its resistance towards degradation by UF_6, in contrast to all other compounds that had been tested. UF_6 is a volatile species that was being used to concentrate ^{235}U for use in the atomic bomb by a gas diffusion process. Consequentially, a huge secret effort was initiated to advance perfluorinated molecule production for use in various aspects of the gas diffusion process, thereby providing the foundation for modern PFAS production.

In 1938, during the interlude between Simmons' initial perfluoroalkane synthesis and the point at which he sent his material to Manhattan project scientists, PTFE was discovered by Plunkett [1]. Plunkett intended to synthesize a new refrigerant using tetrafluoroethylene (TFE) (produced by the Zn-dehalogenation of CF_2ClCF_2Cl), which had been stored in small steel cylinders. Plunkett found that instead of containing the expected TFE gas, these cylinders contained a white powder that was identified to be the derived polymer, PTFE. In its own right, PTFE played an important role in uranium enrichment: it was formed into componentry including reactor linings, gaskets, etc.

The chemistries of perfluoroalkanes developed during this era of rapid discovery eventually became indispensable for peace-time nuclear energy production as well [1], not to mention numerous other important applications highlighted later in this chapter (Sect. 9.6).

9.4 Commercial PFAS Production

Following the expedited research into PFASs during the Manhattan project, a collection of modern synthetic approaches emerged. For commercial production, PFAS small molecules and fluids are typically produced either by direct fluorination of hydrocarbons and hydrocarbon derivatives or by the telomerization/oligomerization of unsaturated or cyclic organofluorine monomers. PFAS plastics, meanwhile, are commonly produced either through the polymerization of fluorinated monomers to yield fluoropolymers or by incorporating poly- or perfluoroalkyl moieties as polymer side-chains [1].

Electrochemical fluorination (ECF) (using HF) and the cobalt trifluoride process (using F_2) both transform hydrocarbons or hydrocarbon derivatives into the analogous PFASs. Both of these technologies were developed during the Manhattan project [1, 64]–In fact, the Simon's ECF process was a wartime secret which was not described publicly until 1949 at which point 3M held exclusive rights to the process and already had pilot-scale production [1].

In the Simons' ECF process, hydrocarbons are dissolved in anhydrous HF and fluorinated at the anode, which is typically nickel, and the cell potential is 4.5–7 V. The electrolyte consists of protonated organic substrate and sometimes additives. During ECF a film forms at the anode which has a different nature in the presence or absence of organic substrate. This film appears to hinder electron transfer from the anode to the bulk electrolyte and substrates must diffuse to the film where they are fluorinated. Mass transport within this film may be rate limiting. The product distribution of ECF is broad and suggests a mechanism involving radical species and cations [1]. The large percentage of branched isomers for PFAAs produced by ECF has been used to fingerprint PFASs produced by ECF [8, 44, 65]. The Simons' ECF process appears to be the most important ECF process within the global economy although the CAVE-Philips process is also described in *Organofluorine Chemistry Principles and Commercial Applications*, which has a narrow product distribution and is useful for volatile substrates [1].

Commercial products directly produced by Simons' ECF are either those with acid functional group (e.g., -CO_2^- or -SO_3^-); or inert fluids like perfluoroalkanes and cyclic alkanes, ethers, and tertiary amines (which are no longer basic due to the inductively withdrawing perfluoroalkyl groups). PFCAs and PFSAs are produced by hydrolysis of the corresponding acyl fluoride (produced according to Eq. 9.1) or sulfonyl fluoride [1]. Acyl fluorides can be used to form esters, amides, and other compounds, while the esters so formed can be hydrogenated. Sulfonyl fluorides are often substituted with an amine, which can be a handle for further derivitization or this step may form the final POSF product [18].

$$C_7H_{15}COCl + 16HF \xrightarrow{ECF} C_7F_{15}COF + HCl + 16H_2 \tag{9.1}$$

ECF was the predominant method of ammonium PFOA production between 1947 and 2002, where the USA and Belgium were the major producers, followed by Italy and then Japan. A smaller fraction of (ca. 10–20%) was produced by TFE telomerization (*vide infra*) in Germany and Japan. Circa 1999, the yearly estimated global ammonium PFOA production was ca. 260 ton, and as of 2006 the estimated historical production was 3600–5700 ton [65]. Comparatively large quantities of POSF have been produced by ECF: From 1949 to 2002, 3M company produced ca. 78% of global POSF through ECF: up to 3665 ton y^{-1}, with major manufacture cites again in USA and Belgium. Between 1970–2002, it was estimated that 96,000 ton of POSF were produced by all methods, alongside another ca. 26,500 ton of unwanted byproducts and waste [63].

Although Simons' ECF is theoretically an efficient approach to hydrocarbon fluorination because it does not require F_2 production, its broad distribution of products is an important consideration. The cobalt trifluoride process using F_2 has better selectivity [1]. In this process, the large exothermicity of carbon fluorination is controlled by introducing the intermediate step of CoF_2 oxidation by F_2 to yield CoF_3 which is then used to fluorinate the organic substrate [1, 64]. Historically, the cobalt trifluoride process has been important for high-boiling point perfluoroalkane production [1].

Beyond ECF, TFE telomerization with iodopentafluoroethane presents another commercially important approach to producing functionalized PFASs [1]. In this process, iodopentafluoroethane is produced according to reaction 9.2:

$$5C_2F_4 + IF_5 + 2I_2 \xrightarrow{catalyst} 5C_2F_5I \tag{9.2}$$

Which is then used to telomerize TFE in a highly exothermic process:

$$C_2F_5I + n\,C_2F_4 \rightarrow C_2F_5(C_2F_4)_nI \tag{9.3}$$

The commercial production of TFE monomer involves the pyrolysis of CHF_2Cl refrigerant that is prepared from methylenetrichloride by halogen exchange [1]:

$$CHCl_3 \xrightarrow{HF,\ SbF3} CHF_2Cl \xrightarrow{700^\circ C} HCl + CF_2 :\rightarrow CF_2 = CF_2 \tag{9.4}$$

Fluorotelomers are the primary feedstock for multifarious products including surfactants, finished products, and surface modifications. The average chain length of fluorotelomeriodide products is influenced by C_2F_5I : TFE ratios and historically procedures were often optimized to generate C8-C9 perfluoroalkyl chain lengths for optimal repellency properties of the final material [1]. Only PFCAs and perfluoroalkylsulfonyl chlorides are directly prepared from these perfluoroalkyl iodides, however preparing perfluoroalkyl ethyl iodides ($C_nF_{2n+1}C_2H_4I$) by facile ethylenation opens a multitude of conventional iodide conversion reactions to produce other PFASs (e.g., FTOHs, FTSHs; FTSCNs) [1, 18].

Polymerization of fluorinated monomers is also the process used for synthesizing oligomeric perfluoropolyether fluids. These are produced by photooxidation of perfluoroolefins, or from polymerization of oxetanes followed by direct fluorination [1].

Beyond small molecule and fluidic PFASs, fluoropolymer plastics represent a massive share of PFAS production [1, 64]. The annual global production of fluorinated plastics in 1988 was estimated to be 55–60,000 ton y^{-1} [1]. Fluoropolymers are most often produced through free-radical polymerization in water and/or fluorinated media under mild conditions. The most important commercial PFAS plastics are homopolymers of TFE, chlorotrifluoroethene, vinyl fluoride, and vinylidene fluoride; in addition to copolymers of the aformentioned with hexafluoropropene, perfluoropropyl vinyl ether, and ethene. PTFE is the most prevelent fluoroplastic. In addition to fluoropolymers that posses fluorinated backbones, fluoroalkyl moieties can be incorporated as side-chains of aliphatic polymers [1, 8].

9.5 Structure and Properties

To better understand how PFASs are suited for specialty applications, which will be described in the next section, one may examine the physicochemical nature of the fluoroalkyl moiety and the resulting properties. These properties all relate to the preeminent electronegativity of fluorine. One intuitive metric of electronegativity, χ, is the average of an elements ionization energy, I, and electron affinity, EA (Eq. 9.5) [61]. In fluorine's bond with carbon, there exists a large degree of polarization with a partial positive charge on carbon and partial negative charge on fluorine: $C^{\delta+} - F^{\delta-}$. In addition, the C-F bond length is short, ca. 1.35 Å. The close proximity and opposite partial charges of C and F give rise to strong electrostatic interactions, making the C-F bond the strongest single bond in organic chemistry [61].

$$
\begin{aligned}
I: &\quad F \rightarrow F^{+} + e^{-} &\quad 1679\ \text{kJ mol}^{-1} \\
-EA: &\quad F + e^{-} \rightarrow \text{F-} &\quad -328\ \text{kJ mol}^{-1} \\
&\chi = \frac{1}{2}(I + E_{ea})
\end{aligned}
\tag{9.5}
$$

In addition to being highly polarized, the C-F bond is poorly polarizable; in conjunction with the strong electrostatic interactions between C and F, this makes fluoride a poor leaving group in organofluorine compounds [45]. Compared to the corresponding chloride, alkyl fluorides are $10^2 - 10^6$ times less reactive towards S_N1 or S_N2 nucleophilic displacement [1].

Table 9.1 Experimental F-CF_xH_{3-x} bond dissociation energies

	F-CH_3	F-CFH_2	F-CF_2H	F-CF_3
BDE (kJ mol^{-1})	460.2 ± 8.4	496.2 ± 8.8	533.9 ± 5.9	546.8 ± 2.1

Adapted from Shi et al. [77]

Table 9.2 Experimental bond dissociation energies of fluorine-substituted ethanes and haloethanes

	BDE (kJ mol^{-1})			
X	CH_3CH_2-X	CF_3CF_2-X	CF_3CH_2-X	CF_3CF_2-X
H	418.8	416.3	446.4	429.7
F	451.5	522.2	457.7	530.5
Cl	350	–	–	346
Br	291	–	–	287
I	231	218	236	219

Adapted from Smart [1]

Table 9.3 Experimental C-C and C-O Bond dissociation energies of fluorine-substituted ethanes and ethers

Ethane	BDE (C-C)[a]	Ether	BDE (C-O)[a]
CH_3CH_3	369	CH3OCH3	348
CH_3CF_3	423.4	CF3OCF3	440.2
CF_3CF_3	413		

[a]BDE in kJ mol^{-1}

Adapted from Smart [1]

Beyond C-F being a phenomenally strong bond in of itself, fluorine substitution influences the strengths of neighboring bonds (Tables 9.1, 9.2, and 9.3) [1]. The stability of the PFASs is limited by the C-C bonding, with longer-chain lengths and branched compounds being less stable. In fact, perfluoroethers have a higher thermal stability than branched perfluoroalkanes [1].

The small size of fluorine (1.47 Å Van der Waals radius) allows substitution for practically any hydrogen atom [61]. Regardless, fluorine is significantly larger than hydrogen (for which the Van der Waals radius is 1.2 Å) [45], this makes fluoroalkyl chains sterically bulkier than the corresponding hydrocarbon. Steric bulk and electrostatic repulsion shield fluorinated carbon centers from nucleophilic attack [64]. The steric protection afforded by the fluoroalkyl moiety is exhibited in the anomalous stability of $(^iC_3F_7)_3C\cdot$ [1].

Collectively the electronic and steric considerations we just described help explain the remarkable thermal stability of the fluoroalkyl moiety, and resilience towards oxidation and reduction.

In addition to conferring remarkable stability, the electronegativity of fluorine lowers fluoroalkane surface energy, attenuating intermolecular forces. In part, this is due to the tight binding of fluorine's valence electrons, making organofluorine a poor hydrogen bond donor [1, 61]. In addition, the low polarizability of the C-F bond and fluorine lone pairs makes London dispersion forces especially weak. As a result, perfluoroalkanes have a low energy of dissolution in both protic solvents and hydrocarbon media: the perfluoroalkyl moiety is both hydro-and lipophobic [45]. This property is widely recognized in the application to nonstick cookware, yet the low surface energy produces many other important properties. For example, compared to the corresponding alkane, the boiling points of perfluoroalkanes are remarkably low especially when then considering their molar mass; e.g., C_8F_{18} has a boiling point of 104 °C and a molecular weight of 438.059 g mol^{-1}, whereas C_8H_{18} has a boiling point of 126 °C and a molecular weight of 114.23 g mol^{-1} [1].

The electronic and steric properties of organofluorines impart further useful properties to the fluoroalkyl moiety: low dielectric constant, high electrical resistivity, low refractive indices, large thermal expansion coefficients, and good compressibility to name a few [1].

9.5.1 PFAS Surfactants

Given that the perfluoroalkyl moiety is both hydro- and lipophobic, the addition of a hydrophilic functionality or lipophilic hydrocarbon chain provides sepparate domains with differing solvation preferences within a single molecule. This causes them to selectively adsorb to aqueous or organic interfaces, respectively; and so they are surface active reagents (surfactants). This surface activity lowers the surface tension at the interface where they adsorb. Higher selectivity for adsorbing to the interface reduces the surface tension to a greater extent, and PFAS surfactants lower surface tension significantly even at low concentrations [18].

Once a certain amount of surfactant has adsorbed to an interface, adding more surfactant to solution no longer reduces the surface tension because the excess surfactant aggregates and forms micelles. The point at which this occurs is called the critical micelle concentration (CMC). The CMCs of PFCAs and PFSAs are low [18, 46].

PFAS surfactants can also be adsorbed onto the surface of textiles and other materials, imparting water and oil repellency. Optimal water repellency is achieved with chain lengths ≥C10 although improvement is minimal above C7 [1].

PFAS surfactants can have a cationic, anionic, amphoteric, or polar hydrophiles. Between the fluorinated segment and hydrophile there can be some sort of linkage, such as the C_2H_4 linkage of FTOHs, which can change solubility, reduce volatility, and decrease the acid dissociation constant of the hydrophile. PFAAs, which do not have such a linkage, are strong acids in part due to the inductively electron withdrawing pefluoroalkyl chain.

PFSAs are notably stronger acids than PFCAs. In fact, there has been some confusion about the acid dissociation constants of medium to long-chain PFCAs [68]. Some measurements may have been complicated due to PFCA aggregation and preferential solvation [10, 42]. More recently, mass-spectrometric and volatility studies indicated that the pK_a of C$\leq$8 PFCAs is around 0.5 or less, at least at sub-mM concentrations [10, 38, 93]. Therefore, at environmentally relevant pH and concentration, they are understood to exist as their conjugate-base anions. This is important for environmental transport because PFCAs in their deprotonated versus neutral states will evaporate from aqueous solution at dramatically different rates. Furthermore, deprotonated PFAAs are water soluble at environmentally relevant concentrations, which governs their transport to a significant extent (*vide infra*, Sect. 9.8.2).

9.6 PFAS Applications

At the heart of the PFAS lifecycle are the many applications exploiting their unique properties. While PFASs are more expensive to produce than hydrocarbon analogues, they are suitable for many stringent specific needs [1]. The following summary of applications is not comprehensive, and certain examples may never have been commercially important, while some uses may have been deprecated.

9.6.1 Perfluoroalkanes

Perfluoroalkanes are essential in the plasma etching processes in microelectronics fabrication. Furthermore they can be used as coolants, refrigerants, cleaning agents, foam blowing agents, as fire fighting agents, as heat transfer agents, as environmental tracers, and as aids in electrical component testing. Due to high oxygen solubility, emulsions of perfluoroalkanes can be used as biomedical oxygen transport agents. Perfluoroalkanes have also played an important role in particle physics experiments [1].

9.6.2 PFAS Surfactants

This is the most notorious class of PFASs. Ecological and epidemiological concerns are well documented for certain members of this class including PFCAs, PFSAs, and FTOHs. These compounds are prevalent in the environment in part because of their large-scale production and the water solubility imparted by their hydrophilic moieties.

By reducing surface tension, PFAS surfactants are powerful wetting and leveling agents. They have been used as wetting agents for etching solutions, solder, and cleaning solutions. And as leveling agents for for inks, paints, and polishes. Furthermore, addition of PFAS surfactants to oils and adhesives has been used to improve penetration [18].

Contrary to their application as whetting agents, when adsorbed to substrate surfaces, PFAS surfactants provide both oil and water repellency. Surface treatment with

PFAS surfactants can also reduce friction, impart alkali and acid corrosion resistance, as well as provide antifogging, antistatic, and antisoiling properties. The low surface energy and inertness of such coatings has been used to improve the biocompatability of biomedical devices [18].

The disparate solvation preferences between the different domains of PFAS surfactants makes them useful emuslifiers and dispersants for other low surface energy compounds such as silicones, fluoroalkanes, and fluorinated monomers. They have been used extensively for the emulsion polymerization of TFE, and they have been used to disperse PTFE in greases. In the inverted paradigm, PFAS surfactants can be used as dispersants within a fluorinated continuous phase. For example, they can be used to disperse medicinals in self-propelling aerosols [18].

Notably, PFAS surfactants can act as either foaming or antifoaming agents and they are constituents in aqueous film forming foams (AFFFs), where low surface energy allows them to easily spread over the surface of burning hydrocarbons. They also have an important application in metal plating baths where they regulate foam formation and improve the quality of finishes [18].

Further applications of PFAS surfactants include cleaning hard surfaces; scale removal in metal pickling solutions, liquid crystal applications, electrode fabrication, application as electrolyte constituents in electrochemical cells and batteries, impregnation into electrical insulators to improve a number of properties, improving water resistance in water-based inks, mineral floatation such as in uranium recovery, promoting petroleum recovery from oil wells, insecticides, and various aspects of leather processing and treatment [1, 18].

9.6.3 Perfluoropolyethers

To an extent, these are also PFAS surfactants due to their ether linkage [18]. Importantly, these are often the basis for fluorinated lubricants and they can be blended with PFAS plastics to prepare greases [1, 64]. Compared to perfluoroalkanes, perfluoropolyethers have facile rotation around the ether linkage, which provides good conformational freedom. This improves their low-temperature properties and they are fluids even at high molecular weights. Although lubrication properties are not better than hydrocarbon-based lubricants, the physicochemical properties of perfluorination are advantageous for many applications [64]. For example, perfluoropolyethers are useful lubricants and working fluids for processes under vacuum such as semiconductor manufacturing [1].

9.6.4 PFAS Plastics

Within the realm of PFAS plastics there are thermoplastics, elastomers, membranes, etc. Fluoropolymers are desirable due to the useful properties imparted by the fluoroalkyl moiety as previously discussed (Sect. 9.5). There are homo- and copolymers of many different monomers, and some PFAS plastics posses fluorinated side-chains instead of a fluorinated backbone [1, 8].

The most commercially important fluoroplastic, PTFE, is a crystalline thermoplastic. PFAS thermoplastics have many important applications include electrical conductor insulation, sealing applications, reactor lining, tubing, protective and anti-stick coatings, biomedical applications, and waterproof materials such as Gore-Tex (at least in the past). PTFE micropowders find application as additives to other media and other plastics. Certain amorphous fluoropolymer thermoplastics have good optical transparency and low refractive indices useful in applications such as contact lenses or optical coatings [1].

Although produced on a much smaller scale than PFAS thermoplastics, the rubber-like properties of fluoropolymer elastomers are highly desired for certain applications. In general, a lack of crystallinity in PFAS elastomers enables conformational freedom while post-polymerization covalent cross-linking anchors polymers together in a 3-dimensional network. Their primary use is as sealing componentry such as O-rings and gaskets, which are used in many industries where the performance characteristics of PFASs are needed. Other applications might include wire coatings, tubing, and personal protective equipment to name a few. Although the C-H bonds in many PFAS elastomers introduce weakness, there are perfluorinated elastomers that rival PTFE in their thermal and chemical stability. In contrast to crystalline perfluoropolymer thermoplastics, PFAS elastomers can swell in the presence of certain solvents [1].

PFAS membranes formed from fluorinated ionomers constitute another category of PFAS plastics that is important for ion-exchange processes in electrochemical applications. It is understood that the formation of ionic domains within the bulk contributes to excellent ion-transport performance. Both PFCA and PFSA-type (e.g., Nafion) PFAS membranes have been developed [1].

9.7 Regulatory Environment

With the advent of modern PFAS production methods and with their numerous applications, it is not surprising that organic fluorine was detected in human blood serum as early as 1968 [4, 85]. While biochemical investigations of PFASs had been pursued for some time [5, 18, 41], it was not until the turn of the century that there were significant attitude changes towards PFASs and regulatory measures began to be put in place.

In 2000, 3M, the major producer of PFOS in the USA and Belgium, announced that it would phase out the production of PFOS and related compounds–excluding some essential production for AFFFs–based on their findings of its presence and persistence in blood and the environment, coupled with the finding that PFOS exposure was linked to mortality in newborn rats [88, 89].

For PFOA, the United States Environmental Protection Agency (US EPA) began a stewardship program for eight major manufactures in 2006. The goals, which have since been achieved, were to attain a 95% reduction in PFOA production relative to a year 2000 baseline, and work towards eliminating PFOA emissions and its presence in products by 2015 [91]. Since June 2020, the EPA was finalizing a new rule requiring their notification and approval for the new import, manufacture, or utilization of certain long-chain PFAS carboxylates, which was in addition to a standing rule regarding sulfonates [91].

The Stockholm convention on persistent organic pollutants currently lists PFOA, PFOS, their salts, and POSFs. Despite their perceived harm, the indispensable utility of these compounds has led to their time-limited exemption from the Stockholm Convention for certain uses. PFOS, its salts, and POSFs are allowed to be used in metal plating, AFFFs, and leaf-cutting ant bait (not time-limited, *vide infra*, Sect. 9.8.3) [74]. PFOA, its salts, and related compounds are allowed for use in semiconductor fabrication, photographic coatings, textiles for protection, medical devices, AFFFs, pharmaceutical products, PTFE and polyvynilidene fluoride production for water treatment membranes, fluorinated ethylene propylene production for electrical wire and cable, and fluoroelastomer production for car interiors [75].

Other regulatory measures and environmental organizations are involved in PFAS management in the US and around the globe.

9.8 PFASs in the Environment and Biota

In order to effectively manage the impact of PFASs on humans and the environment, it is important to characterize how they are released into the environment, how organisms are exposed to them, and what their environmental and biochemical behavior is. Some sources indicate that there are more than 4,700 PFASs [78], yet the majority of research has focussed on a few specific compounds; most notably PFOS and PFOA. Even for closely-related substances like shorter-chain PFCAs and PFSAs, perfluoroalkyl phosphinic acids, and perfluoroether carboxylic and sulfonic acids, there has been only limited studies and some level of regulatory consideration or measures taken [98]. As a result, the following discussion emphasizes PFOS and PFOA.

9.8.1 Environmental Release

Because PFASs are so stable, their historic and ongoing environmental release leads to accumulation [98]. PFASs can be released into the environment by direct or indirect means. PFASs can be directly released at the point of manufacture, according to their application in some secondary manufacturing process, according to their presence within an industrial product or some consumer product, or due to their disposal. Furthermore, some PFASs of concern (such as PFAAs) are released into the environment due to the transformation of a precursor, which is considered indirect environmental release.

9.8.1.1 Transformation of PFAA Precursors

Indirect environmental release of PFAAs such as PFOA and PFOS is very significant because they are the dead-end daughter product of many PFASs following environmental degradation via a variety of mechanisms [8, 109, 111]. In one example, POSF derivatives and FTOHs have been shown to degrade to PFCAs and PFOS when they are oxidized by hydroxyl radical [31], which is an environmentally important species [22]. In fact, this approach can be used to transform unidentified PFASs into more-readily analyzed PFAAs in the total oxidizable precursor assay [31].

PFAS surfactants are the most widely discussed PFAA precursors, yet there are many compounds that can degrade to PFCAs such as hydrofluorocarbons and hydrofluoroethers, partially fluorinated alkanes, and side-chain fluorinated polymers [8]. In 2006, the Organization for Economic Co-operation and Development (OECD) compiled a list of more than 1,000 potential PFCA precursors [60].

Due to the large-scale production of side-chain fluorinated polymers, their role as PFCA precursors is a significant concern. In an investigation of the biodegradation of fluoroacrylate polymer in soil to yeild PFOA, it was estimated that half-lives could range from decades to thousands of years depending on the amount of surface area exposed [70, 103]. Based on this evidence, even conservative estimates indicate significant environmental release of PFCAs due to side-chain fluorinated polymer biodegradation [95].

In contrast to side-chain fluorinated polymers, fluoropolymers lack labile functional groups. On the basis of environmental and biological hazard assessment is has been concluded that fluoropolymers are of low concern and should not be considered according to the same regulations as other PFASs [25]. It should be noted that the data supporting this conclusion was primarily for PTFE, which is perfluorinated, and that other fluoropolymers may not be quite as inert (*vide supra*, Sect. 9.5). Although this will not happen within the environment, thermolysis of perfluoropolymers has been shown to generate PFCAs and this could potentially contribute to environmental PFCA release [96].

9.8.1.2 Appraisal of the Origins PFAAs in the Environment and the Extent of Their Environmental Release

On the basis of both indirect and direct release, between 1951 and 2002, an estimated 1290–14,220 ton of C4-C14 PFCAs were released to the environment [95]. C8 and C9 PFCAs are understood to have been historically produced in the largest quantities and their use and release is better documented compared to other homologues [65, 95]. For example, in the year 2000, an estimated 5–10% of ammonium PFOA production, constituting ca. 20 ton, was estimated to be environmentally released by the largest US ECF plant. Meanwhile, fluoropolymer production is the largest historic utilization of PFCAs and an estimated 60% used therein has been environmentally released–which constitutes ca. 80% of total estimated C8 and C9 PFCA environmental release (although estimates vary). In addition to fluoropolymer production, protective coatings applied as PTFE dispersions containing ammonium PFOA salt are estimated to have caused significant direct environmental release [65]. Furthermore, a multitude of patents incorporating PFCAs suggests that numerous consumer products contained these compounds, constituting an important but poorly documented avenue of direct PFCA human exposure and environmental release [65].

AFFFs usage represents a significant point source of PFAS contamination [57, 58]. There is a potential link between facilities that used AFFFs and the occurrence of PFASs in drinking water across the US [32]. By 2006, the total estimated C8 and C9 PFCA release from AFFFs was 50–100 ton [65]; furthermore, this may be the most prominent mode of direct PFOS dispersal [63]. Beyond direct PFAA environmental release from AFFF usage, these products are known to consist of complex mixtures of PFAS surfactants including cationic and zwitterionic species, many of which are probably PFAA precursors [3].

Although PFOS is ubiquitous in environment and biota, a much greater quantity of its progenitor, perfluorooctanesulfonyl fluoride, has been further derivitized to make other compounds (POSFs). Because POSFs can environmentally degrade to PFOS, a discussion of POSF environmental release is directly relevant to PFOS environmental occurence, although the extent of environmental transformation is unknown. Data from the ECF production of POSFs by 3M has been used to estimate global PSOF production and environmental release [63]. More than 90% of waste (24,000 ton) from the ECF POSF production process is in solid form which was disposed of by land farming (prior to 1998), landfilling and incineration. Between 1970 and 2002 the liquid waste stream was estimated to have caused 650–2600 ton of environmental release to water and air. Presumably, a large portion of land-farmed POSF waste will also eventually make its way into the water shed. Secondary manufacturing was estimated to have released an additional 2600–10,000 ton of POSFs to air and water. Furthermore, release from products such as carpets, packaging, apparel, performance chemicals, and AFFFs is estimated to have caused 4200–42,000 ton to be released to air and water.

Interestingly, pesticide application has historically caused significant release of PFASs to the environment. For example between 2004 and 2015 the application of sulfuramid for leaf cutter ant control in Brazil alone led to the release of 20 ton of PFOS and ca. 150 ton of the POSF N-ethylperfluorooctanesulfonamide, which is the active insecticide [99].

Due to the change in PFAA use practices around the turn of the century that include reduced usage, recapture, and recycling efforts; PFAA environmental release in the US, Western Europe, and Japan has declined dramatically. However, production in India, Poland, China, and Russia has increased substantially meaning that emission was geospatially redistributed while the magnitude of PFAS environmental release globally many be even greater now than in the past (Fig. 9.4) [95].

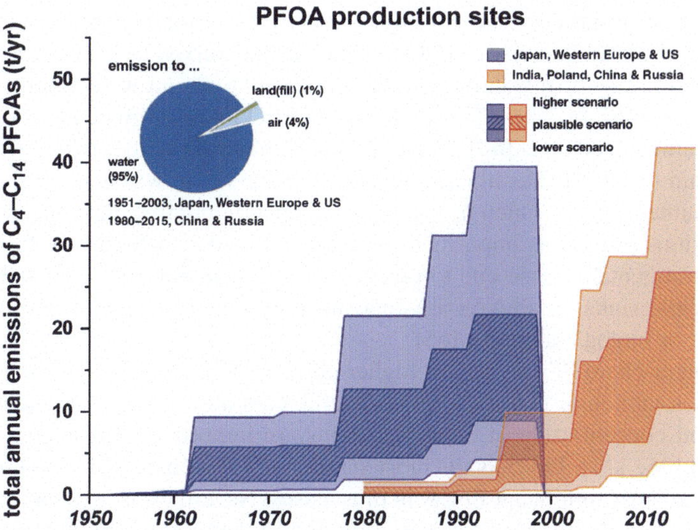

Fig. 9.4 Estimated C4-C14 PFCA emission over time from PFOA production sites in the USA, Western Europe, and Japan (purple); and India, Poland, China, and Russa (orange). The pie chart describes the distribution of emission to different media. Reproduced from Wang et al., with permission from Elsevier [95]

9.8.2 Environmental Transport and Inventory

PFASs may be transported throughout the environment by currents, aerosols, transformation of volatile precursors [20, 98], and by human activities [111]. Various environmental compartments can be expected to harbor PFASs including water, soils, sediments, and atmospheric matrices. PFAS Environmental monitoring typically reports PFOS to be the most frequently detected PFAA [11], although the scope of analytes has historically often been limited and has not included shorter-chain PFCAs, PFAA precursor degradation intermediates, and a variety of other PFAS analytes [111].

The high water solubility of PFAAs means that oceanic currents may be the most significant mechanisms of environmental transport [65]. However, models for the atmospheric oxidation of 8:2 FTOH predicts that it will cause PFCA contamination throughout the northern hemisphere, including remote regions. Monitoring PFCA contamination in the lakes of the Canadian Rockies is consistent with this prediction: between 1985 and 2003 there was an increase in PFCA concentrations corresponding to increased FTOH production. Surprisingly, a greater quantitity of 8:2 FTOH is anticipated to be oxidized and deposited in remote areas such as the Arctic than in source areas because NO_x pollution in metropolitan areas reduces FTOH atmospheric degradation. In urban areas such as these, PFAA point sources might make a larger contribution to local contamination [111].

Based on environmental monitoring, in 2006 it was estimated that in the northern hemisphere the total quantity of PFOA in the oceanic mixed layer was between 110 and 10,000 ton, in fresh waters between 4 and 800 ton, and in sediments between 3 and 340 ton [65]. In 2009, 235–1,770 ton of PFOS was estimated to reside in the global oceanic mixed layer [63]. There are much greater quantities of short-chain and longer-chain (>C9) PFCAs in oceanic surface waters than was estimated based off of quantifiable sources; which might be due to uncertainty in their usage, significant concentrations present as impurities in products based on C8 and C9 PFAAs, and their environmental release due to precursor transformation [95]. The only known environmental sinks for PFAAs are deposition into the deep ocean, sedimentation and burial, or biological uptake [65].

The longer-chain PFAAs have a higher affinity for sediments than shorter-chain homologues, although their adsorption coefficients are modest. Sediment organic content and concentration of divalent cations are important parameters for PFAA absorption [26, 111]. PFOS is the most significant PFSA detected in sediment, and PFOA is the most significant PFCA in most cases. Nevertheless, in some cases TFA was observed in the highest concentration and, notably, this PFCA homologue is often left unmonitored. POSFs typically occur at the same order of magnitude as PFAA concentrations in sediment samples.

Freshwater sediments usually have higher PFAS concentrations than coastal sediments, however there are exceptions. For example, PFOS concentrations of 73–537 ng g^{-1} dry weight were found at a site in the Yangtze River estuary of China where there is intensive human activity. Increased salinity may enhance PFAS adsorption to sediments, which could make estuary sediments an important environmental sink [111].

The temporal profile of PFAS sediment concentrations have tracked well with global trends in production [111]. Concentrations are generally highest in the Asia-Pacific region, which may correspond to an increase in production by nations in that area whereas North America has seen a decrease in sediment concentrations corresponding to regulatory measures.

Due to its largely anthropogenic nature, wastewater treatment plant sludge typically has at least an order of magnitude greater dry weight PFAS concentration than sediments, with a value as high as 7500 ng g^{-1} reported in a sample from Hong Kong. PFCA concentrations were higher in Asian-Pacific countries while PFSAs had a more even global distribution. Differences in a PFAS profiles between intake and effluents at waste water treatment facilities indicate the biodegradation of PFAA precursors. The occurrence of high levels of PFASs in wastewater treatment plant sludge is probably indicative of human exposure [111].

PFAS occurrence in soils is poorly documented compared to occurrence in other solid matrices like sediments and waste water sludge [111]. Elevated levels have been detected near point sources such as facilities that use AFFFs or a fluoropolymer production plant. Amending farmland with waste water sludge (biosolid applications) is a significant mechanisms of PFAS dispersal that can lead to elevated levels in soil. For example, one case study evaluating PFAS levels in farmland that had been amended with biosolids from a wastewater treatment facility known for processing the effluent from a fluorochemical production plant found high levels of PFAAs (6000 ng g^{-1} dry weight).

9.8.3 *Uptake by Plants and Animals*

Knowing that PFASs have been used in so many products and that they are present throughout the environment, it is not surprising that they are ubiquitous in human blood serum. A representative study of nearly 8,000 USA residents found PFOS and C6-C9 PFCAs in >95% of participants [40]. Only PFOS concentrations (1999–2008) noticeably decreased after the turn of the century regulatory measures. PFOA and especially PFOS have been found in human blood serum not only in the US, but in many other counties as well [39]. Furthermore, PFOS (and to a lesser extent PFOA) has been detected wildlife globally, even in remote regions [21].

Drinking contaminated water is thought to be the primary route for human PFAA uptake [83, 99, 111]. On this basis, the US EPA issued a drinking water lifetime advisory limit for the combined concentrations of PFOS and PFOA equal to 70 ppt (ng L^{-1}) [90]. Nonetheless, PFAS exposure routes are poorly understood and there are many possiblitites (Fig. 9.5) [79]. Dietary intake is one possible mechanism of exposure: elevated PFAS levels have been detected in populations with high seafood consumption and PFASs incorporated in nonstick food wrapping leaches into food simulants. Inhalation of volatile precursors is another possible mechanism of exposure, as is dust inhalation [83]. Occupationally exposed workers are known to have an order of magnitude greater PFAS blood serum levels than is typically found, and workers exposed to AFFFs through training exercises have some of the highest recorded values [44, 78].

The bioaccumulation of PFAAs results in a higher biotic concentration compared to the environmental concentration, resulting in an increased risk of adverse health effects. Typically, the bioaccumulation potential of a persistent organic pollutants is screened according to octanol/water partition coefficients given their propensity to partition to lipids. PFAAs conversely have been seen to bind to proteins and are present in much greater concentrations in proetin-rich compartments such as blood and liver than other biological compartments. Empirical determinations of PFAA bioaccumulation potential based in studies of fish and aquatic invertebrates do not meet bioaccumulation regulatory criteria for $\leq$C7 PFCAs [11], whereas PFOA and PFOS are considered bioaccumulative by some regulatory agencies [11, 95] Bioaccumulation potential generally relates to perfluoroalkyl chain length and is greater for PFSAs than analogous PFCAs. It is questionable whether previous evaluations of PFAA bioaccumulation potential are transferable between species considering that the elimination time of PFAAs varies dramatically (*vide infra*, Sect. 9.8.4) [97], and is particularly long in humans.

Although it does not have a regulatory context, the biomagnification potential of PFAAs up food chains has also been explored, which is important for understanding

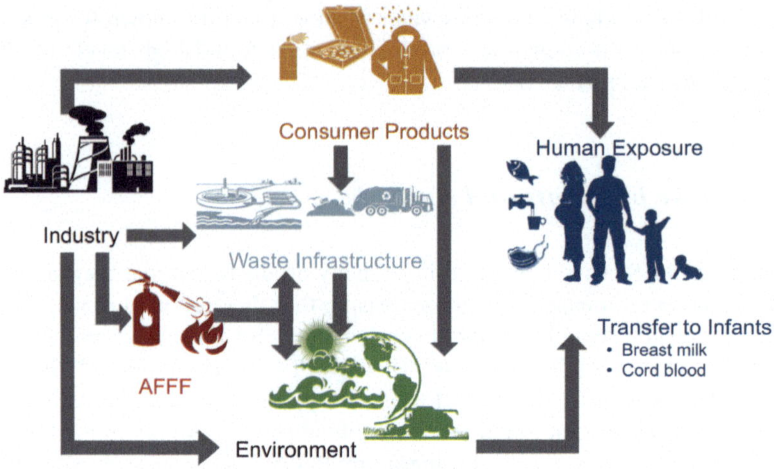

Fig. 9.5 Nonoccupational pathways for human exposure to PFASs. Reproduced from Sunderland et al., with permission from Springer Nature [83]

PFAS occurence in higher trophic levels. PFOS, in particular, has notable biomagnification potential; for example, in polar bears, which are apex predators [11].

Plants, in addition to animals, can take up and concentrate PFASs providing additional exposure routes to livestock and humans. Plants can be exposed to PFASs through irrigation by contaminated water, biosolid application, landfill release, industrial emissions, and by pesticide application. In general, shorter-chain PFAAs have higher bioaccumulation potential in plants, which is thought to come from their smaller size and increased water solubility. Furthermore, shorter-chain PFAAs are more-readily transferred to the above-ground portion of plants, whereas long-chain PFAAs are more prevelent in roots [99].

9.8.4 Epidemiology

The epidemiological concerns for PFASs are largely built on the fact that certain PFASs bioaccumulate and have long residence in humans coupled with extensive studies showing mammalian toxicity at high doses. In addition, some studies have established relationships between PFAS occurence in humans and adverse health effects. Although the levels of some legacy PFASs in human blood serum have declined in recent years, the presence of new PFASs is inadequately characterized [83], and results have indicated that certain next generation PFASs exhibit comparable or heightened toxicities relative to legacy compounds [23, 94].

The phamacokinetics of PFAAs differ greatly with species, and differ by gender for some species [44]. Typically, half-lives ($T_{1/2}$s) for PFAA elimination from human blood serum decrease with chain length, although there are exceptions such as the C6 PFSA, perfluorohexanesulfonic acid (PFHxSA), which has a $T_{1/2}$ much greater than PFOS. Reported $T_{1/2}$ values for PFOA, PFOS, and PFHxSA are 3.8, 5.4, and 8.5 years, respectively. These values are much greater than those known for other mammals including monkeys.

Early epidemiological studies were performed by leading PFAS manufacturers and are not part of the peer reviewed literature. They may be obtained as part of EPA public docket AR-226 and are reviewed in the literature [44, 79, 83]. Many of these early studies on adverse health effects of PFAA occurrence in humans present conflicting results, have a small sample size, or involve confounding factors [44]. More recently, in the most complete longitudinal PFAS epidemiological study of more than 32,000 people living next to a fluorotelomer process PFAS production plant, there was a positive correlation between PFOA in blood serum and kidney cancer, testicular cancer, ulcerative colitis, high cholesterol, and thyroid disease. While PFOA and PFOS presence was correlated with pregnancy induced hypertension and there was a negative correlation between PFOS and birth weight [2, 14, 51, 80, 83]. Children have statistically higher blood serum concentration of PFASs than adults and PFAS occurrence in children has been correlated to immunological, metabolic, kidney-related, thyroid related, and developmental issues [67]. The presence of PFAAs including PFOS and PFOA in umbilical cords indicate these compounds cross the placenta [44].

The adverse impacts of PFAS occurrence in other biota is also a notable concern. A 2012 review summarized that the concentrations of PFASs in surface waters were not acutely toxic to aquatic organisms, yet some compounds (e.g., PFOS) caused long-term negative impacts [15]. The literature on PFAA toxity in mammals including nonhuman primates is extensive and suggests various forms of toxicological harm and several biochemical mechanisms, although most of these studies involve high doses [44, 78, 79]. Due to biochemical differences between species, studies in other mammals cannot be directly evaluated in the context of human health outcome [44, 83], yet they play an important role in understanding PFAS toxicology and biochemical activity. It is not entirely clear what the ecotoxicity of PFASs is at known concentrations in the environment and it may be that there is only mild risk of adverse effects. This matter complicated by inadequate understanding and inventory of of the less-studied PFASs which are often occur simultaneously with PFOA and PFOS [78].

9.9 Preventing PFAS Contamination and Remediating Contaminated Matrices

In light of the adverse health effects associated PFAAs and their ubiquitous presence, their remediation in the environment and removal from waste streams is a primary concern. Due to the obvious mechanism of exposure by drinking water (Fig. 9.6),

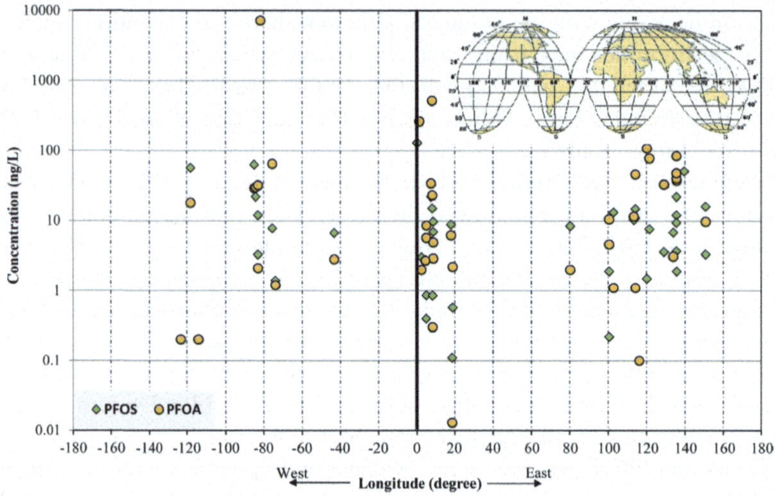

Fig. 9.6 Reported concentrations of PFOA and PFOS in drinking water by latitude. Reproduced from Rahman et al., with permission from Elsevier [66]

and corresponding stringent drinking water standards, treating contaminated water has received the most attention. However, it is also necessary to remediate sediments and soil.

The physicochemical properties of PFASs such as their thermal stability and resistance to oxidative and reductive degradation (*vide supra*, Sect. 9.5) make them difficult to remediate, although certain properties such as surface activity can be taken advantage of, as will be described. Conventional drinking water treatment is ineffective for PFAS removal [66] and so techniques developed for other pollutants are currently being commercially applied (e.g., granular activated carbon), but these may require additional disposal. There is a perceived need to supplement or replace existing technologies which has ignited an explosion of research to that end.

Despite the rapid evolution of PFAS remediation literature, there is still significant opportunity and need for improvements. For example, to date remedial efforts have focused primarily on the long long-chain PFAAs, PFOA and PFOS, while there are thousands of other PFASs, many of which are being used in place of these legacy C8 PFAAs [32, 69]. In addition, most laboratory-scale studies are performed at $\mu g\ L^{-1}$ or greater concentrations and have not been evaluated for $ng\ L^{-1}$ concentrations as are typically found in the environment [19]. In the field, additional challenges must be met. For example, in ex-situ ground water treatment, there can be back diffusion of contaminant slow-mixing zones into treated regions. Meanwhile, precursors residing in another phase or at an interface could cause a time-release of daughter products into water [69].

The optimal treatment approach is probably highly situationally dependent, with influencing factors including location, PFAS speciation and concentration, solution matrix, and what the treatment goals are; for example, whether PFAA transformation or complete mineralization is desired [92]. Treatment-train approaches, where multiple technologies are coupled together, can be used to overcome the difficulties of PFAS remediation [69]. This could include a removal/concentration step used in tandem with a destruction step, technologies effective for either long or short-chain PFAS remediation being coupled, technologies that are effective for either anionic or zwitterionic species being coupled, or a cleanup step being used to deal with harmful byproductsetc. [53, 69]. Herein we focus on individual techniques, many of which can be and have been elaborated on.

9.9.1 PFAS Removal Fom Water

When the need arose to prevent and remediate PFAS water contamination, the immediate response appears to have been containment and/or capture by granular activated carbon (GAC) or ion-exchange resins (IXs) [66, 69]. With good PFOA and PFOS removal, GAC is the golden standard absorptive technology [69]. Once GAC reaches capacity, it can can be reactivated by thermal treatment. Unfotunately, GAC suffers from competitive absorption–e.g., of natural organic matter–and it is less effective for removing of shorter-chain PFAAs [66]. Adsorption efficiency of other PFASs including cationic, zewitterionic, and other anionic species needs further study [69].

In comparison to GAC, some IXs have been shown to be more effective for both long and short-chain anionic PFAAs [66, 69]. In addition, IXs exhibit faster PFAS removal and occupy a smaller footprint. However, IXs are currently incinerated after use and to make them cost-competitive with GAC, resin regeneration may be necessary. Given their high performance, they may also be used as a finishing step after GAC. Problems with IXs include competitive anion binding and PFAS removal could be reduced in high ionic strength matrices [69]. The suitability of IXs for other PFASs–e.g., cationic and zwitterionic PFASs–needs further investigation, and for these other species tandem application of different resins may be necessary.

A number of other adsorbents have been investigated for PFAS removal. For example, the polymeric material, Osorb, which absorbs organics but not water. Then there is a family crosslinked cyclodextrins that have been extensively investigated for PFAS removal and are the basis for the startup, Cyclopure [69, 108]. Electrocoaggulation has also been shown to efficiently remove PFAAs from solution–this technology is already utilized in waste water treatment and has low maintenance costs [55]. A detailed review PFAS adsorption was published by Du et al. [17].

Beyond adsorbents, other PFAS removal approaches are important: most notably reverse osmosis and nanofiltration, which are highly effective for PFAA removal. However these systems are expensive and only typically applied to drinking water treatment. Ground water and other solution matrices can cause problems such as membrane fouling or reduced efficacy [66, 69]. For example, isopropanol, which may be present in wastewater from semiconductor manufacturing, has been shown to reduce PFOS removal [92].

Finally, ozofractionation is a promising PFAS removal technology that takes advantage of the PFAS surface activity. In this system, solution is sparged with small ozone bubbles which remove PFAS surfactants due to their partitioning to the water/gas interface. Other organics including PFAA precursors are oxidized. The PFASs are concentrated in a fraction constituting 0.5–2% of influent volume at the top of the apparatus, which is collected for further disposal. This technology is compatible with certain levels of heterogeneity, and a finishing step can be added for more complete PFAS removal. This technology has demonstrated good metrics for removing mixtures of PFASs from water [69].

9.9.2 PFAS Destruction

9.9.2.1 Bioremediation

PFAA precursors can be biologically transformed into persistent PFAAs [13]. Furthermore, biologically relevant reductants have been shown to defluorinate PFAAs [59], and some microbial communities have been reported to reductively defluorinate certain PFASs [110]–in both cases under anaerobic conditions. Meanwhile, in one case, both PFOA and PFOS were reported to have been biologically mineralized in the presence of Fe^{III} as an electron acceptor and NH_3 as an electron donor [33]. Furthermore, laccase enzyme has been reported to oxidize PFAAs in the presence of hydroxybenzotriazole [52]. To date, there are still few examples of PFAS bioremediation in the literature [13, 33, 69].

9.9.2.2 Chemical and Electrochemical Oxidation

Some chemical oxidants can mineralize PFCAs and there are limited reports of PFSA oxidation [16, 54]. During PFCA oxidation, shorter-chain homologues are produced with concomitant CO_2 and F^- production. Eventually complete mineralization can be achieved. Advanced oxidation processes, which utilize ozone, hydroxyl radical, or O-atoms are incapable of directly oxidizing PFAAs [92]. On the other hand, sulfate radical (2.1–3.1V/NHE), generated by activating peroxydisulfate (persulfate), is capable of mineralizing PFCAs. Reports of PFSA oxidation by sulfate radical are inconsistent [7, 16, 107]. There are various approaches for activating persulfate including light, electron transfer reagents, and thermal energy [104]. Activated persulfate has potential for *in-situ* PFAS ground water remediation [6, 48, 62], although the production of more-mobile short-chain PFCAs from a multitude of potential precursors is a cause for concern [69]. Other oxidants that reportedly oxidize PFCAs and/or PFSAs include NO_2, $KMnO_4$, and $CO_3^{\cdot-}$ [47, 50, 87].

As previously stated, hydroxyl radical is incapable of directly oxidizing PFCAs [35]. This is the principal oxidant in the Fenton reaction, and so it is surprising that there are some reports of successful PFCA oxidation using Fenton's reagent.

However, results are mixed [72, 73, 84, 106], with some results being poorly defined [56, 69]. An extension of the Fenton reaction, the photo-assisted Fenton reaction, has been demonstrated to mineralize PFCAs [84]. Ferric ion-mediated photocatalytic reactions relevant to the photo-assisted Fenton reaction are discussed later.

The adventitious scavenging of chemical oxidants by species in the sample matrix other than the contaminant of concern is a major drawback to PFAS chemical oxidation. PFCAs and PFSAs are so difficult to oxidize that one can expect interference from many other compounds including other organics or inorganic anions. Even hydroxide can interfere with PFCA oxidation by sulfate radical, which limits the appropriate pH to more acidic conditions [7, 92]. Furthermore, oxidation of some compounds likely to be found in PFAS contaminated water–e.g., Cl^-–leads to toxic byproducts.

Electrochemical oxidation presents an attractive remediation approach that is similar in many respects to chemical oxidation. However, in this approach reactivity is provided on the basis of either direct electron transfer from the PFAS to the anode, or by indirect chemical oxidation of the PFAS by complimentary reactive species generated at the anode. Electrochemical oxidation can degrade both PFCAs and PFSAs [69]. Multiple anodes have been investigate with the boron-doped diamond electrode being the best studied. This anode is favorable for its high oxygen evolution overpotential, making it more efficient, but it is expensive [55]. As with chemical oxidation, other species in the matrix may compete with PFAA degradation and form toxic bypoducts [55, 69].

9.9.2.3 Reductive Defluorination

Given the high electronegativity of fluorine, one might expect that PFASs to be prone to reductive defluorination yet this process requires potent electron transfer reagents; most notably the hydrated electron, $e^-_{(aq)}$ (-2.9 V/SHE), which can be produced through numerous methods that ionize molecules such as photolysis, radiolysis, and electrical discharge (*vide infra*); or by direct injection. Other reductants have also been understood to defluorinate PFAAs including Fe^0, $(CH_3)_2CO^{\cdot -}$, $CO_2^{\cdot -}$, and cobalamins [37, 92]. While PFSAs are recalcitrant towards chemical oxidation, they are more-readily reduced than PFCAs. In addition, branched PFAAs are easier to defluorinate than linear isomers [49, 92]. Reductive defluorination of PFAAs leave either hydrofluoroalkanes or olefins, and complete mineralization requires additional treatment [92]. Scavenging of $e^-_{(aq)}$ by dissolved oxygen and other oxidants such as nitrate is a notable barrier to both *in-situ* and *ex-situ* application of this approach [69].

9.9.2.4 Photolysis and Photocatalysis

Both PFOA and PFOS can be directly photolyzed by deep-UV light. Vacuum UV light ($\lambda < 200$ nm) is more effective UVC light ($\lambda < 300$ nm) and there is no evidence that solar radiation ($\lambda > 290$ nm) can directly degrade these PFAAs. Photoytic transformation products are shorter-chain PFCAs as is seen for oxidative degradation. VUV irradiation barely penetrates aqueous solution, and competitive photon absorption by other species in the sample matrix reduces efficiency [92]. The presence of molecules such as Fe^{III} and $PW_{12}O_{40}^{3-}$ have been shown to improve the efficiency of PFCA [29, 30, 102] and PFSA [37] photolysis; in addition, numerous heterogenous semiconductor photocatalysts have been investigated, including TiO_2, Ga_2O_3, In_2O_3, $BiPO_4$, $Bi_3O(OH)(PO_4)_2$, and BiOBr, [71, 100, 105]. Catalyst-mediated photolytic systems typically use longer wavelengths than used for direct photolysis. Significant advances are needed for the large-scale implementation of heterogeneous photocatalysis for water treatment [9, 100], making the near-term application to PFAS remediation is unlikely.

9.9.2.5 Thermolysis

Incineration is capable of destroying the gambit of PFASs. Even CF_4 can be destroyed [92]. Incineration is most efficient when used to destroy solids and becomes more energy intensive when other components within the waste stream must also be thermolyzed [92]. Thermal treatment has been widely employed for PFAS destruction. However, pyrolysis transformation products include fluoroalkanes that are potent greenhouse gases and reactions with co-contaminants can form other harmful byproducts [55]. Nonetheless, combustion conditions may change the product distribution to CO, CO_2, and HF [92], and additived such as $Ca(OH)_2$ can help reduce the production of harmful byproducts [55]. With regards to the real-world thermal treatment of PFASs, a recent report of unpublished data described high soil and surface water PFAS concentrations near a kiln that was being used to destroy AFFF, suggesting significant environmental release during this process [27].

9.9.2.6 Sonochemical Degradation

It was known that sonolysis degrades PFAS since long before major remediation efforts were underway [1]. Sonolysis makes use of sound waves to produce microscopic bubbles by negative pressure, which then collapse with compression; collectively cavitation [69]. Cavitation causes high pressures (ca. 14,000 psi), vapor temperatures exceeding 10,000 K, and it is accompanied by sonoluminescence [55, 92]. Contaminant decomposition occurs by the action of radicals produced in the vapor and by pyrolysis and combustion of molecules at the gas/water interface and in the gas phase. Lower-frequency sonolysis causes higher energy output with cavitation whereas higher frequency promotes a greater quantity of smaller bubbles and hence greater surface area. The surface activity of PFAS surfactants causes them to partition to these bubbles where they are degraded–thus high frequency sonolysis is advantageous for PFAS destruction. PFAS degradation studies have often been performed under argon since higher temperatures and transformation product yeilds are achieved relative to air [55]. Immediate mineralization of PFAAs into CO and CO_2 is observed, and it is thought that the intermediates of pyrolized PFAAs–perfluoroolefins and 1H-perfluoroalkanes–preferentially partition into the vapor phase of gas bubbles where they will be further pyrolized to form C1 fluorocarbon radicals that will react with OH·, O-atom, and H-atom [92]. Using sonolysis, rapid mineralization of PFAS can be achieved. Other species in the sample matrix can inhibit degradation, increase it, or leave efficiency relatively unchanged [55]. Equipment expenses are considerable, and there are significant design challenges anticipated for scale-up, but operating costs should be reasonable [69].

9.9.2.7 Degradation by High-Voltage Electric Discharge

High voltage discharge within water and in the gas phase above water; electrohydraulic discharge and nonthermal plasma, respectively; can be used to generate highly reactive species such as OH·, O-atom, H-atom, and $e^-_{(aq)}$ and free electrons. Discharge may also be accompanied by shockwaves, light emission, and significant heat. In the aqueous phase, the electric field can cause cavitation. Electrohydraulic reactors and nonthermal plasma have been used for PFOS and PFOA degradation [55]. In systems that apply gas-phase discharge, it has been observed that the reactions at the gas/water interface are important, which can be enhanced by bubbling gas through solution [81, 86]. Transformaton products may vary with approach but include fluoroalkanes (which can be potent greenhouse gasses), shorter-chain PFCAs, and oxygen-substituted PFCAs [55, 100, 101]. One study demonstrated that ground water matrix does not significantly impact PFAA degradation efficiency [82].

9.9.3 Remediation Specific to Solid Matrices

There are several options to deal with PFAS contaminated soilds and sediments: they can be excavated and then treated or contained elsewhere, contained or stabilized in place, or the PFAS can be extracted and dealt with. If containment is used, the possibility of future environmental release is a drawback. In soil stabilization, adsorbants are mixed in with contaminated soil to retain the PFASs, preventing their transport to other environmental compartments. Research on the efficacy of this approach are currently in the early stage. Within the PFAS destruction regime, traditional incineration of excavated material is the most obvious choice that has a known track record. A number of other thermal approaches may also be suitable. For example, thermal desorption of PFAS from soil followed by afterburning the gas-phase contaminants. Circa 2018, a large-scale case study was not available for this approach and it had not been evaluated for PFAA precursors. Alternatively, the vapor energy generator system can be used. This process used 1100 °C steam and additional heat is provided by burning syngas ($CO + H_2$) produced by water splitting and CO evolution from the heated organic matter in the soil. This technology has a relatively small footprint and good energy efficiency, and so it may be useful for on-site remediation. Circa 2018 trials were underway to evaluate this technology [69]. Several more-exotic approaches have also been investigated for the remediation of PFAS adsorbed to solids such as ball milling, electron beams, and UV irradiation [36, 69, 92].

9.10 Future Outlook

PFAS stewardship will continue to challenge society on a global level in the foreseeable future. These substances are critical for applications such as AFFFs that are necessary to protect personal from great hazards, and for which there are no suitable alternatives. The need for PFASs must be balanced with the risk that they pose to humans and ecosystems. This risk is still not satisfactorily understood even for the best-studied compounds although significant advances in this area have been made [28]. Meanwhile, poorly understood PFASs already exist in the environment and new PFASs are increasingly being used in place of the legacy compounds and finding their way into the environment [32].

Based on the many uncertainties of PFAS life cycles and impacts, many researchers advocate for a precautionary approach. Some researcher call for regulators to consult with the scientific community during the approval process for new PFASs because the bulk of information on their properties and impacts are provided therein [76]. Other scientists call for a class-based approach to regulation and elimination of all non-essential applications due to the fact that a meaningful understanding of the properties of a single PFAS takes considerable research effort and time, not to mention that once a given PFASs is present in the environment, remediation is difficult at best [43].

In addition to looking forward, existing PFAS problems must be reconciled. Over 6 million US citizens already have drinking water that exceeds EPA guidelines [78], and given environmental transport and transformation considerations, it is possible that PFAA levels have not even reached a steady state. Although many polities have made considerable efforts to mitigate PFAS environmental release, trends in the global economy are worrisome. With existing and future environmental release and human exposure, we are left to wonder who is going to foot the bill of remediation and health care. At least in the US, a precedent has been established for the manufacturer to pay considering they were responsible for identifying the hazards of their product and profited the most from its production [24].

References

1. Banks RE, Smart BE, Tatlow JC (eds) (1994) Organofluorlne chemistry principles and commercial applications. Springer Science + Busniess Media, New York. 9781489912046
2. Barry V, Winquist A, Steenland K (2013) Perfluorooctanoic acid (PFOA) exposures and incident cancers among adults living near a chemical plant. Environ Health Perspect 121 (11–12):1313–1318. ISSN 00916765. https://doi.org/10.1289/ehp.1306615

3. Barzen-Hanson KA, Roberts SC, Choyke S, Oetjen K, McAlees A, Riddell N, McCrindle R, Ferguson PL, Higgins CP, Field JA (2017) Discovery of 40 classes of per- and polyfluoroalkyl substances in historical aqueous film-forming foams (AFFFs) and AFFF-impacted groundwater. Environ Sci Technol 51(4): 2047–2057. ISSN 15205851. https://doi.org/10.1021/acs.est.6b05843
4. Belisle J (1981) Organic fluorine in human serum: natural versus industrial sources. Science, 212(4502):1509–1510. ISSN 00368075. https://doi.org/10.1126/science.7233236
5. Belisle J, Hagen DF (1980) A method for the determination of perfluorooctanoic acid in blood and other biological samples. Anal Biochem 101(2):369–376. ISSN 10960309. https://doi.org/10.1016/0003-2697(80)90202-X
6. Bruton TA, Sedlak DL (2017) Treatment of aqueous film-forming foam by heat-activated persulfate under conditions representative of in situ chemical oxidation. Environ Sci Technol 51(23):13878–13885. ISSN 15205851. https://doi.org/10.1021/acs.est.7b03969
7. Bruton TA, Sedlak DL (2018) Treatment of perfluoroalkyl acids by heat-activated persulfate under conditions representative of in situ chemical oxidation. Chemosphere, 206:457–464. ISSN 00456535. https://doi.org/10.1016/j.chemosphere.2018.04.128, https://doi.org/10.1016/j.chemosphere.2018.04.128%0A, http://linkinghub.elsevier.com/retrieve/pii/S004565351830777X
8. Buck RC, Franklin J, Berger U, Conder JM, Cousins IT, Voogt PD, Jensen AA, Kannan K, Mabury SA, van Leeuwen SPJ (2011) Perfluoroalkyl and polyfluoroalkyl substances in the environment: terminology, classification, and origins. Integrated Environ Assess Manage 7(4):513–541. ISSN 15513793. https://doi.org/10.1002/ieam.258
9. Cates EL (2017) Photocatalytic water treatment: so where are we going with this? Environ Scie Technol 51(2):757–758. ISSN 15205851. https://doi.org/10.1021/acs.est.6b06035
10. Cheng J, Psillakis E, Hoffmann MR, Colussi AJ (2009) Acid dissociation versus molecular association of perfluoroalkyl oxoacids: environmental implications. J Phys Chem A 113(29):8152–8156. ISSN 10895639. https://doi.org/10.1021/jp9051352
11. Conder JM, Hoke RA, Wolf WD, Russell MH, Buck RC (2008) Are PFCAs bioaccumulative? A critical review and comparison with regulatory criteria and persistent lipophilic compounds. Environ Sci Technol 42(4): 995–1003. ISSN 0013936X. https://doi.org/10.1021/es070895g
12. Cormanich RA, O'Hagan D, Bühl M (2017) Hyperconjugation is the source of helicity in perfluorinated n-Alkanes. Angewandte Chemie—Int Edition, 56(27):7867–7870. ISSN 15213773. https://doi.org/10.1002/anie.201704112
13. Coyle C, Ghosh R, Leeson A, Thompson T (2020) US department of defense-funded research on treatment of per- and polyfluoroalkyl substance-laden materials. Environ Toxicol Chem 1–13. ISSN 15528618. https://doi.org/10.1002/etc.4836
14. Darrow LA, Stein CR, Steenland K (2013) Serum perfluorooctanoic acid and perfluorooctane sulfonate concentrations in relation to birth outcomes in the Mid-Ohio Valley, 2005–2010. Environ Health Perspect 121(10):1207–1213. ISSN 00916765. https://doi.org/10.1289/ehp.1206372
15. Ding G, Peijnenburg WJGM (2013) Physicochemical properties and aquatic toxicity of poly- and perfluorinated compounds. Critical Rev Environ Sci Technol 43(6):598–678. ISSN 15476537. https://doi.org/10.1080/10643389.2011.627016
16. Dombrowski PM, Kakarla P, Caldicott W, Chin Y, Sadeghi V, Bogdan D, Barajas-Rodriguez F, Chiang SYD (2018) Technology review and evaluation of different chemical oxidation conditions on treatability of PFAS. Remediation, 28(2):135–150. ISSN 15206831. https://doi.org/10.1002/rem.21555
17. Du Z, Deng S, Bei Y, Huang Q, Wang B, Huang J, Yu G (2014) Adsorption behavior and mechanism of perfluorinated compounds on various adsorbents-a review. J Hazardous Mater 274:443–454. ISSN 18733336. https://doi.org/10.1016/j.jhazmat.2014.04.038
18. Erik K (2001) Fluorinated surfactants and repellents: second edition, 2nd edn. Revised and Expanded. Marcell Dekker, New York, NY. 08247-0472-X
19. Fujii S, Polprasert C, Tanaka S, Lien NPH, Qiu Y (2007) New POPs in the water environment: distribution, bioaccumulation and treatment of perfluorinated compounds—a review paper. J Water Supply: Res Technol—AQUA

20. Galloway JE, Moreno AVP, Lindstrom AB, Strynar MJ, Newton S, May AA, May AA, Weavers LK, Weavers LK (2020) Evidence of air dispersion: HFPO-DA and PFOA in Ohio and West Virginia surface water and soil near a fluoropolymer production facility. Environ Sci Technol 54(12):7175–7184. ISSN 15205851. https://doi.org/10.1021/acs.est.9b07384
21. Giesy JP, Kannan K (2001) Global distribution of perfluorooctane sulfonate in wildlife. Environ Sci Technol 35(7):1339–1342. ISSN 0013936X. https://doi.org/10.1021/es001834k
22. Gligorovski S, Strekowski R, Barbati S, Vione D (2015) Environmental implications of hydroxyl radicals (·OH). Chem Rev 115(24):13051–13092. ISSN 15206890. https://doi.org/10.1021/cr500310b
23. Gomis MI, Vestergren R, Borg D, Cousins IT (2018) Comparing the toxic potency in vivo of long-chain perfluoroalkyl acids and fluorinated alternatives. Environ Int 113(2017):1–9. ISSN 18736750. https://doi.org/10.1016/j.envint.2018.01.011
24. Head R (2019) . PFAS regulations: how we got here and how to recover the costs. NEWWA
25. Head R (2019) . PFAS regulations: how we got here and how to recover the costs. NEWWA
26. Higgins CP, Luthy RG. Sorption of perfluorinated surfactants on sediments. Environ Sci Technol 40(23):7251–7256. ISSN 0013936X. https://doi.org/10.1021/es061000n
27. Hogue C (2020) Incinerators may spread, not break down, PFAS. C&EN Global Enterprise 98(17):15–15. ISSN 2474-7408. https://doi.org/10.1021/cen-09817-polcon4, https://cen.acs.org/environment/persistent-pollutants/Incincerators-spread-break-down-PFAS/98/web/2020/04
28. Holmquist H, Fantke P, Cousins IT, Liagkouridis I, Peters GM (2020) An (Eco) toxicity life cycle impact assessment framework for per- and polyfluoroalkyl substances. Environ Sci Technol 54:6224–6234. https://doi.org/10.1021/acs.est.9b07774
29. Hori H, Hayakawa E, Einaga H, Kutsuna S, Koike K, Ibusuki T, Kiatagawa H, Arakawa R (2004) Decomposition of environmentally persistent perfluorooctanoic acid in water by photochemical approaches. Environ Sci Technol 38(22):6118–6124. ISSN 0013936X. https://doi.org/10.1021/es049719n
30. Hori H, Yamamoto A, Koike K, Kutsuna S, Osaka I, Arakawa R (2007) Photochemical decomposition of environmentally persistent short-chain perfluorocarboxylic acids in water mediated by iron(II)/(III) redox reactions. Chemosphere 68(3):572–578. ISSN 00456535. https://doi.org/10.1016/j.chemosphere.2006.12.038
31. Houtz EF, Sedlak DL (2012) Oxidative conversion as a means of detecting precursors to perfluoroalkyl acids in urban runoff. Environ Sci Technol 46(17):9342–9349. ISSN 0013936X. https://doi.org/10.1021/es302274g
32. Hu XC, Andrews DQ, Lindstrom AB, Bruton TA, Schaider LA, Grandjean P, Lohmann R, Carignan CC, Blum A, Balan SA, Higgins CP, Sunderland EM (2016) Detection of poly- and perfluoroalkyl substances (PFASs) in U.S. drinking water linked to industrial sites, Military fire training areas, and wastewater treatment plants. Environ Sci Technol Lett 3(10):344–350. ISSN 23288930. https://doi.org/10.1021/acs.estlett.6b00260
33. Huang S, Jaffé PR, Defluorination of perfluorooctanoic acid (PFOA) and perfluorooctane sulfonate (PFOS) by acidimicrobium sp. Strain A6. Environ Sci Technol. ISSN 15205851. https://doi.org/10.1021/acs.est.9b04047
34. Jang SS, Blanco M, Goddard WA, Caldwell G, Ross RB (2003) The source of helicity in perfluorinated N-alkanes. Macromolecules 36(14):5331–5341. ISSN 00249297. https://doi.org/10.1021/ma025645t
35. Javed H, Lyu C, Sun R, Zhang D, Alvarez PJJ (2020) Discerning the inefficacy of hydroxyl radicals during perfluorooctanoic acid degradation. Chemosphere 247. ISSN 18791298. https://doi.org/10.1016/j.chemosphere.2020.125883
36. Jiaoqin L, Wenrui X, Chenguang L, Daniel VHJ, Yumeng Q, Wu N, Gadah AB, Qu R, Zunyao W (2020) Kinetics and mechanism analysis for the photodegradation of PFOA on different solid particles. Chem Eng J 383(2019):123115. ISSN 13858947
37. Jin L, Zhang P, Shao T, Zhao S (2014) Ferric ion mediated photodecomposition of aqueous perfluorooctane sulfonate (PFOS) under UV irradiation and its mechanism. J Hazardous Mater 271:9–15. ISSN 0304-3894. https://doi.org/10.1016/j.jhazmat.2014.01.061

38. Johansson JH, Yan H, Berger U, Cousins IT (2017) Water-to-air transfer of branched and linear PFOA: influence of pH, concentration and water type. Emerg Contaminants 3(1):46–53. ISSN 24056642. https://doi.org/10.1016/j.emcon.2017.03.001
39. Kannan K, Corsolini S, Falandysz J, Fillmann G, Kumar KS, Loganathan BG, Mohd MA, Olivero J, Wouwe NV, Yang JH, Aldous KM (2004) Perfluorooctanesulfonate and related fluorochemicals in human blood from several countries. Environ Sci Technol 38(17):4489–4495. ISSN 0013936X. https://doi.org/10.1021/es0493446
40. Kato K, Wong LY, Jia LT, Kuklenyik Z, Calafat AM (2011) Trends in exposure to polyfluoroalkyl chemicals in the U.S. population: 1999–2008. Environ Sci Technol 45(19):8037–8045. ISSN 0013936X. https://doi.org/10.1021/es1043613
41. Kissa Erik (1994) Fluorinated surfactants in blood. J Fluorine Chem 66:5–6. https://doi.org/10.1016/0022-1139(93)02904-S
42. Kutsuna S, Hori H, Sonoda T, Iwakami T, Wakisaka A (2012) Preferential solvation of perfluorooctanoic acid (PFOA) by methanol in methanol-water mixtures: a potential overestimation of the dissociation constant of PFOA using a Yasuda-Shedlovsky plot. Atmos Environ 49:411–414. ISSN 13522310. https://doi.org/10.1016/j.atmosenv.2011.12.009
43. Kwiatkowski CF, Andrews DQ, Birnbaum LS, Bruton TA, Dewitt JC, Knappe DRU, Maffini MV, Miller MF, Pelch KE, Reade A, Soehl A, Trier X, Venier M, Wagner CC, Wang Z, Blum A (2020) Scientific basis for managing PFAS as a chemical class. Environ Sci Technol Lett. ISSN 23288930. https://doi.org/10.1021/acs.estlett.0c00255
44. Lau C, Anitole K, Hodes C, Lai D, Pfahles-Hutchens A, Seed J (2007) Perfluoroalkyl acids: a review of monitoring and toxicological findings. Toxicol Sci 99(2):366–394. ISSN 10966080. https://doi.org/10.1093/toxsci/kfm128
45. Lemal DM (2004) Perspective on fluorocarbon chemistry. J Organ Chem 69(1):1–11. ISSN 00223263. https://doi.org/10.1021/jo0302556
46. Lin IJ (1972) The hydrophile-lipophile balance (hlb) of fluorocarbon surfactants and its relation to the critical micelle concentration (cmc). J Phys Chemi 76(14):2019–2023. ISSN 00223654. https://doi.org/10.1021/j100658a020
47. Liu CS, Shih K, Wang F (2012) Oxidative decomposition of perfluorooctanesulfonate in water by permanganate. Separ Purif Technol 87:95–100. ISSN 13835866. https://doi.org/10.1016/j.seppur.2011.11.027
48. Liu H, Bruton TA, Doyle FM, Sedlak DL (2014) In situ chemical oxidation of contaminated groundwater by persulfate: decomposition by Fe(III)- and Mn(IV)-containing oxides and aquifer materials. Environ Sci Technol 48(17):10330–10336. ISSN 15205851. https://doi.org/10.1021/es502056d
49. Liu J, Hoomissen DJV, Liu T, Maizel A, Huo X, Fernández SR, Ren C, Xiao X, Fang Y, Schaefer CE, Higgins CP, Vyas S, Strathmann TJ (2018) Reductive defluorination of branched per- and polyfluoroalkyl substances with cobalt complex catalysts. Environ Sci Technol Lett 5(5):289–294.ISSN 23288930. https://doi.org/10.1021/acs.estlett.8b00122
50. Li A, Zhang Z, Li P, Cai L, Zhang L, Gong J (2017) Nitrogen dioxide radicals mediated mineralization of perfluorooctanoic acid in aqueous nitrate solution with UV irradiation. Chemosphere 188:367–374. ISSN 18791298. https://doi.org/10.1016/j.chemosphere.2017.08.170
51. Lopez-Espinosa MJ, Mondal D, Armstrong B, Bloom MS, Fletcher T (2012) Thyroid function and perfluoroalkyl acids in children living near a chemical plant. Environ Health Perspect 120(7):1036–1041. ISSN 00916765. https://doi.org/10.1289/ehp.1104370
52. Luo Q, Yan X, Lu J, Huang Q (2018) Perfluorooctanesulfonate degrades in a laccase-mediator system. Environ Sci Technol 52(18):10617–10626. ISSN 15205851. https://doi.org/10.1021/acs.est.8b00839
53. Lu D, Sha S, Jiayue L, Zhuangrong H, Zhang JX (2020) Treatment train approaches for the remediation of per- and polyfluoroalkyl substances (PFAS): a critical review. J Hazardous Mater 386(2019):121963. ISSN 18733336
54. Marek T, Anna B-C, Iwona B, Krzysztof K (2018) Oxidation advanced, processes reduction, treatment for aqueous perfluorooctanoate (PFOA) and perfluorooctanesulfonate (PFOS)—a

review of recent advances. Chem Eng J 336(2017):170–199. ISSN 13858947. https://doi.org/10.1016/j.cej.2017.10.153. November
55. Merino N, Qu Y, Deeb RA, Hawley EL, Hoffmann MR, Mahendra S (2016) Degradation and removal methods for perfluoroalkyl and polyfluoroalkyl substances in water. Environ Eng Sci 33(9):615–649. ISSN 15579018. https://doi.org/10.1089/ees.2016.0233
56. Mitchell SM, Ahmad M, Teel AL, Watts RJ (2013) Degradation of perfluorooctanoic acid by reactive species generated through catalyzed H_2O_2 propagation Reactions. Environ Sci Technol Lett 1(1):117–121. ISSN 23288930. https://doi.org/10.1021/ez4000862
57. Moody CA, Field JA (2000) Perfluorinated surfactants and the environmental implications of their use in fire-fighting foams. Environ Sci Technol 34(18):3864–3870. ISSN 0013936X. https://doi.org/10.1021/es991359u
58. Moody CA, Martin JW, Kwan WC, Muir DCG, Mabury SA (2002) Monitoring perfluorinated surfactants in biota and surface water samples following an accidental release of fire-fighting foam into Etobicoke Creek. Environ Sci Technol 36(4):545–551. ISSN 0013936X. https://doi.org/10.1021/es011001+
59. Ochoa-Herrera V, Sierra-Alvarez R, Somogyi A, Jacobsen NE, Wysocki VH, Field JA (2008) Reductive defluorination of perfluorooctane sulfonate. Environ Sci Technol 42(9):3260–3264. ISSN 0013936X. https://doi.org/10.1021/es702842q
60. OECD Environment Health and Safety Publications—Series on Risk Management (2007) Lists of PFOS, PFAS, PFOA, PFCA, related compounds and chemicals that may degrade to PFCA. Env Jm Mono (2006)15 21 (April 2006):1–157, . ISSN 1609-1914. ENV/JM/MONO(2007)10
61. O'Hagan D (2008) Understanding organofluorine chemistry. An introduction to the C-F bond. Chem Soc Rev 37(2):308–319. ISSN 03060012. https://doi.org/10.1039/b711844a
62. Park S, Lee LS, Medina VF, Zull A, Waisner S (2016) Heat-activated persulfate oxidation of PFOA, 6:2 fluorotelomer sulfonate, and PFOS under conditions suitable for in-situ groundwater remediation. Chemosphere 145:376–383. ISSN 18791298. https://doi.org/10.1016/j.chemosphere.2015.11.097
63. Paul AG, Jones KC, Sweetman AJ (2009) A first global production, emission, and environmental inventory for perfluorooctane sulfonate. Environ Sci Technol 43(2): 386–392. ISSN 0013936X. https://doi.org/10.1021/es802216n
64. Peer K (2013) Modern fluoroorganic chemistry, 2nd edn. Wiley-VCH Verlag GmbH, Weinheim, Germany. 9783527329441
65. Prevedouros K, Cousins IT, Buck RC, Korzeniowski SH (2006) Sources, fate and transport of perfluorocarboxylates. Environ Sci Technol 40(1):32–44. ISSN 0013936X. https://doi.org/10.1021/es0512475
66. Rahman MF, Peldszus S, Anderson WB (2014) Behaviour and fate of perfluoroalkyl and polyfluoroalkyl substances (PFASs) in drinking water treatment: a review. Water Res 50:318–340. ISSN 18792448. https://doi.org/10.1016/j.watres.2013.10.045
67. Rappazzo KM, Coffman E, Hines EP (2017) Exposure to perfluorinated alkyl substances and health outcomes in children: a systematic review of the epidemiologic literature. Int J Environ Res Public Health 14(7):1–22. ISSN 16604601. https://doi.org/10.3390/ijerph14070691
68. Rayne S (2013) Comment on "Letter to the Editor regarding, 'polyfluorinated compounds: past, present, and future. Environ Sci Technol 47(5):2151. ISSN 0013-936X. https://doi.org/10.1021/es400475a
69. Ross I, McDonough J, Miles J, Storch P, Kochunarayanan PT, Kalve E, Hurst J, Dasgupta SS, Burdick J (2018) A review of emerging technologies for remediation of PFASs. Remediation 28(2):101–126. ISSN 15206831. https://doi.org/10.1002/rem.21553
70. Russell MH, Berti WR, Szostek B, Buck RC (2008) Investigation of the biodegradation potential of a fluoroacrylate polymer product in aerobic soils. Environ Sci Technol 42(3):800–807. ISSN 0013936X. https://doi.org/10.1021/es0710499
71. Sahu SP, Qanbarzadeh M, Ateia M, Torkzadeh H, Maroli AS, Cates EL (2018) Rapid degradation and mineralization of perfluorooctanoic acid by a new petitjeanite Bi3O(OH)(PO4)2Microparticle ultraviolet photocatalyst. Environ Sci Technol Lett 5(8):533–538. ISSN 23288930. https://doi.org/10.1021/acs.estlett.8b00395

72. Schröder HF, José HJ, Gebhardt W, Moreira RFPM, Pinnekamp J (2010) Biological wastewater treatment followed by physicochemical treatment for the removal of fluorinated surfactants. Water Sci Technol 61(12):3208–3215. ISSN 02731223. https://doi.org/10.2166/wst.2010.917
73. Schröder HF, Meesters RJW (2005) Stability of fluorinated surfactants in advanced oxidation processes—a follow up of degradation products using flow injection-mass spectrometry, liquid chromatography-mass spectrometry and liquid chromatography-multiple stage mass spectrometry. J Chromatogr A 1082(1 SPEC. ISS.):110–119. ISSN 00219673. https://doi.org/10.1016/j.chroma.2005.02.070
74. Secretariat of the Basel Rotterdam and Stockholm Conventions. Exempted use for PFOS, its salts and PFOSF (2019), 2019
75. Secretariat of the Basel Rotterdam and Stockholm. Time-limited exempted uses for PFOA, its salts and PFOA-related compound (2019)
76. Sheriff I, Debela SA, Kabia OA, Ntoutoume CE, Turay MJ (2020) The phase out of and restrictions on per-and polyfluoroalkyl substances: time for a rethink. Chemosphere 251. ISSN 18791298. https://doi.org/10.1016/j.chemosphere.2020.126313
77. Shi J, He J, Wang HJ (2011) A computational study of C-X (X=H, C, F, Cl) bond dissociation enthalpies (BDEs) in polyhalogenated methanes and ethanes. J Phys Organ Chem 24(1):65–73. ISSN 08943230. https://doi.org/10.1002/poc.1704
78. Sinclair GM, Long SM, Jones OAH (2020) What are the effects of PFAS exposure at environmentally relevant concentrations? Chemosphere 258:127340. ISSN 18791298. https://doi.org/10.1016/j.chemosphere.2020.127340
79. Steenland K, Fletcher T, Savitz DA (2010) Epidemiologic evidence on the health effects of perfluorooctanoic acid (PFOA). Environ Health Perspect 118(8):1100–1108. ISSN 00916765. https://doi.org/10.1289/ehp.0901827
80. Steenland K, Zhao L, Winquist A, Parks C (2013) Ulcerative colitis and perfluorooctanoic acid (PFOA) in a highly exposed population of community residents and workers in the Mid-Ohio Valley. Environ Health Perspect 121(8):900–905. ISSN 00916765. https://doi.org/10.1289/ehp.1206449
81. Stratton GR, Bellona CL, Dai F, Holsen TM, Thagard SM (2015) Plasma-based water treatment: conception and application of a new general principle for reactor design. Chem Eng J 273:543–550. ISSN 13858947. https://doi.org/10.1016/j.cej.2015.03.059
82. Stratton GR, Dai F, Bellona CL, Holsen TM, Dickenson ERV, Thagard SM (2017) Plasma-based water treatment: efficient transformation of perfluoroalkyl substances in prepared solutions and contaminated groundwater. Environ Sci Technol 51(3):1643–1648. ISSN 15205851. https://doi.org/10.1021/acs.est.6b04215
83. Sunderland EM, Hu XC, Dassuncao C, Tokranov AK, Wagner CC, Allen JG (2019) A review of the pathways of human exposure to poly- and perfluoroalkyl substances (PFASs) and present understanding of health effects. J Exposure Sci Environ Epidemiol 29(2):131–147. ISSN 1559064X. https://doi.org/10.1038/s41370-018-0094-1
84. Tang H, Xiang Q, Lei M, Yan J, Zhu L, Zou J (2012) Efficient degradation of perfluorooctanoic acid by UV-Fenton process. Chem Eng J 184:156–162. ISSN 13858947. https://doi.org/10.1016/j.cej.2012.01.020
85. Taves DR (1968) Evidence that there are two forms of fluoride in human serum [16]. Nature 217(5133):1050–1051. ISSN 00280836. https://doi.org/10.1038/2171050b0
86. Thagard SM, Stratton GR, Dai F, Bellona CL, Holsen TM, Bohl DG, Paek E, Dickenson ERV (2017) Plasma-based water treatment: development of a general mechanistic model to estimate the treatability of different types of contaminants. J Phys D: Appl Phys 50(1):2017. ISSN 13616463. https://doi.org/10.1088/1361-6463/50/1/014003
87. Thi LAP, Do HT, Lee YC, Lo SL (2013) Photochemical decomposition of perfluorooctanoic acids in aqueous carbonate solution with UV irradiation. Chem Eng J 221:258–263. ISSN 13858947. https://doi.org/10.1016/j.cej.2013.01.084
88. Tullo A (2000) 3M study raised EPA's concerns. Chem Eng News Arch 78(22):12–13. ISSN 0009-2347. https://doi.org/10.1021/cen-v078n022.p012a

89. Tullo A (2000) How-to manual elucidates lab design, construction process. Chem Eng News Arch 78(21):9–10. ISSN 0009-2347. https://doi.org/10.1021/cen-v078n021.p010a
90. US Environmental Protection Agency (2016) Drinking water health advisories for PFOA and PFOS. URL https://www.epa.gov/ground-water-and-drinking-water/drinking-water-health-advisories-pfoa-and-pfos
91. US EPA (2020) Risk management for per- and polyfluoroalkyl substances (PFAS) under TSCA
92. Vecitis CD, Park H, Cheng J, Mader BT, Hoffmann MR (2009) Treatment technologies for aqueous perfluorooctanesulfonate (PFOS) and perfluorooctanoate (PFOA). Frontiers Environ Sci Eng China 3(2):129–151. ISSN 16737415. https://doi.org/10.1007/s11783-009-0022-7
93. Vierke L, Berger U, Cousins IT (2013) Estimation of the acid dissociation constant of perfluoroalkyl carboxylic acids through an experimental investigation of their water-to-air transport. Environ Sci Technol 47(19):11032–11039. ISSN 0013936X. https://doi.org/10.1021/es402691z
94. Wang Y, Chang W, Wang L, Zhang Y, Zhang Y, Wang M, Wang Y, Li P (2019) A review of sources, multimedia distribution and health risks of novel fluorinated alternatives. Ecotoxicol Environ Safety 182(July):109402. ISSN 10902414. https://doi.org/10.1016/j.ecoenv.2019.109402
95. Wang Z, Cousins IT, Scheringer M, Buck RC, Hungerbühler K (2014) Global emission inventories for C4-C14 perfluoroalkyl carboxylic acid (PFCA) homologues from 1951 to 2030, Part I: production and emissions from quantifiable sources. Environ Int 70:62–75. ISSN 18736750. https://doi.org/10.1016/j.envint.2014.04.013
96. Wang Z, Cousins IT, Scheringer M, Buck RC, Hungerbühler K (2014) Global emission inventories for C4-C14 perfluoroalkyl carboxylic acid (PFCA) homologues from 1951 to 2030, part II: the remaining pieces of the puzzle. Environ Int 69:166–176. ISSN 18736750. https://doi.org/10.1016/j.envint.2014.04.013
97. Wang Z, Cousins IT, Scheringer M, Hungerbuehler K (2015) Hazard assessment of fluorinated alternatives to long-chain perfluoroalkyl acids (PFAAs) and their precursors: Status quo, ongoing challenges and possible solutions. Environ Int 75:172–179.ISSN 18736750. https://doi.org/10.1016/j.envint.2014.11.013
98. Wang Z, Dewitt JC, Higgins CP, Cousins IT (2017) A never-ending story of per- and polyfluoroalkyl substances (PFASs)? Environ Sci Technol 51(5):2508–2518. ISSN 15205851. https://doi.org/10.1021/acs.est.6b04806
99. Wang W, Rhodes G, Ge J, Yu X, Li H (2020) Uptake and accumulation of per- and polyfluoroalkyl substances in plants. Chemosphere 261. ISSN 18791298. https://doi.org/10.1016/j.chemosphere.2020.127584
100. Wang L, Sun L, Yu Z, Hou Y, Peng Z, Yang F, Chen Y, Huang J (2017) Synergetic decomposition performance and mechanism of perfluorooctanoic acid in dielectric barrier discharge plasma system with Fe3O4@SiO2-BiOBr magnetic photocatalyst. Molecular Catalysis 441:179–189. ISSN 24688231. https://doi.org/10.1016/j.mcat.2017.08.014
101. Wang X, Wang P, Liu X, Hu L, Wang Q, Xu P (2020) Enhanced degradation of PFOA in water by dielectric barrier discharge plasma in a coaxial cylindrical structure with the assistance of peroxymonosulfate. Chem Eng J 389(February):124381. ISSN 1385-8947. https://doi.org/10.1016/j.cej.2020.124381
102. Wang Y, Zhang P, Pan G, Chen H (2008) Ferric ion mediated photochemical decomposition of perfluorooctanoic acid (PFOA) by 254 nm UV light. J Hazardous Mater 160(1):181–186, 2008. ISSN 03043894. https://doi.org/10.1016/j.jhazmat.2008.02.105
103. Washington JW, Ellington JJ, Jenkins TM, Evans JJ, Yoo H, Hafner SC (2009) Degradability of an acrylate-linked, fluorotelomer polymer in soil. Environ Sci Technol 43(17):6617–6623. ISSN 0013936X. https://doi.org/10.1021/es9002668
104. Wojnárovits L, Takács E (2019) Rate constants of sulfate radical anion reactions with organic molecules: a review. Chemosphere 220:1014–1032. ISSN 18791298. https://doi.org/10.1016/j.chemosphere.2018.12.156

105. Xu B, Boshir M, Zhou JL, Altaee A, Wu M (2017) Photocatalytic removal of per fl uoroalkyl substances from water and wastewater: mechanism , kinetics and controlling factors. Chemosphere 189:717–729. https://doi.org/10.1016/j.chemosphere.2017.09.110
106. Xuechun Wang, Yuan Zhuang, Jia Zhang, Laizhou Song, and Baoyou Shi. Pollutant degradation behaviors in a heterogeneous Fenton system through Fe/S-doped aerogel. *Science of the Total Environment*, 714: 1–12, 2020. ISSN 18791026. https://doi.org/10.1016/j.scitotenv.2019.136436
107. Yang S, Cheng J, Sun J, Hu Y, Liang X (2013) Defluorination of aqueous perfluorooctanesulfonate by activated persulfate oxidation. PLoS ONE 8(10):6–15. ISSN 19326203. https://doi.org/10.1371/journal.pone.0074877
108. Yang A, Ching C, Easler M, Helbling DE, Dichtel WR (2020) Cyclodextrin polymers with nitrogen- containing tripodal crosslinkers for efficient PFAS adsorption. ACS Mater Lett 2:1240–1245. https://doi.org/10.1021/acsmaterialslett.0c00240
109. Young CJ, Mabury SA (2010) Atmospheric perfluorinated acid precursors: chemistry, occurrence, and impacts. In: De Voogt P (eds) Reviews of environmental contamination and toxicology 208:2–98. Springer. ISBN 978-1-4419-6879-1. https://doi.org/10.1007/978-1-4419-6880-7
110. Yu Y, Zhang K, Li Z, Ren C, Chen J, Lin Y-H, Liu J, Men Y (2020) Microbial cleavage of C-F bonds in two C 6 per- and polyfluorinated compounds via reductive defluorination. Environ Sci Technol. ISSN 0013-936X. https://doi.org/10.1021/acs.est.0c04483
111. Zushi YZ, Masunaga S, Lee H, Mabury SA (2014) Transformation products of emerging contaminants in the environment. Wiley, West Syssex, United Kingdom. ISBN 9781118339596. https://doi.org/10.1002/9781118339558

Index

M. Shukla et al. (eds.), *Emerging Materials and Environment*, Challenges and Advances in Computational Chemistry and Physics 37, https://doi.org/10.1007/978-3-031-39470-6

The manufacturer's authorised representative in the EU is Springer Nature Customer Service Centre GmbH, Europaplatz 3, 69115 Heidelberg, Germany. If you have any concerns regarding our products, please contact ProductSafety@springernature.com

Printed and bound by CPI Group (UK) Ltd, Croydon, CR0 4YY
01/10/2026
02225896-0010